Green Chemistry and Water Remediation

Advances in Green and Sustainable Chemistry

Series Editors

Bela Torok

Timothy Dransfield

Green Chemistry and Water Remediation

Research and Applications

Edited by

Sanjay K. Sharma
Green Chemistry and Sustainability Research Group,
Department of Chemistry, JECRC University,
Jaipur, Rajasthan, India

ELSEVIER

Elsevier
Radarweg 29, PO Box 211, 1000 AE Amsterdam, Netherlands
The Boulevard, Langford Lane, Kidlington, Oxford OX5 1GB, United Kingdom
50 Hampshire Street, 5th Floor, Cambridge, MA 02139, United States

Library of Congress Cataloging-in-Publication Data
A catalog record for this book is available from the Library of Congress

British Library Cataloguing-in-Publication Data
A catalogue record for this book is available from the British Library

ISBN: 978-0-12-817742-6

Publisher: Susan Dennis
Acquisitions Editor: Anneka Hess
Editorial Project Manager: Mona Zahir
Production Project Manager: Sruthi Satheesh
Cover Designer: Alan Studholme

Typeset by TNQ Technologies

This book is a sincere gratitude to
All the Warriors
fighting against COVID-19
across the world.

Stay safe, stay healthy, may God bless us all.

Contents

5. Removal of dyes and pigments from industrial effluents

Ahmet Gürses, Kübra Güneş and Elif Şahin

6. Membrane processes for removal of polyaromatic hydrocarbons from wastewater

S.R. Barman, U. Roy, P. Das and A. Mukhopadhayay

7. Hybrid application of advanced oxidation processes to dyes′ removal

Süheyda Atalay and Gülin Ersöz

8. Wastewater Treatment Plants as emerging source of antibiotic resistance

Neha Sharma and Sanjay K. Sharma

9. Green chemistry and its applications in hospital wastewater and its treatment

Reetu Rani and Suman Singh

10. Recent developments in sonochemical treatments of contaminated wastewaters

Efraím A. Serna-Galvis and Ricardo A. Torres-Palma

11. Modified titania, A green material for wastewater remediation

Sofia Sandhu, Rehmat Brar, Siddhant Gulati and Vasundhara Singh

Contributors

Süheyda Atalay, Department of Chemical Engineering, Ege University, İzmir, Turkey

S.R. Barman, Department of Environmental Science, Asutosh College, Kolkata, West Bengal, India; Department of Environmental Science, University of Calcutta, Kolkata, West Bengal, India

Rehmat Brar, Punjab Engineering College, Chandigarh, Punjab, India

P. Das, Department of Chemical Engineering, Jadavpur University, Kolkata, West Bengal, India

Hasan Demir, Osmaniye Korkut Ata University, Osmaniye, Turkey

Gülin Ersöz, Department of Chemical Engineering, Ege University, İzmir, Turkey

Siddhant Gulati, Punjab Engineering College, Chandigarh, Punjab, India

Kübra Güneş, Chemical Education Ataturk University, Erzurum, Turkey

Suresh Gupta, Department of Chemical Engineering, Birla Institute of Technology and Science (BITS), Pilani, Rajasthan, India

Ahmet Gürses, Chemical Education Ataturk University, Erzurum, Turkey

Suryadi Ismadji, Department of Chemical Engineering, Widya Mandala Surabaya Catholic University, Surabaya, East Java, Indonesia

Yi-Hsu Ju, Graduate Institute of Applied Science and Technology, National Taiwan University of Science and Technology, Taipei City, Taiwan; Taiwan Building Technology Center, National Taiwan University of Science and Technology, Taipei City, Taiwan

Anil Kumar K, Department of Chemical Engineering, Birla Institute of Technology and Science (BITS), Pilani, Rajasthan, India

N. Kusumawati, Analytical Chemistry Laboratory, Mathematics and Natural Science Faculty, Universitas Negeri Surabaya, Surabaya, Indonesia

A. Mukhopadhayay, Department of Environmental Science, University of Calcutta, Kolkata, West Bengal, India

Jindrayani Nyoo Putro, Department of Chemical Engineering, National Taiwan University of Science and Technology, Taipei City, Taiwan

Smita Raghuvanshi, Department of Chemical Engineering, Birla Institute of Technology and Science (BITS), Pilani, Rajasthan, India

E. Rahmadyanti, Water and Irrigation Laboratory, Engineering Faculty, Universitas Negeri Surabaya, Surabaya, Indonesia

Reetu Rani, CSIR-Central Scientific Instruments Organisation, Chandigarh, India; Academy of Scientific and Innovative Research (AcSIR), Ghaziabad, Uttar Pradesh, India

U. Roy, Department of Chemical Engineering, Jadavpur University, Kolkata, West Bengal, India

Elif Şahin, Chemical Education Ataturk University, Erzurum, Turkey

Sofia Sandhu, Punjab Engineering College, Chandigarh, Punjab, India

Shella Permatasari Santoso, Department of Chemical Engineering, Widya Mandala Surabaya Catholic University, Surabaya, East Java, Indonesia

Efraím A. Serna-Galvis, Grupo de Investigación en Remediación Ambiental y Biocatálisis (GIRAB), Instituto de Química, Facultad de Ciencias Exactas y Naturales, Universidad de Antioquia UdeA, Medellín, Colombia

Neha Sharma, Department of Microbiology, Maharishi Markandeshwar Medical College and Hospital, Maharishi Markandeshwar University, Solan, Himachal Pradesh, India

Sanjay K. Sharma, Green Chemistry and Sustainability Research Group, Department of Chemistry, JECRC University, Jaipur, Rajasthan, India

M.M. Sianita, Analytical Chemistry Laboratory, Mathematics and Natural Science Faculty, Universitas Negeri Surabaya, Surabaya, Indonesia

Suman Singh, CSIR-Central Scientific Instruments Organisation, Chandigarh, India; Academy of Scientific and Innovative Research (AcSIR), Ghaziabad, Uttar Pradesh, India

Vasundhara Singh, Punjab Engineering College, Chandigarh, Punjab, India

Felycia Edi Soetaredjo, Department of Chemical Engineering, Widya Mandala Surabaya Catholic University, Surabaya, East Java, Indonesia

Ricardo A. Torres-Palma, Grupo de Investigación en Remediación Ambiental y Biocatálisis (GIRAB), Instituto de Química, Facultad de Ciencias Exactas y Naturales, Universidad de Antioquia UdeA, Medellín, Colombia

Garima Verma, Department of Chemical Engineering, Birla Institute of Technology and Science (BITS), Pilani, Rajasthan, India

About the editor

Professor Sanjay K. Sharma is a Fellow of the Royal Society of Chemistry (UK) and an internationally acclaimed author and editor of many books, research journals, and hundreds of articles from the last two decades. He has also been appointed as *Series Editor* by Springer, UK, for their prestigious book series *SpringerBriefs in Green Chemistry for Sustainability*, where he has been involved in the editing of 39 different titles of various international contributors so far.

Presently, Professor Sharma is working as Professor of Chemistry & Associate Dean (Research), JECRC University, Jaipur, India, where he is engaged in teaching chemistry, environmental chemistry, and green chemistry to undergraduate and postgraduate students, pursuing his research interest in the domain area of green chemistry with special reference of water pollution, corrosion inhibition, and biopolymers. In addition, he takes care of the PhD program as well as research activities of the university.

Dr. Sharma has 20 books on chemistry from international publishers, including Springer, Wiley, the Royal Society of Chemistry, etc., and over 100 research papers of national and international repute to his credit.

He is also a member of the American Chemical Society (USA) and International Society for Environmental Information Sciences (ISEIS, Canada), and is also a life member of various international professional societies, including the International Society of Analytical Scientists, Indian Council of Chemists, International Congress of Chemistry and Environment, Indian Chemical Society, etc.

He was conferred the *Life Time Achievement Award* by the Association of Chemistry Teachers, Mumbai, and SRM University, Chennai, in 2018 for his contribution to chemistry education and research.

He is a resource person for various prestigious national and international organizations, consulting on green chemistry, water sustainability, manuscript writing, and scientific ethics.

The following are his popular books:

1. *Green Chemistry for Dyes Removal from Waste Water: Research Trends and Applications* (from Wiley-Scrivener, USA)
2. *Heavy Metals in Water: Presence, Removal and Safety* (from the Royal Society of Chemistry, UK)
3. *Bio-surfactants: Research Trends and Applications* (from CRC Taylor & Francis, USA)
4. *Waste Water Reuse and Management* (from Springer, UK)
5. *Advances in Water Treatment and Pollution Prevention* (from Springer, UK)
6. *Green Corrosion Chemistry and Engineering* (from Wiley, Germany)
7. *Handbook of Applied Biopolymer Technology: Synthesis, Degradation and Applications* (from the Royal Society of Chemistry, UK)
8. *Handbook on Applications of Ultrasound: Sonochemistry and Sustainability* (from CRC Taylor & Francis, USA)
9. *Green Chemistry for Environmental Sustainability* (from CRC Taylor & Francis, USA)

He loves to answer readers' emails if they write him at: sk.sharmaa@outlook.com.

Preface

When you care for nature, be assured that nature cares for you, more than a mother would do for her child.

Gurudev Sri Sri Ravishankar

Nothing is possible on Earth without water. This is why water is not just a commodity, it is a necessity and an indispensable element of life for all the living beings present on the planet. Therefore, if we don't value it, then we do not value life either … seriously.

However, at the same time, preserving Earth's water is a big challenge for all of us. Available water resources have been contaminated with various types of pollutants and hazardous substances, which cause several water-borne diseases resulting in casualities all around the world. It is a shocking fact that water-borne diseases kill more people each year than war and all other forms of violence combined. Therefore, water treatment and its remediation are vital to avoid disease and conamination.

Researchers all over the world are doing hard work in the area of water remediation to ensure its drinking quality by adopting different techniques and remedies, which is a big service to mankind and the environment, and this book, entitled *Green Chemistry and Water Remediation: Research and Applications*, is a modest contribution in this direction.

This book consists of 11 comprehensive chapters on various topics contributed by internationally recognized scientists and researchers.

Because green chemistry is involved in the life cycle (design, production, usage, and disposal) of every process and production, it gives insight into the removal or reduction of the harmful effects of treatment techniques by minimizing energy consumption and using renewable feedstock. Applications of green chemistry principles in water remediation are the focus of this book.

Chapter 1 provides an overview of research studies on water remediation within the frame work of 12 principles of green chemistry, bibliometric features, and characteristics of related research papers. Bibliometric analysis will be performed on the literature cited in the Web of Science from 1999 to 2018. This bibliometric analysis covers mainly year-wise and geographical-wise distribution, authorship pattern and citation, etc. All these data will be organized, analyzed, tabulated, and presented by using simple statistical methods, where readers will find recent research applications on water remediation with supporting statistical data. This chapter gives an updated literature review of bibliometric analysis related to the theme of this book.

The groundwater of many countries is highly contaminated with fluoride and this is why it is a big challenge worldwide. Chapter 2 focuses on the use of an adsorption technique that uses nanoadsorbents. In this chapter, the use of modified Hummers' method for synthesizing nanomagnesium oxide (n-MgO) and nanocomposites (n-MgO-coated graphene oxides) has been beautifully discussed, which is further used in the efficient removal of fluoride ions from aqueous solution.

Chapter 3 tells us about the batik industry of Indonesia, the UNESCO-recognized humanitarian legacy for oral and nonmaterial cultures of native Indonesia. This chapter deals with sustainability issues related to batik and its treatment, which is a beautiful add-on to this book. It not only emphasizes local problems but also tries to discuss them from a global perspective.

As we know, biosorption has been employed as an alternative wastewater treatment process to reuse wastewater for better economic prospects and environmental consideration, and Chapter 4 tells us about its detailed applications and uses for the removal of dyes. Agricultural and plant wastes, fungus/fungi, and bacteria are some very promising and effective biosorbents in this regard. This chapter also explains the equilibrium and kinetic studies involved, which guide researchers to understand the mechanism more accurately and systematically. Discussion related to the thermodynamics of the adsorption of dyes on biosorbents is another highlight of this chapter.

Chapter 5 also deals with the removal of dyes and pigments from industrial effluents. It effectively explains chemical treatment methods, including coagulation—flocculation, electroflotation, electrocoagulation process, advanced oxidation process (AOP), ozonation process, Fenton process, photocatalytic processes, physical treatment methods, including the adsorption process, irradiation process, ion exchange, membrane processes, membrane filtration, and biological methods, including aerobic treatment, anaerobic treatment, combined anaerobic—aerobic treatment, decolorization by bacteria, and decolorization by fungi and yeast. All these methods are quite promising and result oriented in the removal of dyes from industrial effluent. A chapter worth reading.

Polyaromatic hydrocarbons (PAHs) are an assembly of organic micropollutants that have been linked to cancer and other health hazards in humans. They are also known mutagenic agents and have been listed as priority pollutants by various government agencies around the world. Chapter 6 gives a comprehensive account of their hazardous nature and removal through nanostructured industrial membranes. The efficiency of this technique depends upon the nanostructure of these membranes, which increases their permeability and pollutant-trapping ability in the removal process of PAHs from wastewater.

AOPs are known to be environmentally friendly processes for the mineralization of refractory pollutants such as dyes. Chapter 7 presents a detailed summary of the potential hybrid processes described in the literature for the treatment of dye wastewater. Combination of the processes is considered in two different titles: a combination of AOPs and a combination of physicochemical methods with AOPs.

Antimicrobial agents, such as antibiotics, are being used extensively in humans and veterinary medicine, and often as part of routine agricultural production, which has significantly contributed to the emergence of antimicrobial resistance. Besides others, antibiotics are emerging pollutants, which have found their way into industrial effluents and are rendered to exhibit recalcitrance and resistance. Chapter 8 gives us an insight into bacterial diversity-based analysis utilizing tools of metagenomics, transcriptomics, and metabolomics. The application of wastewater treatment plants (WWTPs) has also been discussed comprehensively to acquaint readers with the latest updates in this area of research.

Due to the presence of high concentrations of hazardous substances such as pharmaceutically active compounds, toxic chemicals, radioisotopes, and pathogens, hospital effluents are of major concern for environmental and public health. Hospital wastewater is discharged directly to municipal WWTPs without any pretreatment and causes many serious health-related implications. Conventional physicochemical wastewater treatment methods fail in terms of high costs of treatments and the generation of harmful by-products. In such a situation, green chemistry techniques can be employed as a viable solution for the treatment of a huge amount of wastewater generated in hospitals. Chapter 9 deals with these promising green techniques. A wonderfully explained chapter.

Water contamination is hazardous for the environment and life on the Earth. Ultrasound-based treatment methods have very significant importance in this regard. Chapter 10 contains critical information about current developments in sonochemical treatments of contaminated water. Sonocatalysis using novel carbonaceous materials (e.g., carbon nanotubes, graphene oxide, and biochar) discussed in this chapter is certainly a highlight. A short, crisp, and impactful chapter.

Titanium dioxide first came under the spotlight when it was found that the oxidizing power of the photogenerated holes in TiO_2 is sufficient to decompose water and air pollutants. In a very short period, TiO_2 emerged as one of the most promising photocatalysts, finding its usage from conventional areas like pigments in foodstuffs, paint, and cosmetics to advanced sectors such as catalysis, photovoltaic cells, sensors, and biomedical treatment due to its broad functionality, longer stability, and nontoxicity. Chapter 11 focuses on reaction mechanisms involved in bandgap modification of TiO_2 with the doping of metals and nonmetals and application of this modified titania in reactive dye degradation. In addition, it details a discussion about the synthesis of modified titania in terms of visible light absorbance and reactivity via bandgap and facet engineering. A very impressive chapter.

With a bunch of so diverse chapters on a variety of topics related to the theme of this book, I believe that this book can be used as an important document to inspire researchers in the area of remediation of water and treatment of wastewater at large.

I once again gratefully acknowledge all the contributors to this book, without whom it could not have been completed and published. I express my highest gratitude and thankfulness to all of them.

Since water is life and its every drop counts, it is our moral as well as ethical duty to save water and serve our nature.

Jai Gurudev!

Sanjay K. Sharma, FRSC
May 13, 2020
Jaipur, India

Acknowledgments

Writing an acknowledgment is the opportunity to express one's gratitude to one's family, friends, colleagues, and well-wishers for extending their never-ending cooperation and support during the journey of any creation or achievement…in this case it's a book

However, this time is a different time. Unfortunately, this book is being published in midst of the COVID-19 pandemic, which is not only a challenge for the whole of humankind, but also a threat for the coming generations for years to come…therefore, first of all…my sincere and heartfelt condolences to the people who died struggling with this deadly coronavirus. Though they did not win, they are given a tough fight in this battle.

At the same time, I salute the doctors, nurses, paramedical staff, police, administrators, and teachers, who are still fighting this devil to save lives and the livelihoods of billions of people around the world, and proving themselves true warriors in the real sense. God bless them.

Now, I want to express special thanks to all the contributors to this book for contributing their work, without which this book would not have been possible. My sincere thanks.

I would like to thank my wife Dr. Pratima Sharma for standing beside me throughout my career. She has been my strength and motivation for continuing to improve my knowledge and move my career forward. She deserves the highest appreciation for being beside me all the way. Her love and sacrifices are endless…I am truly blessed to have a companion like her. Thank you, Pratima.

I also thank my son Kunal and daughter Kritika for always making me smile. I know they are mature now, studying at a university, pursuing computer engineering, and understanding why I spent so much time in front of my computer. They are behind all the reasons for my smile every time. Love you, kids.

I would like to thank the team at Elsevier for providing me with all possible help, assistance, and cooperation during the journey of this book. Working with them is just like working with a family. I am happy to have such a fantastic team. Love you, guys!

I thank all my friends, colleagues, and every person at the JECRC University for all their encouragement, cooperation, and support. It is really a beautiful group of people…I am lucky to have the opportunity of working with them. Thanks a lot, JECRC.

Last, but not least, I am thankful to my students and readers because they are the actual strength behind every teacher, every author, and every book. I am sure they will find this book useful.

May God prevail the humanity and humankind on this planet.

Jai Gurudev!

Sanjay K. Sharma, FRSC
May 13, 2020
Jaipur, India

Chapter 1

Green chemistry and water remediation: bibliometric study and research applications

Hasan Demir[a], **Sanjay K. Sharma**[b]
[a]*Osmaniye Korkut Ata University, Osmaniye, Turkey;* [b]*Green Chemistry and Sustainability Research Group, Department of Chemistry, JECRC University, Jaipur, Rajasthan, India*

1. Introduction

The World Health Organization (WHO) states that the daily water requirement per person is between 50 and 100 L for basic needs and a few health concerns. Humans regardless of their status and societal conditions have the right to have a sufficient and safe quantity of water for their basic needs. 13.6% of the world population cannot access safe drinking water (Ki-moon, 2015; Maestu, 2015) and 2.6 billion people are short of basic sanitation. Before the 21st century, the world's total consumption of water was estimated as 600 km^3/year. In the first decade of the 21st century, global water consumption jumped to 5300 km^3/year. Future estimation of the world's population sees it stabilizing after 2060 with approximately 4 billion people living in urban areas indicating an associated high-water consumption demand (Jain and Singh, 2003). 1.8 billion people will face water scarcity and 66% of the population could be living with the threat of water scarcity by 2025 (Proverb, 2011).

Many industries cause water pollution above the permitted limits. Water resources are threatened by industrial and domestic waste, diffuse pollution from livestock and agriculture farming, nonsustainable use of water infrastructure, etc. (Helmer et al., 1997). The management of water quality, involving raw water catchment management, treatment technology, and maintenance of the distribution network, has become ever-more important in recent years. Management, new technologies, and methodologies are important for providing efficient and sustainable water quality. The data indicate the importance of water remediation using environmentally friendly technologies

Green Chemistry and Water Remediation: Research and Applications.
https://doi.org/10.1016/B978-0-12-817742-6.00001-3

and methodologies. These improvements should provide recycling and reduction of chemicals and materials, waste minimization, and less energy consumption (Ghernaout et al., 2011). These attitudes correspond with the concept of green chemistry. The concept of green chemistry can be defined by 12 principles, which are preventing waste generation, increasing conversion of reactions, designing less hazardous chemical syntheses, designing safer chemicals and products, using safer solvents and reaction conditions, increasing energy efficiency, using renewable feedstocks, avoiding chemical derivatives, using catalysts, designing chemicals and products that degrade after use, analyzing in real-time to prevent pollution, and minimizing the potential for accidents (Anastas and Eghbali, 2010).

This chapter comprehends bibliometric features and characteristics of research papers related to water remediation. The Web of Science was surveyed for research articles about the topic of water remediation indexing in SCI, SCI-expanded, and emerging-SCI between 1999 and 2018. The applied query was: (TS = water remediation) AND DOCUMENT TYPES: (Article) AND PUBYEAR > 1998 AND PUBYEAR < 2019. The search was made on the July 12, 2019. 14,125 papers were found from this search. They were examined in MS Excel and the number of research papers was reduced to 11,224 by searching for the word "water" in the abstracts of research papers. The bibliometric analysis of 11,224 research papers was processed according to year-wise and geographical-wise distribution, authorship pattern and citation, etc. All these data were organized, analyzed, tabulated, and presented by using simple statistical methods in MS Excel, BibExcel, and Pajek software (Peset and Gonzalez, 2012). Research topics such as sorbents in sorption techniques, extraction, photocatalytic degradation, sedimentation, microorganism, membranes, and immobilization are explained in this chapter.

1.1 Scientific categories of research papers

The scientific categories and interaction between these categories are illustrated in Fig. 1.1. The water remediation topic was naturally placed in the environmental sciences and ecology category. However, engineering with 4661 mentions and chemistry with 2621 mentions indicated that researchers were investigating the solution of remediation of water quality under engineering and chemistry science as shown in Fig. 1.1A. The other water remediation research papers were studied under material science (1010), physics (406), microbiology (173), electrochemistry (155), and polymer science (144) categories. Fig. 1.1B indicates the interaction of scientific categories with each other. The broad concept of scientific categories such as environmental sciences, ecology, and water resources plays a bridging role in the interactions of categories.

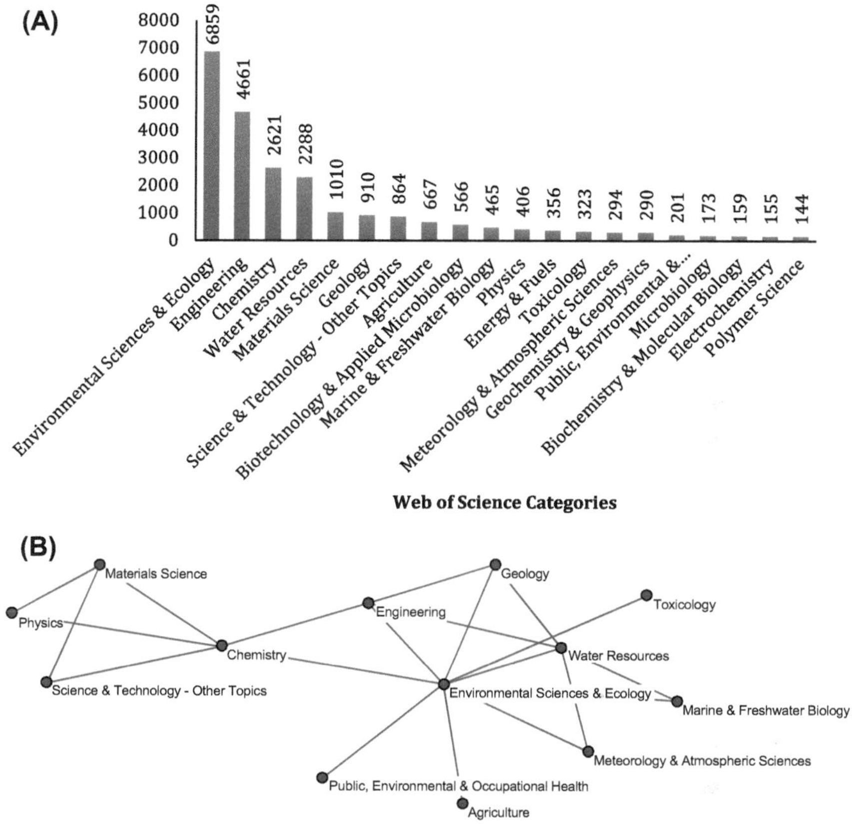

FIGURE 1.1 (A) Top 20 Web of Science categories of published papers and (B) interaction with each other.

1.2 Year-wise distributions

Fig. 1.2 presents the distribution of 11,224 published research papers according to years. In 1999, a minimum of 172 research papers were published. After 2004, the number of published research papers started to increase with increasing years. A pie chart indicates that 70% of these research papers were published after 2010. This phenomenon is because water remediation had become an important issue in the world and gained the attention of researchers.

1.3 Country-wise distribution of contributions

Table 1.1 gives the number of contributions made by 111 countries. Fig. 1.3 illustrates the distribution of contributions made by countries on the world heat map. The distribution of contributions is presented as green (dark gray in printed version) to red colors [grey in printed version] indicating low to high number

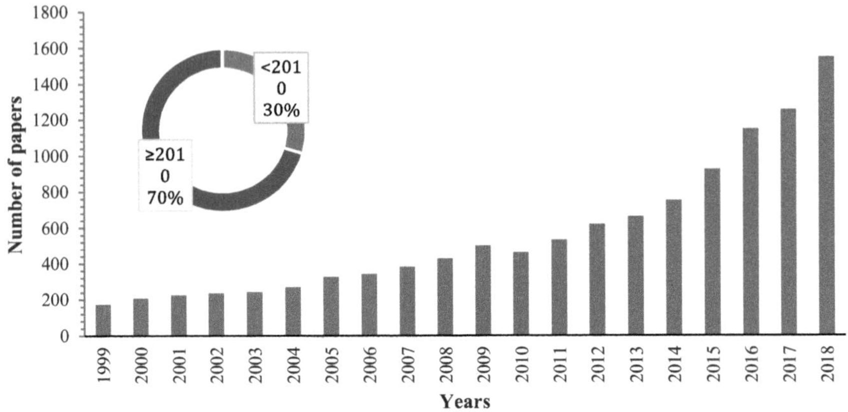

FIGURE 1.2 Year-wise distribution of the number of articles.

TABLE 1.1 Country-wise distribution of contributions.

No.	Country	Contribution
1	Algeria	26
2	Argentina	98
3	Australia	531
4	Austria	46
5	Bangladesh	31
6	Belarus	1
7	Belgium	115
8	Bolivia	5
9	Bosnia	3
10	Brazil	248
11	Brunei	1
12	Bulgaria	10
13	Burkina Faso	2
14	Cameroon	1
15	Canada	686
16	Chad	9
17	Chile	63
18	China	2280
19	Colombia	26

TABLE 1.1 Country-wise distribution of contributions.—cont'd

No.	Country	Contribution
20	Costa Rica	1
21	Croatia	15
22	Cuba	11
23	Cyprus	11
24	Denmark	120
25	Ecuador	6
26	Egypt	108
27	Emirates	24
28	England	467
29	Estonia	10
30	Ethiopia	1
31	Finland	69
32	France	415
33	Georgia	73
34	Germany	474
35	Ghana	9
36	Greece	123
37	Hong Kong	77
38	Hungary	38
39	Iceland	2
40	India	674
41	Indonesia	15
42	Iran	186
43	Iraq	16
44	Ireland	70
45	Israel	55
46	Italy	452
47	Japan	264
48	Jordan	22

Continued

TABLE 1.1 Country-wise distribution of contributions.—cont'd

No.	Country	Contribution
49	Kazakhstan	6
50	Kenya	6
51	Kosovo	1
52	Kuwait	10
53	Laos	2
54	Latvia	6
55	Lebanon	18
56	Libya	2
57	Lithuania	6
58	Macedonia	1
59	Malaysia	94
60	Malta	1
61	Mexico	166
62	Moldova	1
63	Monaco	1
64	Montenegro	2
65	Morocco	16
66	Namibia	1
67	Netherlands	102
68	New Zealand	65
69	Nicaragua	1
70	Nigeria	55
71	Norway	75
72	Oman	13
73	Pakistan	129
74	Peru	23
75	Philippines	10
76	Poland	124
77	Portugal	171

TABLE 1.1 Country-wise distribution of contributions.—cont'd

No.	Country	Contribution
78	Qatar	16
79	Romania	55
80	Russia	77
81	Rwanda	0
82	Saudi Arabia	117
83	Scotland	102
84	Senegal	1
85	Serbia	49
86	Sierra Leone	1
87	Singapore	61
88	Slovakia	27
89	Slovenia	35
90	South Africa	132
91	South Korea	433
92	Spain	462
93	Sri Lanka	15
94	Sweden	158
95	Switzerland	96
96	Syria	4
97	Taiwan	211
98	Tanzania	1
99	Thailand	73
100	Togo	2
101	Trinidad and Tobago	1
102	Tunisia	25
103	Turkey	157
104	United States	3134
105	Uganda	2
106	Ukraine	18

Continued

TABLE 1.1 Country-wise distribution of contributions.—cont'd

No.	Country	Contribution
107	Uruguay	3
108	Venezuela	10
109	Vietnam	14
110	Wales	106
111	Zimbabwe	8

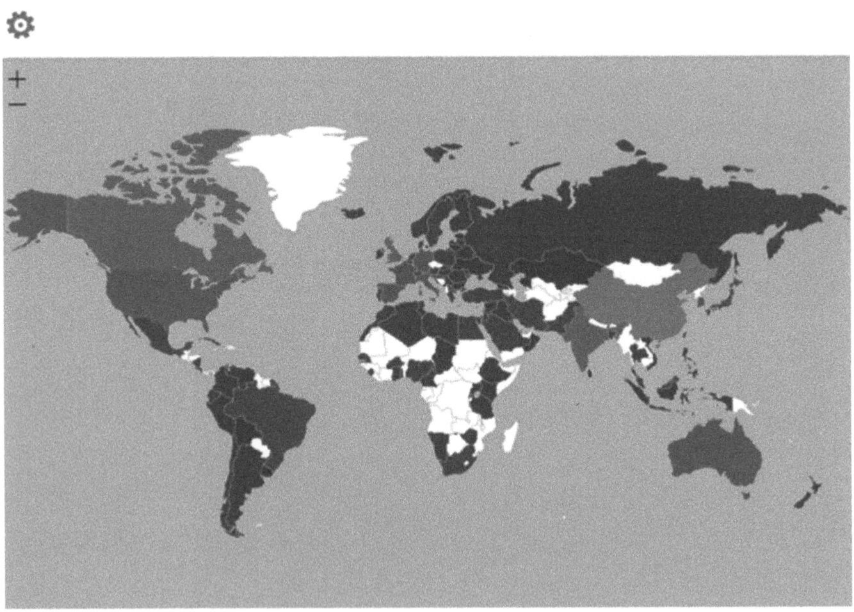

FIGURE 1.3 Geographical-wise distribution of contributions on the Earth map.

of contributions. As is seen in Fig. 1.3, 57% of countries of the world gave a contribution to the water remediation topic. This phenomenon also indicated that water remediation concerns all countries in the world and the countries have shown necessary attention to this topic. The maximum contributions were made by the United States (3134) and China (2280). Canada, India, Australia, and Germany made contributions of 686, 674, 531, and 474, respectively.

1.4 Authorship pattern

Authorship patterns and the collaboration of authors are illustrated in Fig. 1.4. 11,224 research papers were studied and/or published by 49,913 authors. This number includes repetition by the authors. The average number of authors per paper was calculated as 4.4. Solely, 2.9% of the articles were studied and published with a single author. This situation indicated that the importance of collaboration and the water remediation topic require cooperative study. 14.2%

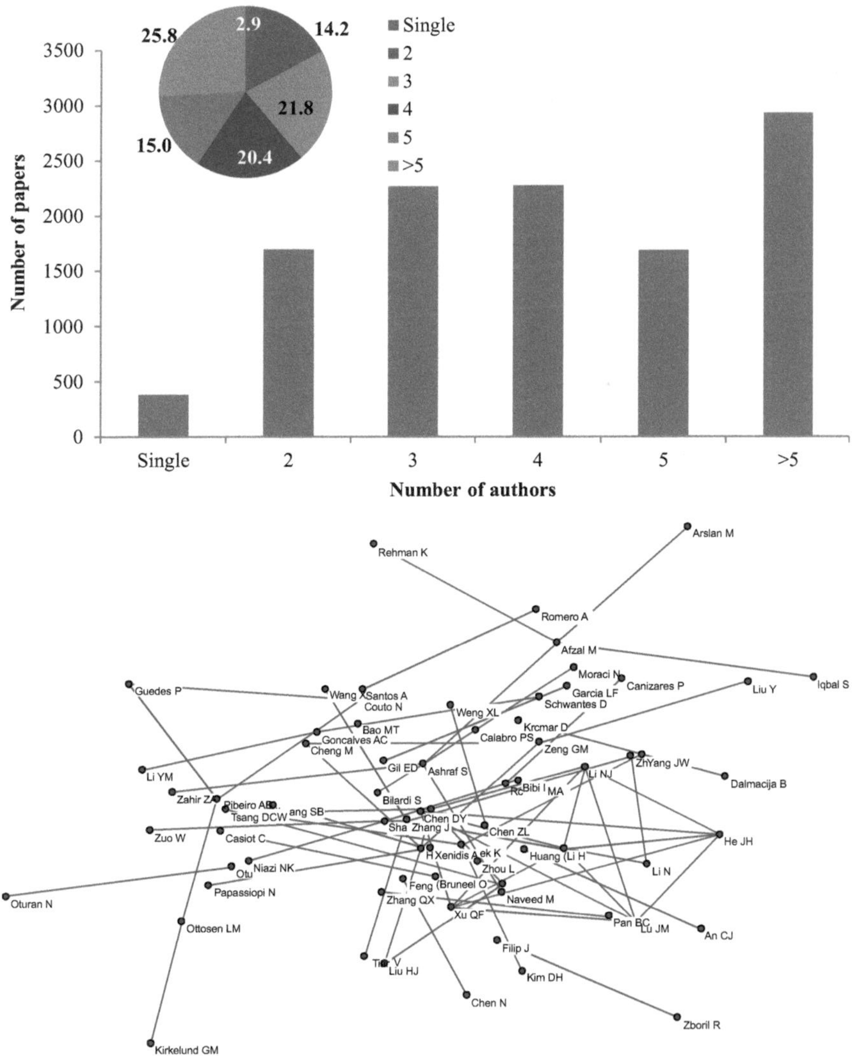

FIGURE 1.4 Authorship pattern of papers.

of published papers were studied by two authors and 2266 papers (21.8%), 2273 papers (20.4%), and 2928 papers (25.8%) were published by three, four, and more than five authors, respectively. Fig. 1.4B also presents the Pajek diagram of relations of the authors who published at least five papers together. The authors Filip J. with Zboril R. and Goncalves A.C. with Schwantes D. published seven papers with cooperation.

Table 1.2 lists the degree of collaboration. The degree of collaboration was evaluated by following Eq. (1.1) (Subramanyam, 1983).

$$C = \frac{N_m}{N_m + N_s} \tag{1.1}$$

where C denotes a degree of collaboration and N_m indicates the number of multiple authors. The number of single authors is denoted with N_s. The degree

TABLE 1.2 Degree of collaboration.

Year	Single author	Multiple authors	C
2018	35	1514	0.98
2017	29	1228	0.98
2016	27	1120	0.98
2015	30	894	0.97
2014	16	735	0.98
2013	16	647	0.98
2012	13	607	0.98
2011	18	513	0.97
2010	14	447	0.97
2009	20	479	0.96
2008	20	406	0.95
2007	20	361	0.95
2006	21	320	0.94
2005	10	316	0.97
2004	19	249	0.93
2003	15	227	0.94
2002	16	219	0.93
2001	14	211	0.94
2000	16	190	0.92
1999	10	162	0.94

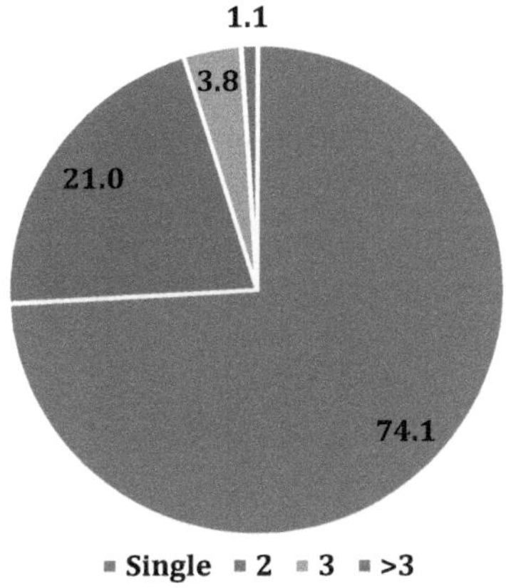

FIGURE 1.5 Number of international collaboration-wise contributions.

of collaboration ranges from 0.92 to 0.98. The average degree of collaboration was calculated as 0.96. After 2011, the average degree of collaboration almost reached equilibrium (0.98) value.

Fig. 1.5 shows percentages of an international collaboration of published papers. 25.9% of all published papers were studied and written by international collaboration and 74.1% of research papers were studied/published in a single country.

Year-wise single, collaboration, and percentage of international collaboration are presented in Fig. 1.6. International collaboration, increases with the progression of time. The maximum percentage of international collaborations was observed as 31.1% in 2018. In 2018, 482 papers were published with international collaborations.

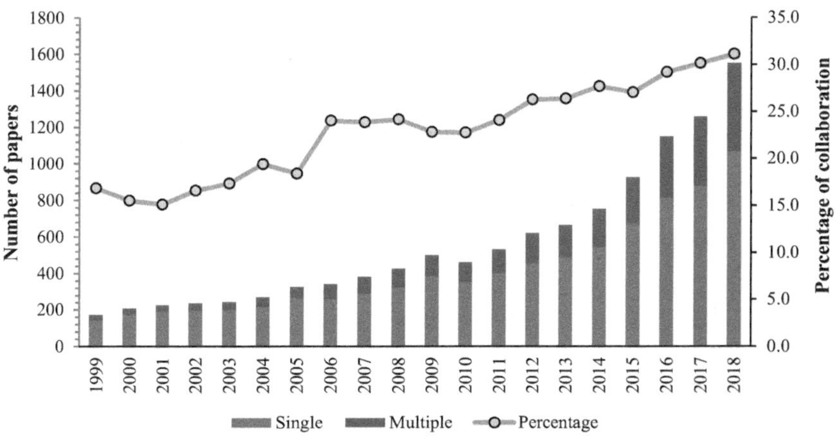

FIGURE 1.6 Year-wise single, collaboration, and percentage of international collaboration.

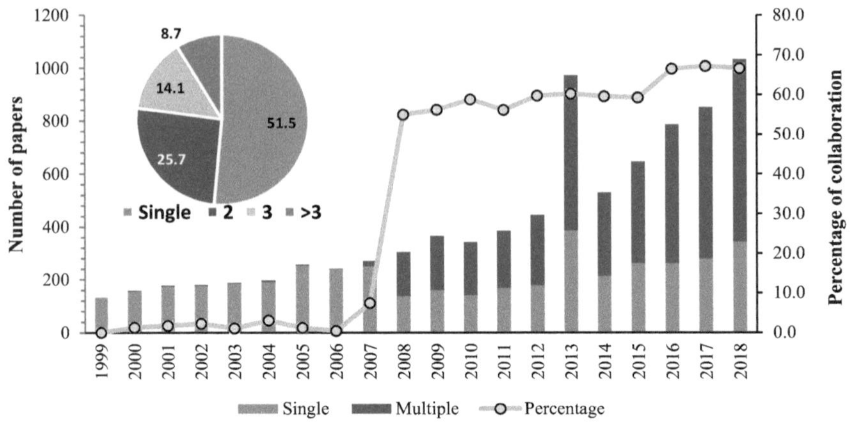

FIGURE 1.7 Year-wise single, collaboration, and percentage of national collaboration.

The research papers published from a single country were reinvestigated by removing single-author publications. Fig. 1.7 reveals national collaboration information. 8456 of 11,224 papers were studied that had a single country and more than one author. In Fig. 1.7, 51.5% of papers were published from a single address. In other words, the papers were studied and/or published by more than one author but the affiliation of papers had only one address. This indicates that the papers may have been studied with professor/student collaboration, which is a very common collaboration in an academy, and/or a collaboration of colleagues who are in the same department and/or institute. The rest of the papers were studied with interdisciplinary cooperation since the affiliation of 48.5% of papers involved more than one address. This result indicated that more than one department/institute or both together made a collaboration for research activities. After 2006, interdisciplinary studies started and rapidly became widespread. In the last 3 years, the percentage of interdisciplinary collaboration reached 66.6%.

Table 1.3 shows the written languages of papers. All research papers were written in 19 languages: 11,063 papers were written in English and only 1.4% of papers were written in other languages. The increment of international collaboration with respect to years influenced the languages of published papers.

1.5 The frequency of references, paper length, and citations

A histogram of the number of references in each paper is shown in Fig. 1.8. The research papers were written using approximately 41 references on average. Distribution in the number of references is a left-skewed distribution, which means most of the papers were written under the average of references: 58.1% of all research papers were written using equals and fewer

TABLE 1.3 The written language of papers.

Languages	Number of papers
English	11,063
Chinese	22
Portuguese	27
Japanese	3
French	9
German	19
Spanish	40
Polish	15
Czech	6
Russian	3
Croatian	2
Hungarian	1
Italian	3
Korean	4
Serbo-Croatian	3
Turkish	1
Serbian	1
Estonian	1
Ukrainian	1

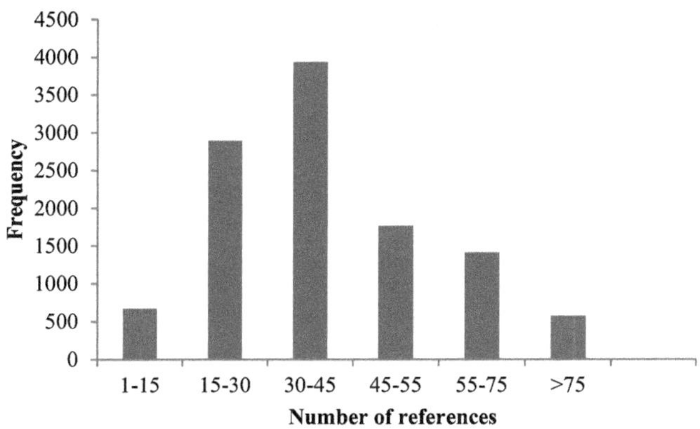

FIGURE 1.8 Histogram of the number of references in each paper.

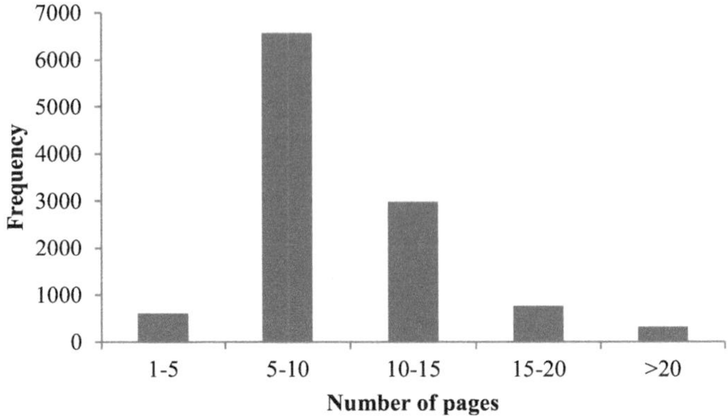

FIGURE 1.9 Frequency of the number of pages in each paper.

than 41 cited references and 35% of papers (3933) were written using 30–45 cited references.

Fig. 1.9 shows the distribution of the number of pages in each research paper. 58.5% of research article length (6569 papers) varied from 5 to 10 pages and 2977 research papers had a length between 10 and 15 pages. The length of 11,224 research papers was 114,866 pages. The average length of research papers was 10 pages and 63.9% of research papers were written under and equal to 10 pages.

The distribution of frequency in the number of citations is presented in Fig. 1.10. The total number of citations from 1999 to 2018 was 238,836. An average number of citations was evaluated as 11,942 per year. The average

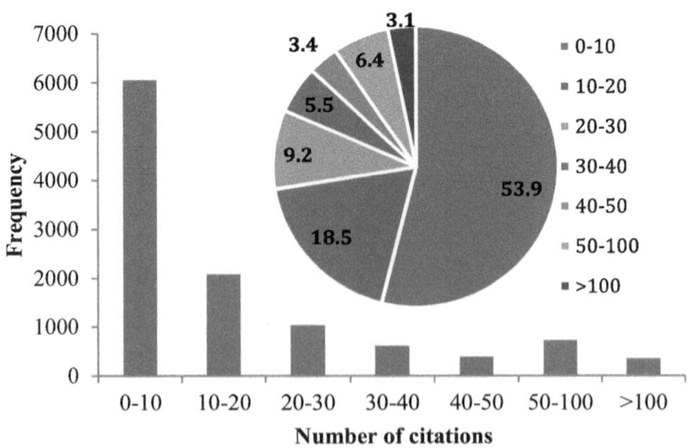

FIGURE 1.10 The number of papers against the number of citations distribution.

number of citations per paper was calculated as 21. It should be kept in mind that the number of citations can be described as live data, which means that these data will continue to increase. 53.9% of papers (6050) were cited less than 10 times and 1072 research papers (9.5%) were cited as equal to and higher than 50 times. High citation quantity can reflect the quality of research.

1.6 Top publishers and journals

Fig. 1.11 illustrates the top publishers that published papers related to green chemistry. The number of published papers from each publisher as shown in Fig. 1.11 was 68.1% of the total number of published papers. In other words, every two or three papers were published by these publishers. 3983 research papers were published by Elsevier and 1502 and 610 research papers were published by Springer and Wiley, respectively.

Fig. 1.12 presents the top 20 journals that published papers related to the topic of water remediation. 11,224 research papers were published in approximately 800 different journals and 34.8% of research papers were published in analyzed journals as shown in Fig. 1.12. The majority of research papers (468) were published in the Journal of Hazardous Materials and 412 and 407 papers were issued by Environmental Science & Technology and Chemosphere, respectively.

1.7 Research funding and supporters

Fig. 1.13 shows the year-wise funded and nonfunded number of papers and the percentage of funding: 6784 research papers were financially supported over

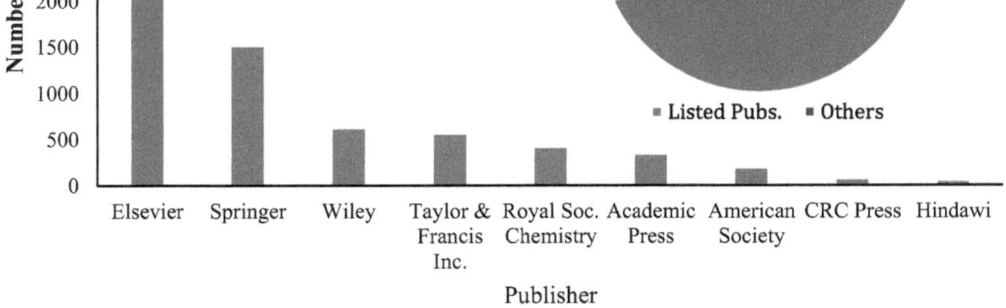

FIGURE 1.11 Number of papers published by publisher.

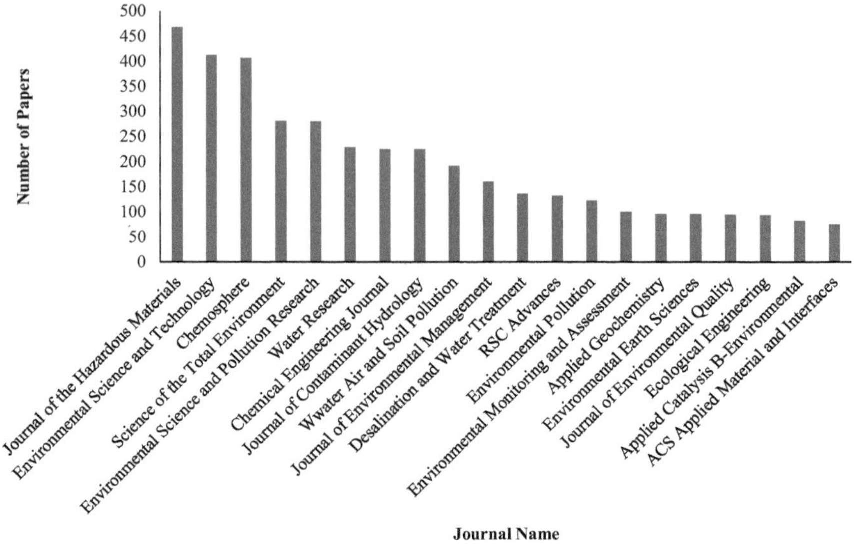

FIGURE 1.12 Top 20 journals.

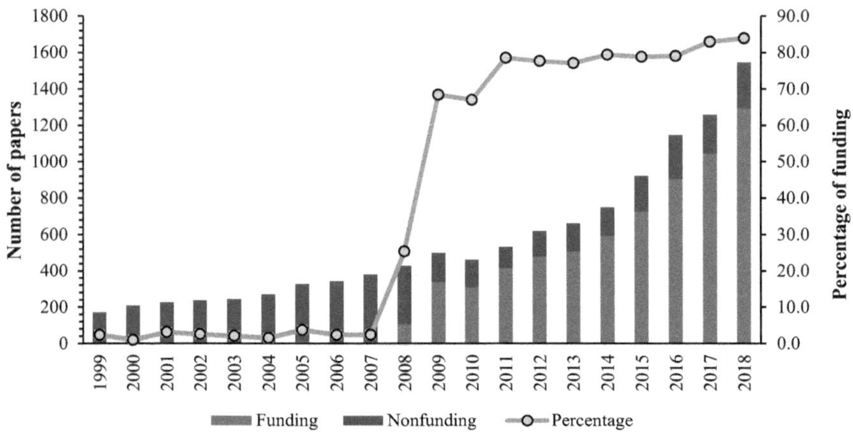

FIGURE 1.13 Year-wise funded and non-funded number of papers and percentage of funding.

the years. The percentage of funded papers jumped from 25.4% to 68.4% after 2008. The maximum percentage of funded papers was observed at 83.9% in 2018. The increment of the funded project may increase national and international collaborations and the number of published papers. Nowadays, an increase in analytical equipment cost, prices of consumables, and labor costs reveal the importance of project grants.

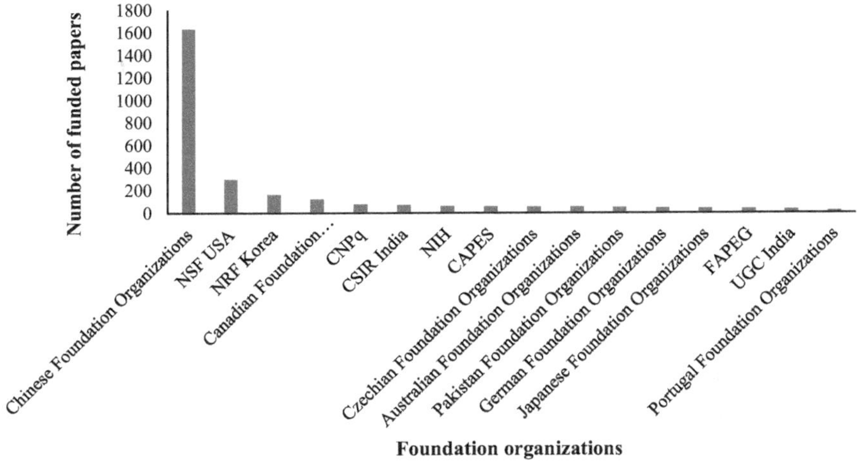

FIGURE 1.14 Foundation organizations.

Project grants were supplied by national and international organizations, some of which are shown in Fig. 1.14. 25% of papers were supported by listed funding organizations as shown in Fig. 1.14. The National Natural Science Foundation of China was acknowledged 1630 times in research papers and 297, 162, and 123 research papers thanked the National Science Foundation, NSF (USA), NRF of Korea, and the Canadian Foundation of organizations.

1.8 Water remediation techniques

Waste, drinking, and groundwater can be treated by various physical and chemical water methods such as adsorption, absorption, coagulation, flocculation, sedimentation, filtration, disinfection, UV treatment, photochemical/electrochemical processes, extraction, ion exchange, microorganism treatment, membrane technology, enzymes, etc. Nowadays, most of the physical and chemical water treatments are approached using green chemistry principles. The definitions and details of physical and chemical treatments are given in the following chapters. In this chapter, 11,224 research papers were examined according to the types of physical and chemical water treatments. Fig. 1.15 illustrates the frequency of water treatment methods studied in 11,224 research papers. This figure was achieved by evaluating the data of the papers' keywords. The adsorption process was mentioned in the keywords 1651 times, which is the maximum value. The sorption process consists of both adsorption and absorption processes and for that reason the amount of sorption was not added to the number of adsorption or absorption terms. The oxidation process was investigated in 944 research papers. Photocatalytic degradation and extraction processes were researched in 466 and 575 papers, respectively. Microorganisms were studied in 163 research papers. Visible-light techniques

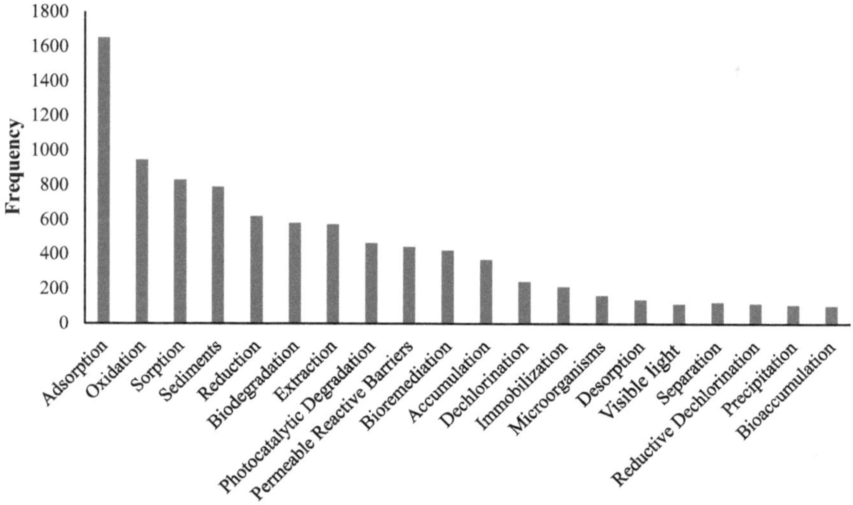

FIGURE 1.15 Frequency of water treatment methods.

were studied in 115 papers. In 422 research papers, water was treated with bioremediation techniques.

Fig. 1.16 presents relations and interactions between words placed in the titles of 11,224 research papers. Dark lines indicate a high number of interactions of words. In other words, dark lines show that the word couples have a high number of citations in the title of research papers. The word water interacts with quality, drinking, soil, remediation, contaminated, and treatment as seen in Fig. 1.16. Fig. 1.16 gives information about techniques and materials used during water remediation such as zero-valent iron, visible light,

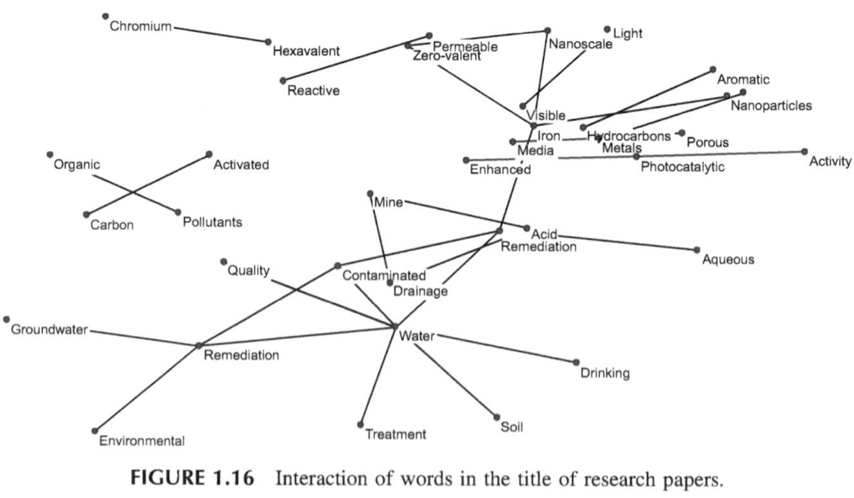

FIGURE 1.16 Interaction of words in the title of research papers.

permeable membrane, nanoparticles/nanotechnology, activated carbon, and photocatalytic treatment.

1.8.1 Sorbents in sorption techniques

The biosorption process is a new technology aimed at recovering valuable resources and removing various toxic/nontoxic pollutants from aqueous media. The biosorption process can be applied for removing heavy metals, pesticides, dyes, pharmaceuticals, volatile organic compounds, aromatic hydrocarbons, nutrients, trace metals, organics, etc. (Asere et al., 2019). Mechanisms and diversity of biosorption techniques are explained in the following chapters. The advantages of biosorption techniques are:

- Low cost
- High efficiency
- Minimal chemical and/or biological waste
- Ability to regenerate biosorbents
- The ability for removing metals as conventional sorbents (Park et al., 2010).

Biosorbents are one of the main components used in biosorption processes. Table 1.4 presents the classification and types of biosorbents. The microorganisms, biopolymers, minerals, and sorbents obtained from waste materials are: and soil/rocks can be classified under biosorbents. Investigation and production of new biosorbents are currently being studied with enthusiasm by researchers. For that reason, the sorbents listed in Table 1.4 can be expanded with new materials. Listed materials were found by surveying 11,224 research and water remediation review papers (Asere et al., 2019; Das and Charumathi, 2012; Fazal et al., 2018; Gaur et al., 2018; Wang and Wang, 2019; Park et al., 2010; Dasgupta et al., 2017; Premarathna et al., 2019; Salama et al., 2019; Varghese et al., 2019).

Fig. 1.17 reveals the frequency of sorbents that were studied in 11,224 research papers. In 8574 research papers, various sorbents were studied, which means that one of the biosorption techniques was applied to these research papers. A zero-valent iron was the most-studied sorbent in 919 research papers among those analyzed. Titanium dioxide, activated carbon, and organic matter were studied in 500, 483, and 468 research papers.

1.8.2 Extraction

Fig. 1.18A illustrates the frequency of materials removed by the extraction method in research papers. Fig. 1.18A presents a general classification of materials such as metals, heavy metals, acids, etc. and compound names such

TABLE 1.4 Types of biosorbents.

Classification name	General sorbents name	Sorbents
Biopolymers	Cellulose	
	Alginate	Brown algae, *Azotobacter vinelandii* and several *Pseudomonas* species
	Chitosan/chitin	
	Ligands	
Minerals	Clays (amorphous form)	Aluminum silicates calcined or natural; montmorillonite, Fuller's earth, smectite, bentonite
	Iron	Goethite (α-FeOOH), hematite (α-Fe_2O_3),magnetite (Fe_3O_4), ferrihydrite ($Fe_5HO_8 \cdot 4H_2O$), maghemite (γ-Fe_2O_3), lepidocrocite
	Hydroxylapatite and struvite	
	Zeolite	
	Metal–organic frameworks	
	Activated alumina	
	Active carbon	
Waste materials/by-products	Sludge	
	Ash and coals	Pyrite ash, fly ash, biochar (pyrolysis of carbonaceous materials)
	Red mud	Bauxite residue
	Agricultural and local waste	Wheat straw, tea wastes, sugarcane bagasse, sawdust, orange waste, coconut fiber, luffa fiber, rice husk, straw
	Biochar	
	Industrial waste	Fermentation wastes, food/beverage wastes, activated sludges, anaerobic sludges, etc.

TABLE 1.4 Types of biosorbents.—cont'd

Classification name	General sorbents name	Sorbents
Microorganisms	Algae	Green algae (*Maugeotia genuflexa*), *Chlorella vulgaris, Chlorella pyrenoidosa, Oscillatoria tenuisin, Chlorella alga, Scenedesmus* sp., *Dunaliella salina, Spirulina maxima, Rhodococcus* genus, *Caepidium antarcticum, Desmarestia* sp., *Focus* sp., *Ascophyllum nodosum*
	Fungi (yeast, molds, mushrooms)	*Lessonia nigrescens*, fungus *Penicillium purpurogenum, Aspergillus niger Phanerochaete chrysosporium, Cunninghamella* sp., *Candida tropicalis, Candida albicans, Candida krusei, Debaryomyces polymorphus, Saccharomyces cerevisiae, Candida* sp., *Trichosporon beigelii, Schizosaccharomyces pombe, Kluyveromyces marxianus, Candida lipolytica, Candida utilis,* etc.
	Bacteria	*Brevibacillus* sp., *Klebsiella pneumoniae, Acetobacter liquefaciens, Pseudomonas desmolyticum, Bacillus* sp., *Zoogloea* sp., *Flavobacterium*
Soil and rocks	Rocks	Pumice
		Red scoria
	Soil	Termite mound
		Laterite soil
		Red soil
		Sewage irrigated soil
Photocatalysis	Nanoparticles	Ga-doped ZnO nanopagodas, ZnO@PbS/graphene oxide, Ag/$AgVO_3$ one-dimensional hybrid nanoribbons, TiO_2 nanoparticles

as iron, lead, cadmium, citric acid, etc. Extraction is a technique that is used for the purification, fractionation, and/or removal of products from a matrix. Liquid–liquid, ion–pair, gas–solid, supercritical fluid, and liquid–solid extraction can be regarded as types of extraction methods based on types of

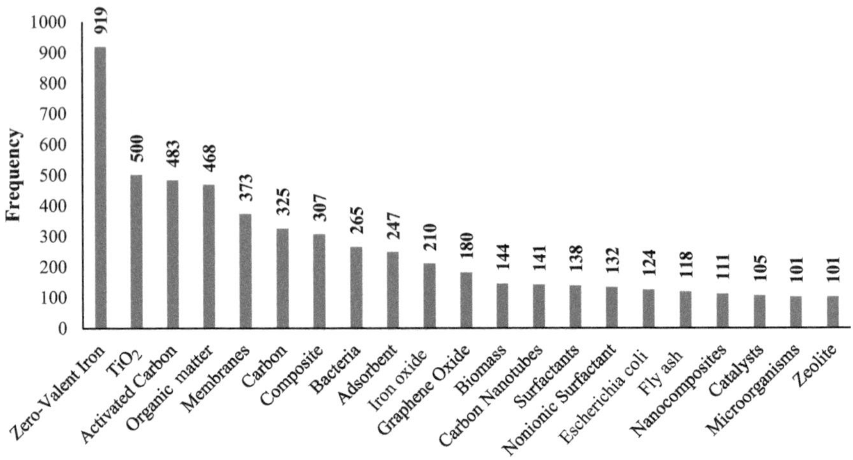

FIGURE 1.17 Most-studied sorbents in analyzed research papers.

phases (Cooke et al., 2000). The extraction methods can also be classified according to types of processes (continuous or batch), and types of assisted techniques (microwave and ultrasound) (Cooke et al., 2000; Tiwari, 2015). In analyzed research papers, lead (74 contributions), cadmium (54), copper (48), zinc (42), and chromium (18) are the most studied toxic metals by removing extraction methods. Removing phenanthrene and polychlorinated biphenyls is studied in 18 and 9 research papers, respectively.

Fig. 1.18B also shows the distribution of a number of papers studying extraction methods and international collaboration percentages. The removal of toxic materials from their matrix with extraction methods gradually increases with increasing years. In 2018, 67 research papers were relevant to extraction methods; 75.1% of research papers used extraction methods published from a single country address; and 24.9% of research papers used extraction methods published from two and more country addresses denoting international collaboration. Low international collaboration percentage may be caused by the advantages of extraction methods, which are easy to apply and low cost.

1.8.3 Photocatalytic degradation

The photocatalytic degradation of pollutants using photocatalysts such as TiO_2, ZnO nanowires, Ag–ZnO, $BaTiO_3$, $KNbO_3$, nanoparticles, $SrBi_4$, Ti_4O_{15}, etc. has recently emerged as a promising water remediation method (Chen et al., 2019; Petronella et al., 2019). The photocatalysts are stimulated

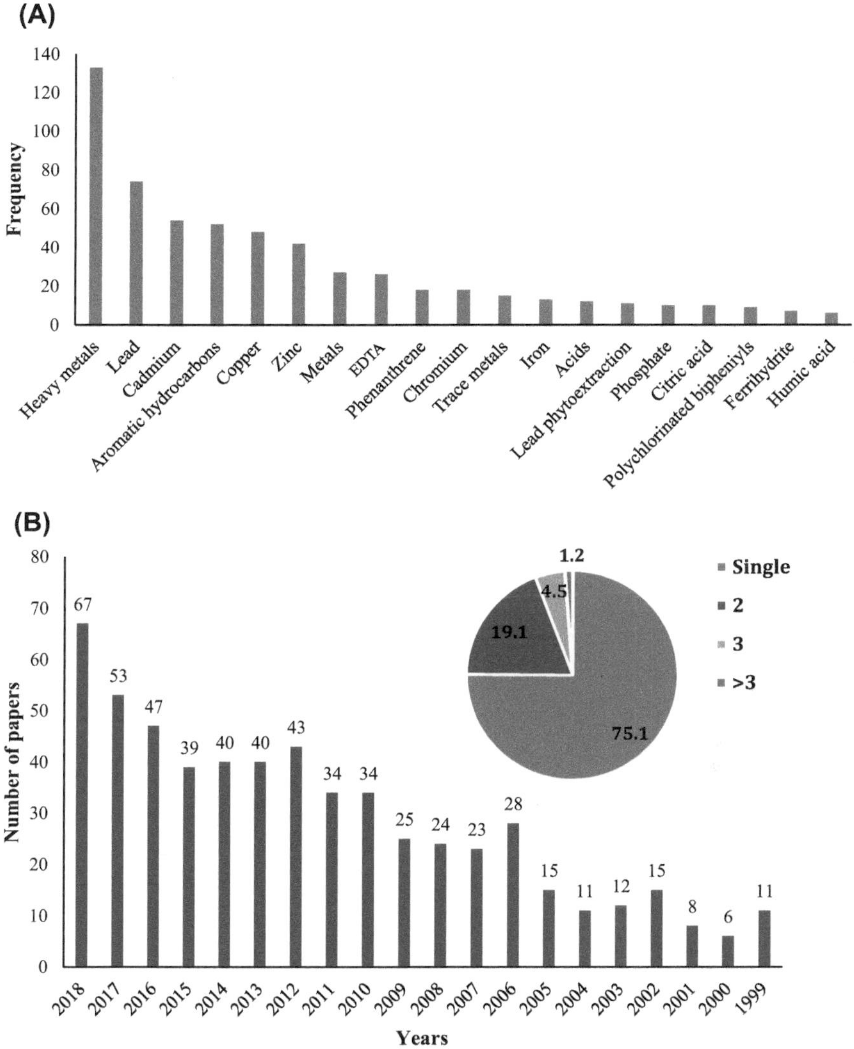

FIGURE 1.18 (A) Frequency of studied materials by extraction techniques and (B) distribution of the number of papers using extraction techniques according to years.

and/or activated with electromagnetic radiation from the sun or artificial light. Irradiation causes the generation of electron–hole pairs (e^-/h^+); h^+ generates •OH radicals onto the surface of photocatalysts and the e^- reacts with $•O_2^-$ to finally produce •OH radicals (Petronella et al., 2019; Tiwari et al., 2019). The

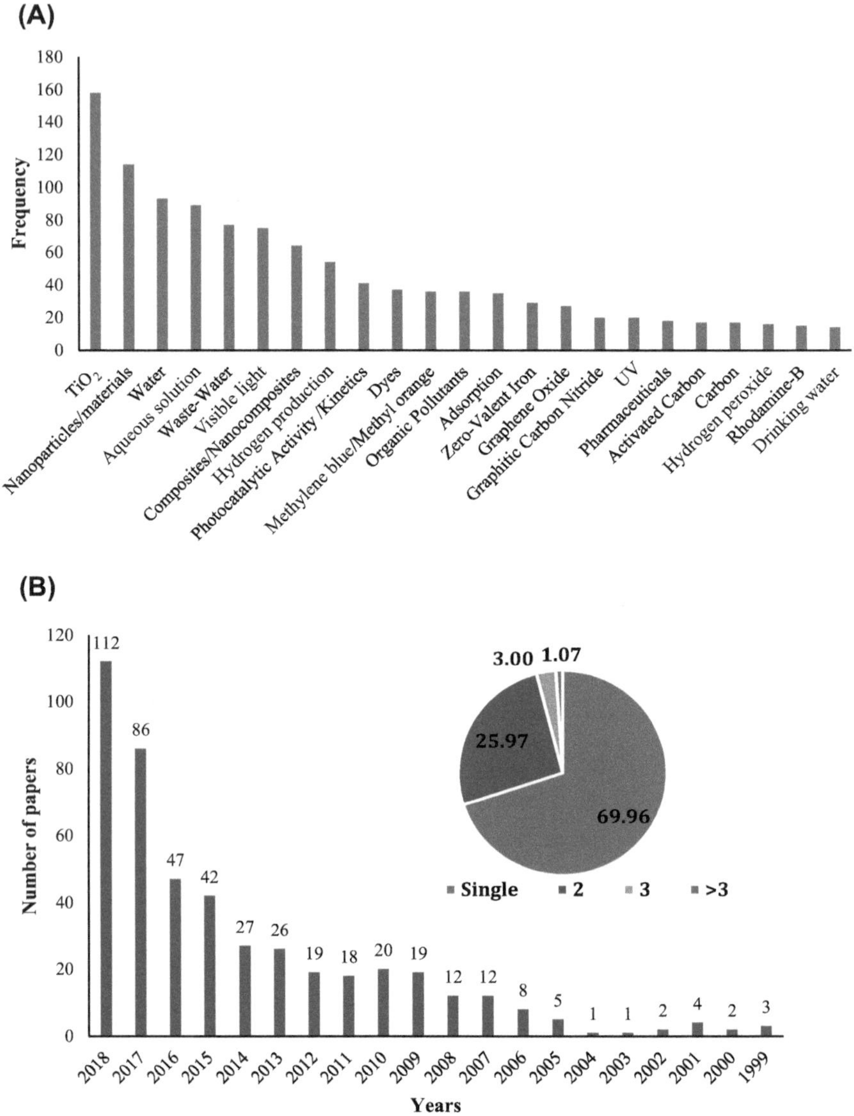

FIGURE 1.19 (A) Frequency of studied materials by photocatalytic degradation. (B) Distribution of the number of papers studying photocatalytic degradation according to years.

efficiency of a photocatalytic treatment depends on the generation of reactive oxygen species and the amount of trapping/sorbing pollutants onto the surface of photocatalysts. In Fig. 1.19A, TiO_2 (158), zero-valent iron (29), graphene oxide (27), and graphitic carbon nitride (20) were used as photocatalysts for remediation of organic effluents. In general, nanoparticles and/or nanomaterials and nanocomposites were also studied as photocatalysts in 114 and 64 research papers, respectively. Photocatalytic treatment is applied to an aqueous solution in 89 papers, wastewater treatment in 77 papers, dye removal in 37 papers, and drinking water treatment in 14 papers.

67.4% of 466 photocatalytic topic research papers were published in the last 5 years. A considerable number of research papers (112) were published in 2018. These data predict that photocatalytic treatment is a popular alternative method to conventional water remediation techniques acknowledged by the scientific community. Approximately 70% of published research papers were addressed from a single country and 30% of research papers were studied with international collaborations.

1.8.4 Flocculation, coagulation, and sedimentation

Flocculation, coagulation, and sedimentation treatments can be used with the help of chemicals to accelerate pollutants, colloid formation, precipitation, and settlement (Homem and Santos, 2011). 789 research papers were related to flocculation, coagulation, and sedimentation treatment. Fig. 1.20A reveals the application area of these treatments and the removed pollutants. Treatments were researched for marine/bay/harbor remediation in 109 papers, soil remediation in 93 papers, and groundwater remediation in 49 papers. The most-studied pollutant removal was polycyclic aromatic hydrocarbons with 107 research papers. Removal of iron, sulfur, polychlorinated biphenyls, cadmium, and lead was researched in 77, 46, 41, 28, and 27 research papers, respectively.

47% of 789 research papers relevant to flocculation/coagulation/sedimentation treatments were published in the last 5 years as shown in Fig. 1.20B. The 71.9% of research papers relevant to flocculation/coagulation/sedimentation treatments were addressed from a single country and 28.1% of 789 research papers were published with international collaboration.

1.8.5 Microorganisms

The scientific community published 163 research papers on water remediation with the topic microorganisms. Microorganisms have been used in recent years for water remediation using fungi, bacteria, and algae (Ejder-Korucu et al., 2015; Gaur et al., 2018 Reeve and Fallowfield, 2018). Fig. 1.21A illustrates the frequency of some keywords that were used to identify these published 163 research papers. The word bacteria was used in 41 research papers for identifying the study. Marine bacteria *Pseudomonas* sp., *Shewanella*, *Escherichia coli*, and other strains were studied in 14, 10, 10, 6, and 6 research papers,

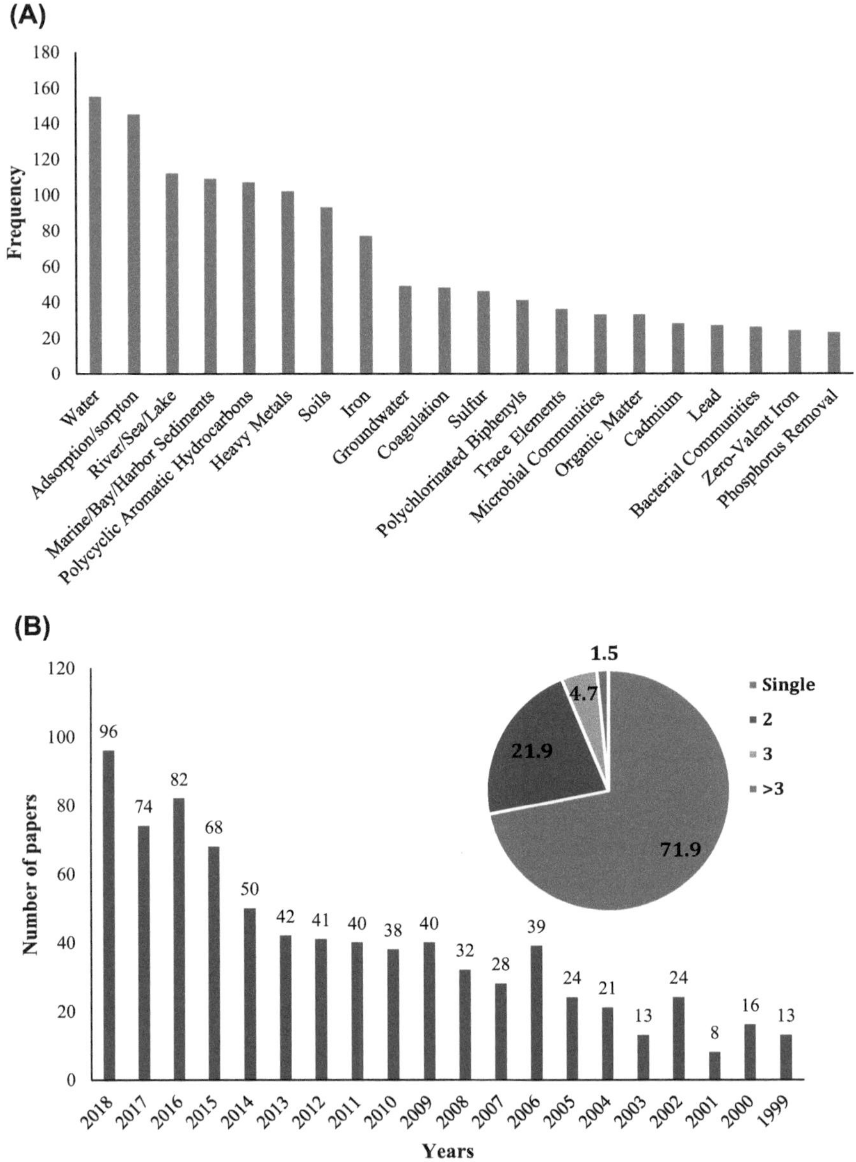

FIGURE 1.20 (A) Frequency of subtopics of flocculation/coagulation/sedimentation treatments and (B) distribution of the number of papers according to years.

respectively. The use of microorganisms in water remediation has several advantages, e.g., low cost and environmentally friendliness completely mineralized end products, etc. (Ejder-Korucu et al., 2015). Hydrocarbons, removal of oil/petroleum effluents, heavy metals, cadmium, copper, and iron

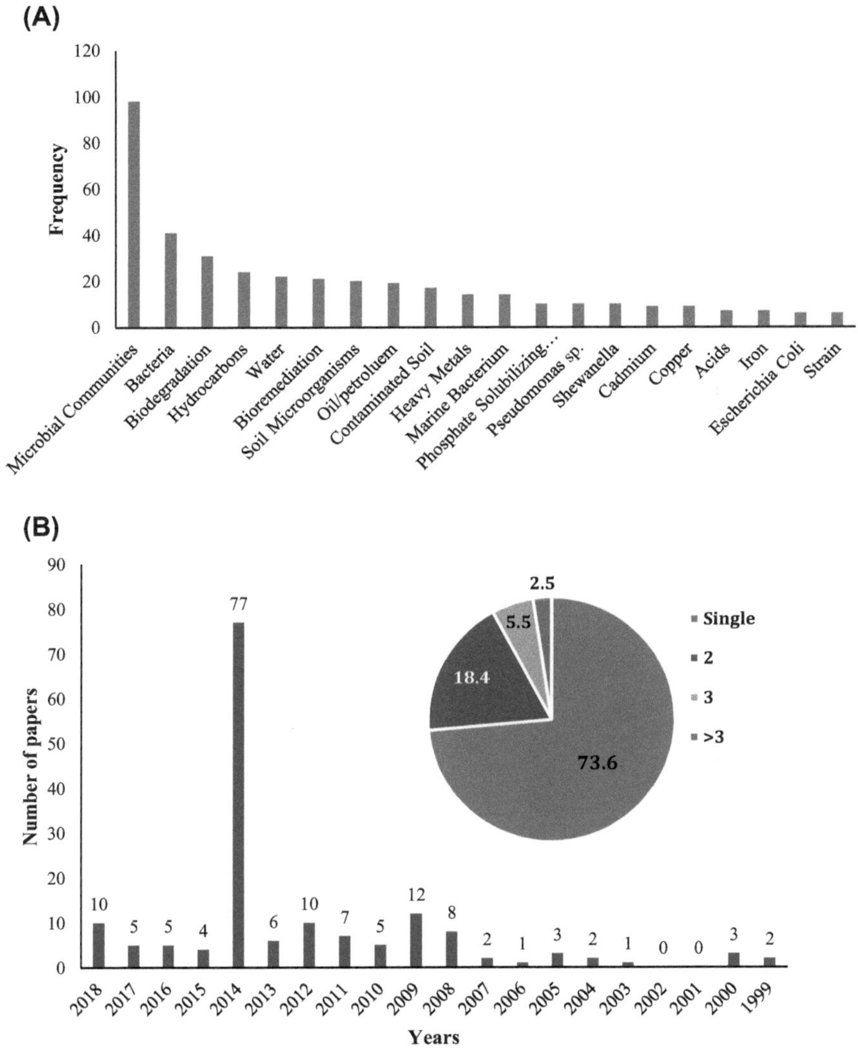

FIGURE 1.21 (A) Frequency of subtopics of bioremediation with microorganism treatments and (B) distribution of the number of papers according to years.

were studied under bioremediation with the topic of microorganisms as shown in Fig. 1.21A.

Fig. 1.21B illustrates the distribution of the number of papers according to years and international collaboration percentage. The maximum number of microorganism research papers published in 2014 was 77. In the last 5 years, 62% of 163 research papers were published and 73.6% of 163 research papers showed single country contributions. The international collaboration percentage was 26.4%.

1.8.6 Permeable reactive barrier/membranes

For approximately 170 years, the scientific community has known that membranes can be separated into two species. However, membranes have some disadvantages: limited life and high cost of replacements. After several

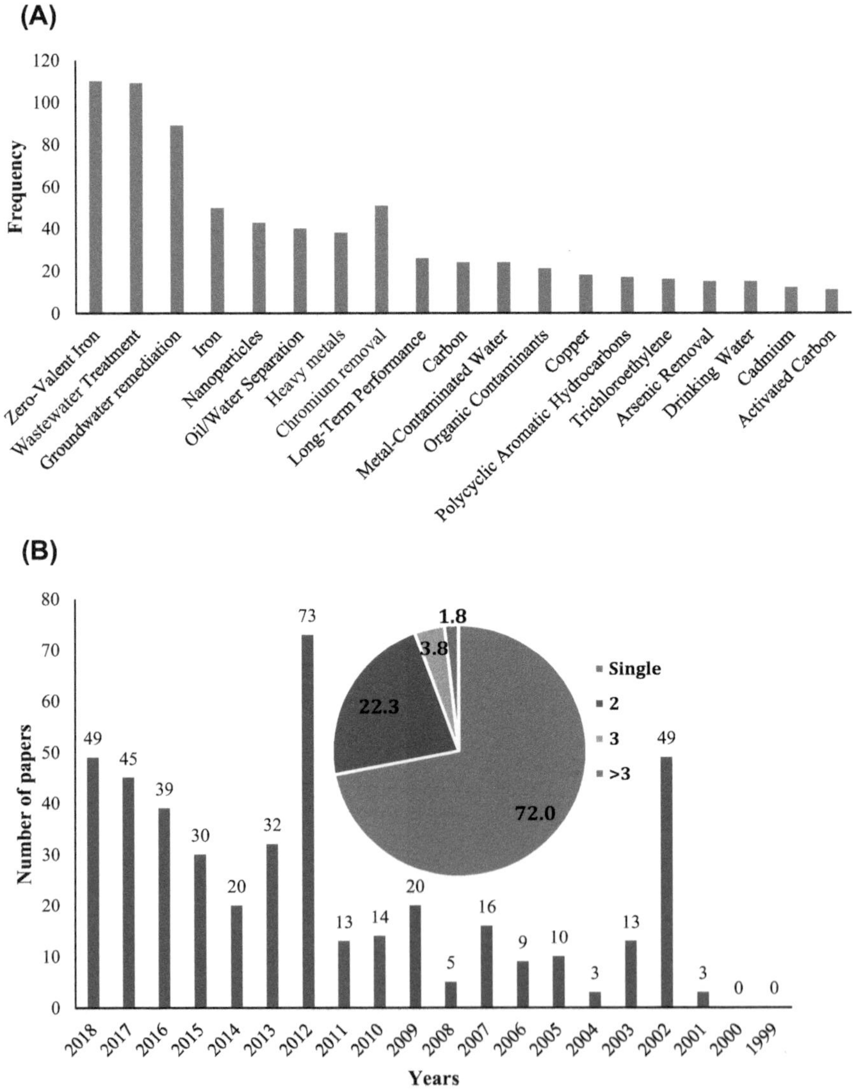

FIGURE 1.22 (A) Frequency of subtopics of membrane treatments and (B) distribution of the number of papers according to years.

modifications in membrane technologies from the 1960s membranes are now widely used (Ejder-Korucu et al., 2015; Cooke et al., 2000). Fig. 1.22A shows the frequency of the studied remediation environment and removal of pollutants in 443 research papers relating to the permeable reactive barrier and membrane processes. Membrane processes were applied to wastewater treatment (in 109 research papers), groundwater remediation (in 89 papers), metal-contaminated water (in 24 papers), and drinking water remediation (in 15 papers). Membranes were used in chromium removal in 51 papers, oil/water separation in 40 papers, heavy metals removal in 38 research papers, and organic contaminants in 21 papers. Zero-valent iron (in 110 papers), nanomaterials (in 40 papers), carbon (in 24 papers), and activated carbon (in 11 papers) were studied for assisting the membrane processes. These assisting materials were able to sequester numerous pollutants (Li et al., 2018). In 26 research papers, the long-term performance of membranes was investigated.

The maximum published research papers relevant to membranes were observed in 2012 with 73 papers as shown in Fig. 1.22B. 41% of 443 research papers relevant to membrane processes were published in the last 5 years. Moreover, all these research papers were published after 2000 and 72% of 443 research papers were single country contributions. The percentage of international collaboration was 28% as shown in the pie chart in Fig. 1.22B.

1.8.7 Encapsulation/immobilization

Immobilization methods are used to increase the performance of the catalyst, prevent loss, and separate nanomaterials sufficiently from water. Iron oxides, TiO_2, fungi (i.e., *Trametes hirsuta*), bacteria, algae, enzymes, etc. can be immobilized onto solid matrices (Atalay and Ersöz, 2015; Gaur et al., 2018; Asere et al., 2019; Taghipour et al., 2019). The solid matrices are activated carbon, zeolite, clay, sand, glass, stainless steel (Atalay and Ersöz, 2015), sodium alginate, polyurethane foam, alumina (Gaur et al., 2018), cellulose beads (Asere et al., 2019), etc. Organic pollutants, dyes, polyaromatic hydrocarbons, pesticides, and toxic compounds can be removed by immobilized active materials onto several solid matrices. Fig. 1.23A presents the frequency of removal of toxic compounds, immobilized catalysts/active materials, and remediated environments. The studied toxic compounds in 212 research papers related to immobilization methods are lead (in 76 papers), cadmium (in 43 papers), zinc (in 31 papers), copper (in 22 papers), chromate (in 10 papers), and arsenate (in 6 papers).

Fig. 1.23B shows the distribution of the number of research papers according to year and percentage of international collaboration. The number of papers increases with increasing years. In the last 5 years, 54.7% of 212 research papers were published. The percentage of international collaboration is 29.7%.

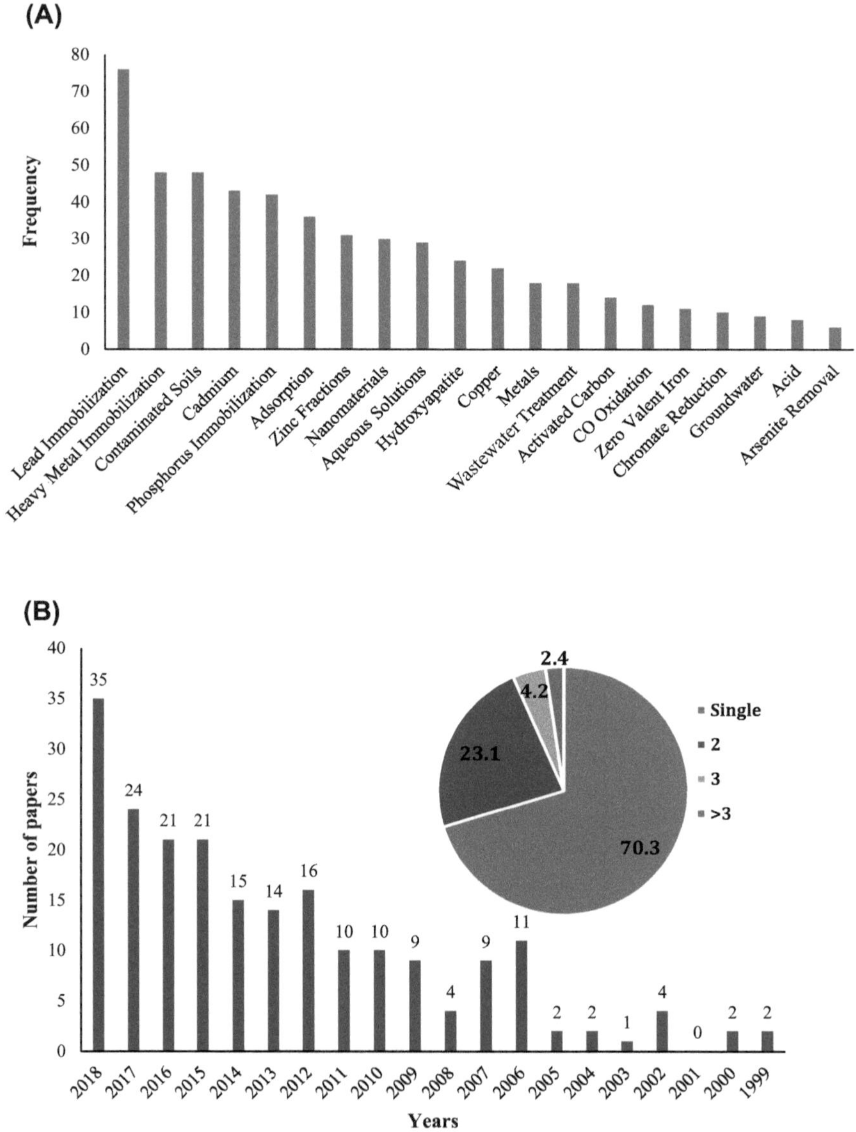

FIGURE 1.23 (A) Frequency of subtopics of encapsulation and immobilization and (B) distribution of the number of papers according to years.

2. Conclusions

11,224 research papers related to water remediation treatment studies were surveyed successfully on the core collection of the Web of Science. Bibliometric analysis was used for investigating authorship pattern, national and international collaboration, and funded/not-funded research. In general, during research, the collaborations were constituted as professor/student, colleagues, and interdisciplinary. To understand the types of collaboration from the affiliation of papers is very difficult. However, interdisciplinary and international collaborations were noted in this chapter.

- 49,913 authors from 111 countries made contributions to water remediation with respect to the green chemistry concept.
- After 2010, 70% of research papers were published.
- 97.1% of research papers were studied by two or more authors with national and/or international collaborations.
- The average length and cited references of research papers were 10 pages and 41 references, respectively.
- The percentage of funded papers reached a maximum value of 83.9% in 2018.
- The national Natural Science Foundation of China supported 1630 research papers.
- The adsorption process was mentioned in keywords 1651 times and the oxidation process was investigated in 944 research papers. Photocatalytic degradation, extraction processes, bioremediation techniques, and microorganisms were researched in 466, 575, 422, and 163 papers, respectively.
- Various sorbents were studied in 8574 research papers, which means that one of the biosorption techniques was applied to these research papers.
- In analyzed research papers, lead (74 contributions), cadmium (54), copper (48), zinc (42), and chromium (18) were the most-studied toxic metals by removing extraction methods.
- TiO_2 (158), zero-valent iron (29), graphene oxide (27), and graphitic carbon nitride (20) were used as photocatalysts for remediation of organic effluents.
- Microorganisms, which are marine bacteria, *Pseudomonas sp.*, *Shewanella*, *E. coli*, and strains, were studied in 14, 10, 10, 6, and 6 research papers, respectively.
- Membrane processes were applied to wastewater treatment (in 109 research papers), groundwater remediation (in 89 papers), metal-contaminated water (in 24 papers), and drinking water remediation (in 15 papers).
- The studied toxic compounds in 212 research papers related with immobilization methods were lead (in 76 papers), cadmium (in 43 papers), zinc (in 31 papers), copper (in 22 papers), chromate (in 10 papers), and arsenate (in 6 papers).

References

Anastas, P., Eghbali, N., 2010. Green chemistry: principles and practice. Chem. Soc. Rev. 39 (1), 301–312. https://doi.org/10.1039/b918763b.

Asere, T.G., Stevens, C.V., Du Laing, G., 2019. Use of (modified) natural adsorbents for arsenic remediation: a review. Sci. Total Environ. 676 https://doi.org/10.1016/j.scitotenv.2019.04.237. Elsevier B.V.: 706–20.

Atalay, S., Ersöz, G., 2015. Advanced oxidation processes for removal of dyes from aqueous media. In: Sharma, S.K. (Ed.), Green Chemistry for Dyes Removal from Wastewater. Scrivener Publishing, pp. 83–110.

Chen, F., Huang, H., Guo, L., Zhang, Y., Ma, T., 2019. The role of polarization in photocatalysis. Angew. Chem. Int. Ed. https://doi.org/10.1002/anie.201901361.

Cooke, M., Hill, E., Surrey, T., Poole, C.F., 2000. Encyclopedia of separation science (Michael Cooke, Colin F. Poole & Ian D. Wilson). Encycl. Separ. Sci. 4593.

Das, N., Charumathi, D., 2012. Remediation of synthetic dyes from wastewater using yeast-an overview. Indian J. Biothecnol. 11, 369–380.

Dasgupta, N., Ranjan, S., Ramalingam, C., 2017. Applications of nanotechnology in agriculture and water quality management. Environ. Chem. Lett. 15, s591–605. https://doi.org/10.1007/s10311-017-0648-9.

Ejder-Korucu, M., Gürses, A., Doğar, Ç., Metin, A., Sharma, S.K., 2015. Green chemistry for dyes removal from wastewater. In: Sharma, S.K. (Ed.), The British Journal of Psychiatry, vol. 112. Scrivener Publishing, New Jersey. https://doi.org/10.1192/bjp.112.483.211-a.

Fazal, T., Mushtaq, A., Rehman, F., Khan, A.U., Rashid, N., Farooq, W., Rehman, M.S.U., Xu, J., 2018. Bioremediation of textile wastewater and successive biodiesel production using microalgae. Renew. Sustain. Energy Rev. 82 (3), 3107–3126. https://doi.org/10.1016/j.rser.2017.10.029.

Gaur, N., Narasimhulu, K., Pydisetty, Y., 2018. Recent advances in the bio-remediation of persistent organic pollutants and its effect on environment. J. Clean. Prod. 198 https://doi.org/10.1016/j.jclepro.2018.07.076. Elsevier Ltd: 1602–31.

Ghernaout, D., Ghernaout, B., Mohamed, W.N., 2011. Embodying the chemical water treatment in the green chemistry-A review. Desalination 271 (1–3). https://doi.org/10.1016/j.desal.2011.01.032. Elsevier B.V.: 1–10.

Helmer, R., Hespanhol, I., United Nations, Environment Programme, Sanitation Collaborative Council, 1997. Water Pollution Control - A Guide to the Use of Water Quality Management Principles. United Nations Environment Programme.

Homem, V., Santos, L., 2011. Degradation and removal methods of antibiotics from aqueous matrices - a review. J. Environ. Manag. 92 (10) https://doi.org/10.1016/j.jenvman.2011.05.023. Elsevier Ltd: 2304–47.

Jain, S.K., Singh, V.P., 2003. Introduction to water resources systems. In: Jain, S.K., Singh, V.P. (Eds.), Water Resources Systems Planning and Management, p. 858. https://doi.org/10.1016/S0167-5648(03)80055-2.

Ki-moon, B., 2015. "The Human Right to Water," no. April 2011: 1–8.

Li, J., Dou, X., Qin, H., Sun, Y., Yin, D., Guan, X., 2018. "SC." Water Research. Elsevier Ltd. https://doi.org/10.1016/j.watres.2018.10.025.

Maestu, J., 2015. A 10 Year Story – The Water For Life Decade 2005-2015 and Beyond - 2015.

Park, D., Yun, Y.S., Park, J.M., 2010. The past, present, and future trends of biosorption. Biotechnol. Bioproc. Eng. 15, 86–102. https://doi.org/10.1007/s12257-009-0199-4.

Peset, F., Gonzalez, L.M., 2012. Building Networks with Bibexcel, Pajek, Vosviewer and GPSVisualizer. http://personales.upv.es/mpesetm/.

Petronella, F., Truppi, A., Dell'Edera, M., Agostiano, A., Lucia Curri, M., Comparelli, R., 2019. Scalable synthesis of mesoporous TiO_2 for environmental photocatalytic applications. Materials 12 (11). https://doi.org/10.3390/ma12111853.

Premarathna, K.S.D., Rajapaksha, A.U., Sarkar, B., Kwon, E.E., Bhatnagar, A., Ok, Y.S., Vithanage, M., 2019. Biochar-based engineered composites for sorptive decontamination of water: A review. Chemical Eng. J. 372, 536–550. https://doi.org/10.1016/j.cej.2019.04.097.

Proverb, K., 2011. Water Quality. United Nations Environment Programme.

Reeve, P.J., Fallowfield, H.J., 2018. Natural and surfactant modified zeolites: a review of their applications for water remediation with a focus on surfactant desorption and toxicity towards microorganisms. J. Environ. Manag. 205, 253–261. https://doi.org/10.1016/j.jenvman.2017.09.077.

Salama, E.S., Roh, H.S., Dev, S., Khan, M.A., Abou-Shanab, R.A.I., Chang, S.W., Jeon, B.H., 2019. Algae as a green technology for heavy metals removal from various wastewater. World J. Microbiol. Biotechnol 35 (75), 2–19. https://doi.org/10.1007/s11274-019-2648-3.

Subramanyam, K., 1983. Bibliometric studies of research collaboration: a review. J. Inf. Sci. 6, 33–38. https://doi.org/10.1177/016555158300600105.

Taghipour, S., Mossa Hosseini, S., Ataie-Ashtiani, B., 2019. Engineering nanomaterials for water and wastewater treatment: review of classifications, properties and applications. New J. Chem. 43 (21), 7902–7927. https://doi.org/10.1039/c9nj00157c.

Tiwari, B.K., 2015. "Trends in analytical chemistry ultrasound: a clean , green extraction technology. Trends Anal. Chem. 71 https://doi.org/10.1016/j.trac.2015.04.013. Elsevier B.V.: 100–109.

Tiwari, A., Shukla, A., Tiwari, D., Suk Soon Choi, H.-G.S., Lee, S.-M., 2019. Titanium dioxide nanomaterials and its derivatives in the remediation of water: past, present and future. Appl. Chem. Eng. 30 (3), 261–279. https://doi.org/10.14478/ace.2019.1035.

Varghese, A.G., Paul, S.A., Latha, M.S., 2019. Remediation of heavy metals and dyes from wastewater using cellulose-based adsorbents. Environ. Chem. Lett. 17, 867–877. https://doi.org/10.1007/s10311-018-00843-z.

Wang, J., Wang, S., 2019. Preparation, modification and environmental application of biochar: A review. J. Cleaner Prod. 227, 1002–1022. https://doi.org/10.1016/j.jclepro.2019.04.282.

Chapter 2

Defluoridation studies using graphene oxidebased nanoadsorbents

Anil Kumar K, Garima Verma, Smita Raghuvanshi, Suresh Gupta
Department of Chemical Engineering, Birla Institute of Technology and Science (BITS), Pilani, Rajasthan, India

1. Introduction

Water is known to be important resource on the planet and is getting polluted by all means of sources such as anthropogenic activities and by industries (Zhang et al., 2019). The indiscriminate release of pollutants from the industries such as semiconductor manufacturing, coal-fired power stations, electroplating, glass and ceramic production, aluminum smelters, beryllium extraction plants, and brick and iron works cause a huge impact on aquatic organisms and human beings (Bhatnagar et al., 2011). Hence, there is a need to treat the pollutants before they are released in the environment. Globally, the generation of wastewater was estimated as 12 billion tonnes in the year 2002 which might reach up to 19 billion tonnes by the year 2025 (Asokan Pappua and Saxenaa, 2018). These pollutants contain majorly organic and inorganic pollutants, which are considered to be harmful pollutants in water. The presence of inorganic pollutants such as heavy metal ions and fluoride ions is considered to be the most electronegative element and are considered to be nonbiodegradable (Verma and Kaur, 2016). Out of all, fluoride is considered to be essential mineral in the body which helps in the formation of carrier-resistant dental enamel and hard tissue mineralization process at lower concentration (0.7 ppm) (Zhang et al., 2015). However, beyond this concentration can cause serious health disorders and other deleterious effects such as dental and skeletal fluorosis, cardiovascular effects, etc. (Ayoob and Gupta, 2007; Hussain et al., 2004). The effluents from the industries contain high amount of

Green Chemistry and Water Remediation: Research and Applications.
https://doi.org/10.1016/B978-0-12-817742-6.00002-5

fluoride ions and is mixed with waterbodies. Due to its nonbiodegradable and nonsoluble properties, it is difficult to remove fluoride from the water. Therefore, it is exigent to develop a method to remove fluoride from the wastewater (Verma and Kaur, 2016).

Several methods have been developed such as membrane separation, precipitation-coagulation, electrocoagulation, nanofiltration, catalytic ozonation, electrodialysis, reverse osmosis, and ion exchange (Bhatnagar et al., 2011). However, these methods have certain pros and cons. Few processes such as chemical precipitation, coagulation, flocculation, and electrochemical coagulation are only effective when the pollutant concentration in aqueous media is more than 50 mg/L. It takes huge amount of chemicals to precipitate the metal ions and formation of large amount of sludge, and treating the formed sludge becomes a big concern to the environmentalists (Devi et al., 2014). Another famous technique that gained attention is Nalgonda technique where alum and lime are rapidly mixed in fluoride water and allowed to settle for 2−3 h after which fluoride is removed with flocs. The Nalgonda technique is mostly preferred at domestic and community levels; as it demands lower energy, simple design, and operations. The limitation reported for this method is the need of high amount of alum for achieving lower concentrations of fluoride in drinking water (Bhatnagar et al., 2011; Gunatilake, 2015). The higher operational costs and fouling problems are reported in ion exchange and reverse osmosis processes. Among all these methods, adsorption process is a surface-based mechanism in which the components in the gas or liquid state are assembled on the surface of an adsorbent layer. It occurs in three ways, by physical adsorption method, chemical adsorption method, or both. In general, physical adsorption is due to the weak forces such as electrostatic force of attraction and van der Waals forces which exist between the molecule (adsorbate) and the surface of adsorbent. In chemical adsorption, a strong covalent bond exists in between adsorbate and adsorbent. The robustness of adsorption technique over other techniques is due to its ability to handle large volumes of water, simple operation, lower energy requirements, higher efficiency, reusability of adsorbents, and its ability to adsorb multiple metal ions at a time (Dahi et al., 1996).

Many adsorbents have been developed and are used for the water remediation especially for the fluoride removal process, such as chitosan, carbonaceous materials, biomass, activated carbon, activated alumina, clay, chitosan zeolites, etc. (Hussain et al., 2004; Devi et al., 2014; Gunatilake, 2015; Dahi et al., 1996). These adsorbents have shown significant adsorption for fluoride ions. However, limitations associated with these adsorbents include (1) higher cost for the production, (2) less eco-friendly, (3) expensive process for the regeneration, (4) pH dependent, (5) large contact time for adsorption, and (6) smaller number of active sites/surface area. Therefore, there is a need to find an alternative adsorbent material for the removal of fluoride ions (Ali, 2012). Nanoparticles have proven themselves as good adsorbents due to their unique features such as small size, large surface area, large number of active sites, and ease of separation (Sánchez et al., 2011). The large surface area to mass ratio of nanoparticles increases their adsorption capability. Moreover, the surface of nanomaterials is reactive mainly due to higher density of lower coordinated atoms at the surface edges (Tamrakar et al., 2019). These unique properties and surface functionalization can be used to adsorb and degrade the pollutants in wastewater.

1.1 Possible mechanism for fluoride ion adsorption on nanoparticles

Several phenomena have been explained for the fluoride adsorption on adsorbent's surface and are illustrated below:

1. Adsorption of fluoride ions which is carried out on the surface of adsorbent is mainly because of physical adsorption, where the fluoride ions diffuse out from the bulk solution to adsorbent surface so-called external mass transfer, which is held by intermolecular forces/electrostatic forces (the force of attraction created between two charged particles) created between the molecules/atom present on the surface of adsorbent and the fluoride ions.
2. As the fluoride ions approach the surface of adsorbent, anchoring groups/functional groups present on the surface of adsorbents are solely responsible for the ion exchange process, i.e., −OH groups present on the metal oxide surface; −COOH, −CO, etc., impregnated on the surface through modification using functionalizing agent. The fluoride is highest electronegative element in the periodic table, which can easily exchange the lesser electronegative atom such as H_2, etc. (Kuang et al., 2017).
3. The micropores present in the adsorbents can successively trap the ions diffusing through it.

The synthesis methods such as sol-gel process, precipitation method, solvo thermal method, pulsed laser ablation, flame spray pyrolysis, mechanical alloying/milling, gas condensation, etc., are used for the preparation of these nanoparticles (Sharma et al., 2009).

The nanoadsorbents have got huge attention for fluoride ion adsorption from wastewater. Li et al. (2001) reported 13.5 times higher capacity nanoadsorbents of carbon nanotubes along with Al_2O_3 than that of AC-300 carbon and four times higher than that of γ-Al_2O_3. Aligned carbon nanotubes and nano-aluminum oxide hydroxide (n-AlOOH) showed maximum fluoride ion adsorption capacity of 4.5 mg/g and 3.259 mg/g, respectively, at pH 7.0 (Wang et al., 2009). Nano-magnesium oxide (surface area of 96.42 m^2/g) showed the fluoride adsorption capacity of 0.6 g/L (Devi et al., 2014). Apart from these studies, various other adsorbents are reported in the literature for their application in fluoride adsorption (Bhatnagar et al., 2011).

Recovery of nanoadsorbents after remediation process is extremely difficult. This is due to the higher ionic activity and higher dispersive nature of nanoadsorbents in the aqueous media. In order to overcome this problem, researchers have developed the nanocomposites, in which the oxides of different metal/adsorbents are mixed together to make a composite or particles are grown on the surface of other particles to make a core shell on the other adsorbent.

Carbon nanostructures have been extensively studied due to their excellent properties and numerous applications. The techniques such as mechanical exfoliation ("Scotch-tape" method) of bulk graphite and by epitaxial chemical vapor deposition method lead to the formation of single-layer transferable graphene nanosheets. However, these routes were found to be less effective for the large-scale manufacturing. Chemical method means such as combustion synthesis and sol-gel techniques are the practical approaches for the bulk-scale synthesis of graphene materials.

The most common approach is the graphite exfoliation method, in which strong oxidizing agents are used to obtain graphene oxide (GO). The determination of the exact structure of GO is difficult, the contiguous aromatic lattice of graphene is interrupted by epoxides, alcohols, ketone carbonyls, and carboxylic groups. GO contains a range of reactive oxygen functional groups, which renders it a good candidate for use in the various applications through chemical functionalization. The graphene honeycomb lattice is composed of two equivalent sublattices of carbon atoms bonded together with σ bonds.

Each carbon atom in the lattice has a π orbital that contributes to a delocalized network of electrons. The determination of exact structure of GO is difficult (Zhu et al., 2010). GO is graphite which has been oxidized to intersperse the carbon layers with oxygen molecules, and then got reduced to separate the carbon layers completely into individual or few layers of graphene. When the oxidizing agents react with graphite, the interplanar spacing between the layers of graphite is increased. The completely oxidized compound can then be dispersed in a base solution such as water, and then GO is produced. GO is produced by the oxidative treatment of graphite which is one of the principle methods developed by Brodie, Hummers, or Staudenmaier (Park and Ruoff, 2009).

Teutli-Sequeira et al. (2015) studied the adsorption of fluoride ions over GO and nanocomposite (GO + ZrO_2) and observed the tremendous increase of adsorption of fluoride ions from 13.8 to 46.0 mg/g. This may be due to the introduction of ZrO_2 nanoparticles in the graphene structure which makes less agglomeration and one molecule of ZrO_2 can alone bind five ions of fluoride. The protonated hydroxyl ions present on the surface of ZrO_2 could facilitate the adsorption of fluoride ions in water. However, the use ZrO_2 is restricted because it is considered as toxic compound. Hence, recent trends have been observed to find the economical and nontoxic materials which can effectively remove fluoride ions from water.

During a decade ago, use of aluminum, magnesium, silicon, etc., based on metal oxide gained huge interest toward water treatment. Magnesium oxide (MgO) adsorbent has been found to be an attractive defluoridizing agent because of its higher adsorption capacity, nontoxic nature, limited solubility in water, and applicability in wide range of pH. Nanoscale MgO may act as a prominent adsorbent for fluoride as it can provide higher surface area. As the crystallite size becomes smaller, the adsorption efficiency of MgO will increase. The surface area of MgO is 92.46 m^2/g with the sorption capacity of 0.6 g/L (Devi et al., 2014).

Minju et al. (2015) explored the use of magnesium-based core shell on iron oxide nanoparticles. They have achieved about 98% removal capacity using MgO-coated magnetite (Fe_3O_4) nanoparticles. This is due to the fact that huge surface of MgO, high isoelectric point, and small crystallite size facilitate the formation of less particle size. Advantage of creating magnetite core facilitates the easy separation of adsorbent after use, when subjected to external magnetic field.

By considering the above facts, this chapter demonstrates the development and use of magnesium oxide nanoparticles (n-MgO) and magnesium oxide–coated GO (nanocomposite) for the removal of fluoride ions from the aqueous media.

2. Materials and methods

2.1 Materials and chemicals required

The chemicals utilized in this chapter were sodium nitrate (Molychem, product code: 18620), potassium permanganate (Molychem), hydrochloric acid (10%) (Molychem, product code: 14860), sulfuric acid (Molychem), hydrogen peroxide (30%) (Rankim, product code: H0120), ethanol (99.9%), hexane (CDH, 99%, 910387), magnesium chloride 6-hydrate ($MgCl_2 \cdot 6H_2O$) (Rankim, 99%, 5833), sodium hydroxide (NaOH), and ethanol (99.9%).

2.2 Synthesis and characterization of adsorbent

2.2.1 Preparation of n-MgO using coprecipitation method

The precursor solution magnesium was prepared by dissolving 0.2 M of magnesium nitrate ($MgNO_3 \cdot 6H_2O$) in 100 mL of deionized water by use of magnetic stirrer. 0.5 M sodium hydroxide (NaOH) solution was added dropwise to the magnesium nitrate precursor solution with the continuous stirring. After few minutes, white precipitate started to appear, which indicated the formation of magnesium hydroxide particles in it. The solution was continued to stir for 2 h, the formed precipitate was filtered, then washed by methanol and water. For washing, centrifugation was done at 5000 rpm for 5 min. The sample was taken out in ptri dish and was dried at 60 °C temperature for overnight. The dried sample was crushed into fine powder using mortar and pestle and then the powder was calcined for 2 h at 300 °C followed by 2 h at 500 °C.

2.2.2 Synthesis of magnesium oxide–coated graphene oxide (nanocomposite)

Sonication method was utilized for the coating of MgO nanoparticles on the surface of nano-GO. In a 50 mL beaker, 0.04 g of GO and 0.07 g of MgO nanoparticles were mixed in 20 mL of ethanol. The solution was then sonicated in ultrasonic liquid processor (Johnson Plastosonic, model MM1010) till the whole sample was fully dispersed (Sridhar et al., 2013). Later, the same sample was kept for drying in an oven for 6 h at 60 °C.

2.2.3 Characterization techniques

n-MgO and nanocomposites were primarily grounded and sieved to make less particle size using pestle and mortar and was subjected for the characterization of phase purity and determination of crystallinity of the compound by using X-ray diffractometer (XRD). The functional groups present on the surface for the adsorption of fluoride were analyzed using Fourier transferred infrared spectroscopy (FTIR) (PerkinElmer India Pvt Ltd., India). In the FTIR analysis, air was taken as background. Since GO is in sheet form, it was kept directly at the tip of the probe. For the fine powder, pellet was formed using potassium bromide (KBr) mixed with finely powdered sample. The surface area of the adsorbent was analyzed using Brunauer–Emmett–Teller (BET) surface area analyzer. SEM-EDX was used to analyze the chemical characterization of the sample. It identified the metals and materials present on the sample. The size of a particle was analyzed using transmission electron microscopy (TEM). The size and the zeta potential were estimated from DLS (dynamic Light Scattering) of Malvern ZEN3600. Thermogravimetric Analysis (TGA) was performed for determining the thermal stability of the sample using TGA 4000 (PerkinElmer India Pvt Ltd., India). A very small amount of the sample approximately 1.5–2.0 g was taken in the platinum pan. The sample was heated and the temperature range for the process was kept from 25 to 800 °C for about 4 h with the rate of increase of 10–20 °C/min.

3. Batch adsorption experiments

The batch studies were conducted in 100 mL conical flask. Initially, 10 ppm and 5 ppm solution of fluoride ion was made by using NaF salt stock solution at pH range of 6–7. The effect of important parameters such as contact time, adsorbent dosage, initial fluoride concentration, and pH of the solution was studied. 50 mL of sample was taken, and the contact time was varied from 0 to 4 h by keeping the adsorbent dosage (0.03 g of n-MgO and 0.05 g of the nanocomposite), initial fluoride concentration (5 and 10 ppm), rotational speed (90–100 rpm), and pH constant. The adsorbent dosages were varied in the range of 0.01–0.2 g for n-MgO and 0.01–0.5 g for nanocomposite by maintaining the contact time (2 h), initial fluoride concentration (5 and 10 ppm), rotational speed (90–100 rpm), and pH (6–7) as constant. The initial fluoride ion concentration was varied from 5 to 100 ppm and the adsorbent dosages of 0.03 g of n-MgO and 0.05 g of the nanocomposite were kept constant. The effect of pH was studied in the range of 3–12 and the initial pH of the solution was maintained by 0.1 N HCl and 0.1 N NaOH. The samples were filtered using Whatman filter paper and followed by nano syringe filter after carrying out the adsorption study. 10 mL of filtered sample was taken in a beaker and 10 mL of TISAB was added in it. The concentration of fluoride ion in the samples was measured using fluoride ion electrode.

4. Results and discussions

4.1 Characterization of adsorbents

4.1.1 Crystallinity

The X-ray diffraction (XRD) is one of the most widely used techniques for finding the crystallinity of the material. It is used to measure the average spacings between layers or rows of atoms and helps to determine the orientation of a single crystal or grain. The XRD patterns of the n-MgO before sorption with fluoride ions were also recorded. Fig. 2.1 represents the X-ray diffractograms of n-MgO sample which showed peaks corresponding to (1,1,1), (2,0,0), (2,2,0), and (2,2,2) planes, indicating the pure phase of MgO and the presence of cubic face-centered MgO.

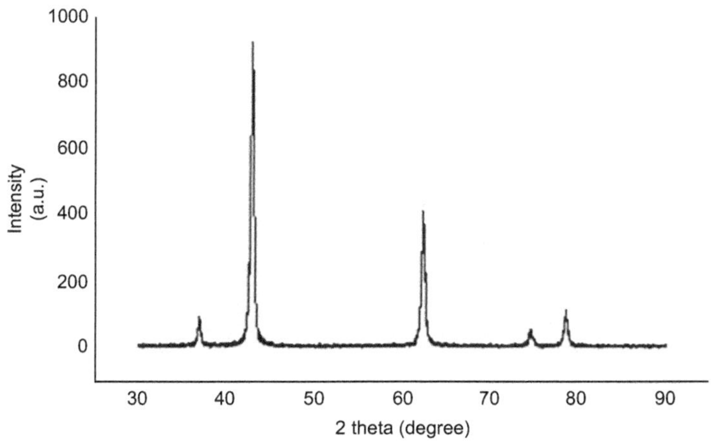

FIG. 2.1 XRD spectra of n-MgO.

The XRD spectra of nanocomposite (n-MgO—coated GOs) are presented in Fig. 2.2. The nanocomposite showed a broad peak centered on 18 °C, which revealed the presence of the amorphous phase of the MgO nanoparticles, with almost no crystalline structure. As the growth of MgO took place at room temperature, most of the grown MgO nanoparticles are naturally amorphous, without any crystalline phase. The nominal shift of the peak of the GO from 10 to 25 °C was due to the sodium hydroxide treatment of MgO nanoparticles, possibly removing a small amount of oxide groups on the surface of the GOs.

4.1.2 Morphology

The SEM image and EDX analysis of n-MgO is shown in Fig. 2.3. The morphology as shown in Fig. 2.3 indicates the well-defined structure of n-MgO

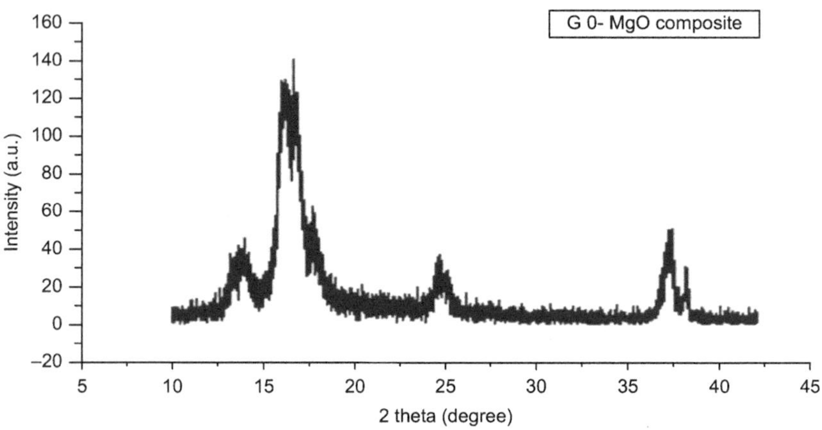

FIG. 2.2 XRD spectra of nanocomposite (n-MgO—coated GOs).

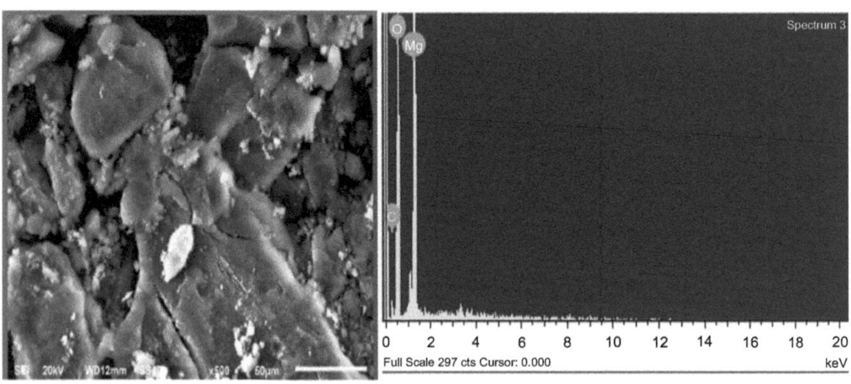

FIG. 2.3 SEM-EDX spectra of n-MgO.

with pores. EDX analysis confirmed the presence of Mg and O on the n-MgO. The presence of carbon (2.06%) impurity on the surface of n-MgO was due to the absorption of carbon dioxide from atmosphere.

SEM-EDX spectra of nanocomposite (n-MgO—coated GOs) confirmed the dense nanoparticles of MgO on the surface of GOs (Fig. 2.4). The presence of carbon, oxygen, and magnesium denotes the formation of composite. The SEM image showed the pores of MgO. This indicated that the GO have no influence on the crystallinity of the grown MgO because of the relatively defective structure of the GOs. EDX analysis has also established the fact that the MgO nanoparticles were successfully synthesized on the surface of the GO.

TEM analysis was conducted to further characterize structures and to estimate the size of developed nanoadsorbents. TEM micrographs of n-MgO and extended images of the same are shown in Fig. 2.5, respectively. The image shows the denser behavior of n-MgO. The particle size of n-MgO was found to be less than 20 nm by TEM analysis.

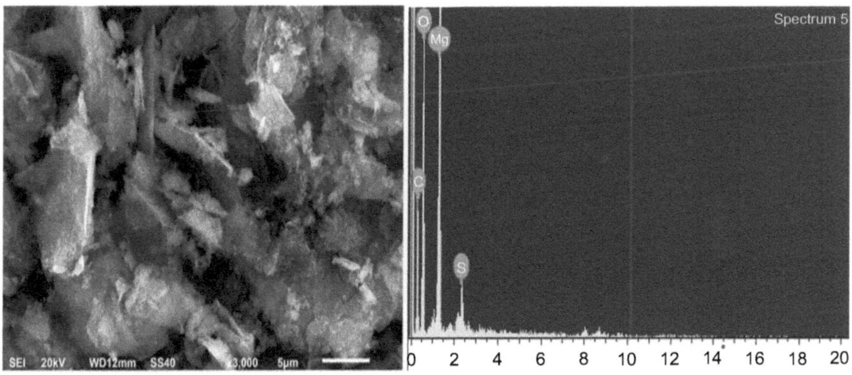

FIG. 2.4 SEM-EDX spectra of nanocomposite (n-MgO—coated GOs).

FIG. 2.5 TEM images of n-MgO.

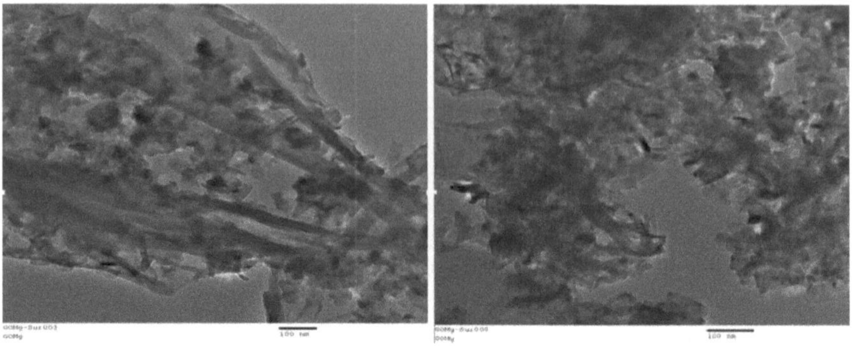

FIG. 2.6 TEM images of nanocomposite (n-MgO—coated GOs).

TEM analysis of nanocomposite (n-MgO—coated GOs) reveals densely distributed MgO nanoparticles on the surface of GOs (Fig. 2.6). The particle size of nanocomposite was found to be less than 90 nm by TEM analysis.

4.2 Surface compositions

FTIR spectroscopy identifies the functional groups present in a molecule and each specific chemical bond often shows a unique vibrational energy absorption band. The FTIR spectra of n-MgO before and after fluoride adsorption are shown in Fig. 2.7A and B, respectively. The broad peak at 3450 cm^{-1} corresponded to the water molecules absorbed from the atmospheric moisture. The peak at 1639 cm^{-1} was due to the bending vibration of —OH group. The peak obtained at 3703 cm^{-1} corresponded to —OH stretching vibration bonded with Mg. There was an increase in the intensity of the peak at 3703 cm^{-1} after fluoride adsorption compared with pure n-MgO as shown in Fig. 2.7B. The slight broadening of the band at 3450 cm^{-1} in the fluoride-sorbed n-MgO was observed due to pure n-MgO which could be due to the formation of O—H— F bond.

FTIR spectra of nanocomposite (n-MgO coated with GOs) before and after fluoride adsorption are given in Fig. 2.8A and B, respectively. The intensities of the peaks are directly related to the amount of sample present. The presence of absorbed water in nanocomposite was shown by a broad peak which appeared at 3392 cm^{-1}. The peak at 1111 cm^{-1} indicated the presence of C—O bond and at 1631 cm^{-1} attributed to the C=C bond. The band at 3392 corresponded to —OH stretching vibration bonded with Mg. There was an increase in the intensity of the peak at 3410 cm^{-1} after fluoride adsorption compared with pure nanocomposite as shown in Fig. 2.8B. The appearance of peak at 3697 cm^{-1} in the fluoride sorbed nanocomposite compared with pure nanocomposite was due to the formation of O—H— F bond.

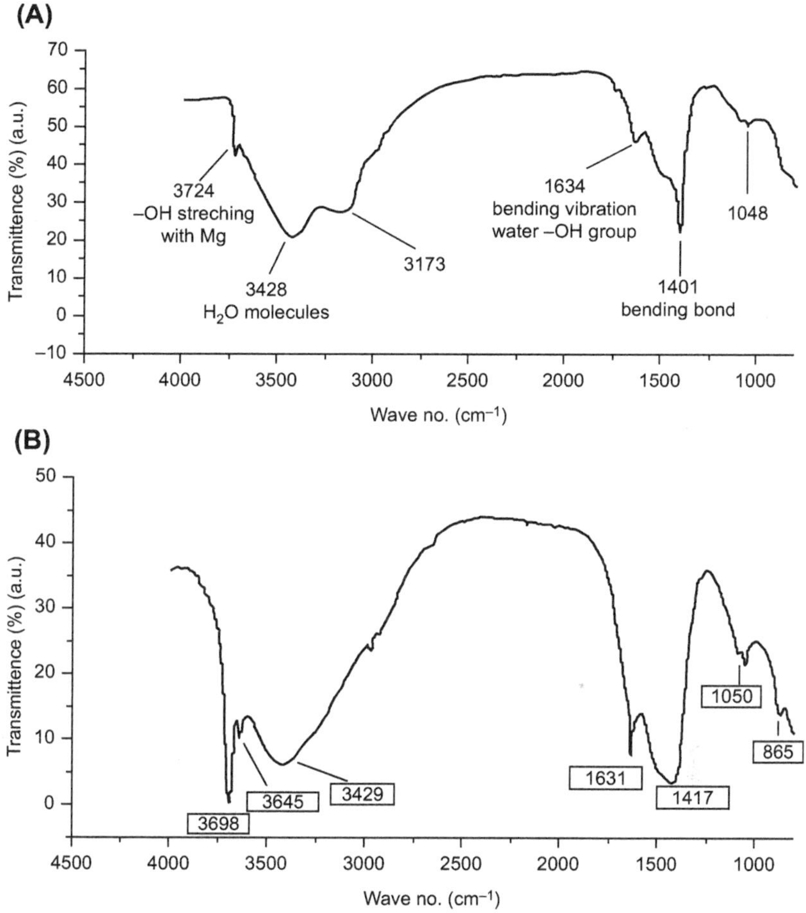

FIG. 2.7 (A) FTIR spectra of n-MgO before fluoride adsorption. (B) FTIR spectra of n-MgO after fluoride adsorption.

4.3 Experimental results and discussion

The experimental results obtained to study the effect of various parameters such as effect of pH, contact time, initial fluoride concentration, and mass of adsorbent for the removal of fluoride ions using n-MgO and nanocomposite are discussed in the following sections.

4.3.1 Effect of pH

The effect of pH was studied for a wide range of 3–12 at an initial fluoride concentration of 10 mg/L. The results obtained for the removal of fluoride at different pH using n-MgO are shown in Fig. 2.9A. The change in percentage

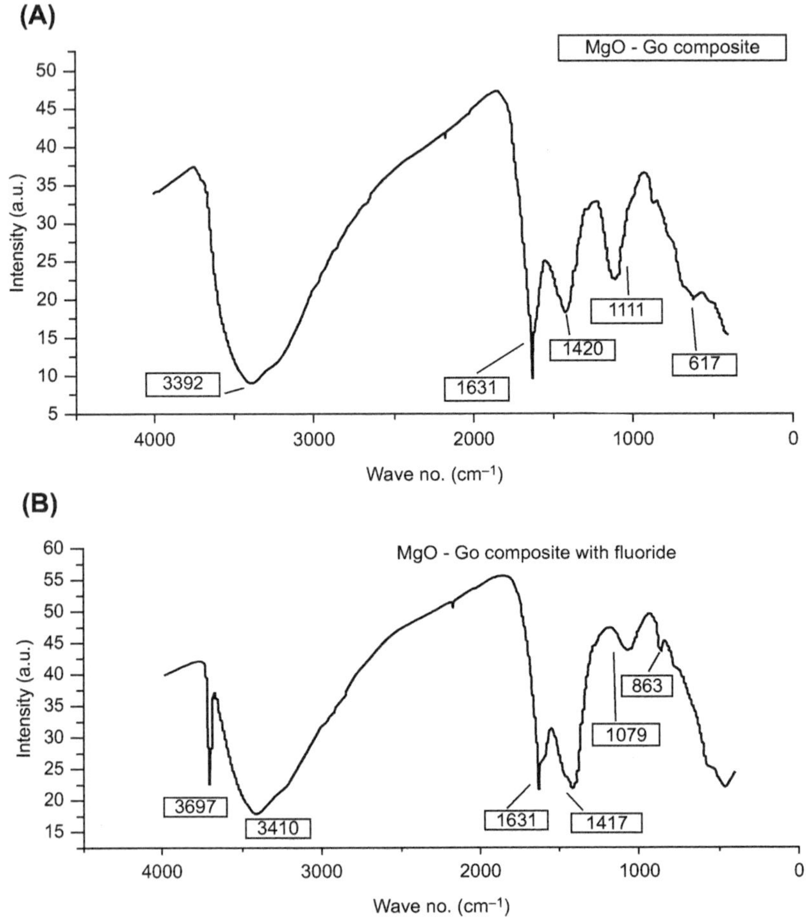

FIG. 2.8 (A) FTIR spectra of nanocomposite before fluoride adsorption. (B) FTIR spectra of nanocomposite after fluoride adsorption.

removal of fluoride was insignificant with the change in solution pH. This may be due to the fact that MgO is basic in nature and therefore the treated water remains basic that is near about pH value of 9–11, irrespective of initial pH. Similar results were reported by Maliyekkal et al. (2008) for defluoridation of water by magnesium oxide. At pH value of 12, there was sudden decrement in the fluoride removal which indicated that pH removal is sensitive above the initial pH value of 11 (Fig. 2.9A).

The pH study was performed for nanocomposites by keeping the adsorbent dosage constant as 0.05 mg/50 mL and the initial fluoride concentration as 10 mg/L (Fig. 2.9B). The maximum fluoride removal was observed in between pH range of 7–11. This indicated that fluoride removal was more in the basic

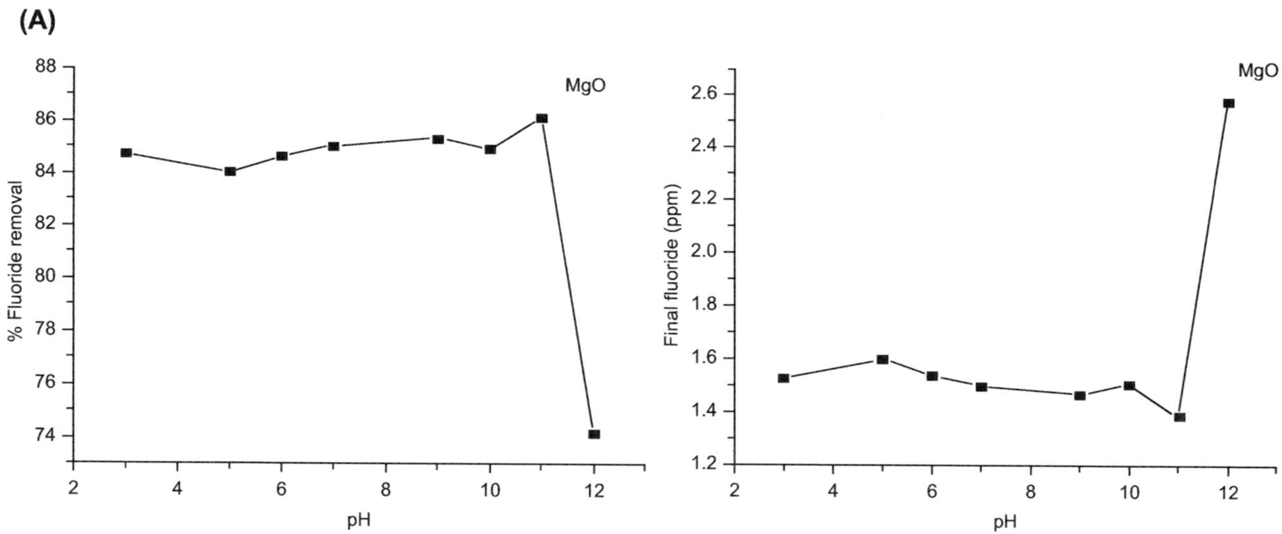

FIG. 2.9 (A) Effect of pH on adsorption of fluoride using n-MgO. (B) Effect of pH on adsorption of fluoride using nanocomposite.

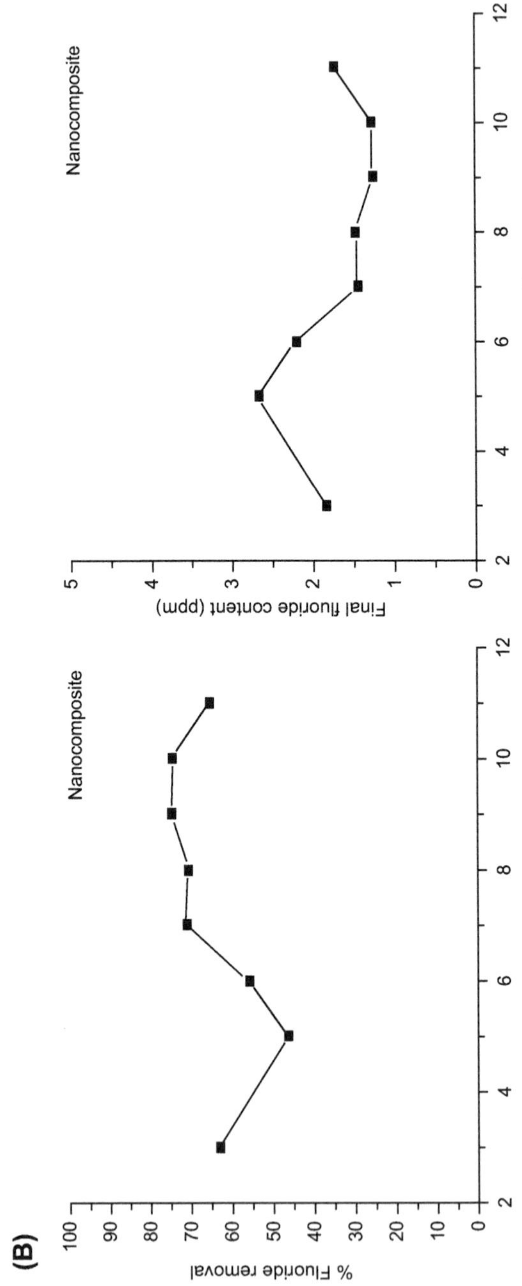
(B)
% Fluoride removal
0
10
20
30
40
50
60
70
80
90
100
2
4
6
8
10
12
pH
Nanocomposite
Final fluoride content (ppm)
0
1
2
3
4
5
2
4
6
8
10
12
pH
Nanocomposite

FIG. 2.9 —cont'd

medium using nanocomposite as adsorbent. The fluoride removal capacity of the adsorbent tends to decrease in the acidic medium. At basic pH, about 70% of fluoride was removed from drinking water.

4.3.2 Effect of initial fluoride concentration

The effect of initial fluoride concentration on the percentage removal of fluoride was studied by varying initial fluoride concentration and keeping the other parameters such as adsorbent dose (0.03 g/50 mL) and time (2 h) as constant. The effect of initial fluoride concentration on n-MgO and nanocomposite is shown in Fig. 2.10A and B, respectively. It was observed that the percentage of fluoride removal decreased as initial fluoride concentration was increased for both adsorbents. This may be due to the fact that at higher fluoride concentration, the binding capacity of the adsorbent reaches to saturation point which resulted in a decrease of overall percentage removal of fluoride.

4.3.3 Effect of adsorbent dosage

The effect of adsorbent dosage on fluoride removal using n-MgO and nanocomposite is shown in Fig. 2.11A and B, respectively. The other parameters such as equilibrium time (2 h), volume of sample (50 mL), and the initial fluoride concentration (10 mg/L) were kept constant. The fluoride removal was increased rapidly with increase in the adsorbent dosage for both adsorbents. The permissible limit of fluoride was reached at only 0.03 g/50 mL and 0.05 g/50 mL adsorbent dosage for n-MgO and nanocomposite, respectively. The increase in the fluoride removal with increase in adsorbent dosage was due to the increase in available active sites of adsorbent.

4.3.4 Effect of contact time

The effect of contact time is studied by carrying out the kinetic studies for the adsorption of fluoride ion on developed adsorbents. 10 mg/L and 5 mg/L of initial fluoride concentration was maintained for n-MgO and nanocomposite adsorbents, respectively, in this chapter. The other parameters of adsorbent dosage (0.03 mg/50 mL for n-MgO and 0.05 mg/mL for nanocomposite) were kept constant. Fig. 2.12A and B shows the change in percentage fluoride removal using n-MgO and nanocomposite, respectively, with time. It was observed that with the lapse of time, the percentage fluoride removal was increased for both the adsorbents. Initially, the rate of adsorption was very rapid; later, it got reduced and reached to the equilibrium value. Within the 4-h interval, percentage removal of fluoride was observed as 92.5% and 75.8% for n-MgO and nanocomposite, respectively. This chapter revealed that the equilibrium time for both adsorbents was found to be 2 h.

This technology is based on the usage of n-Mgo a n-MgO-coated GO as an important method for the removal of fluoride ions.

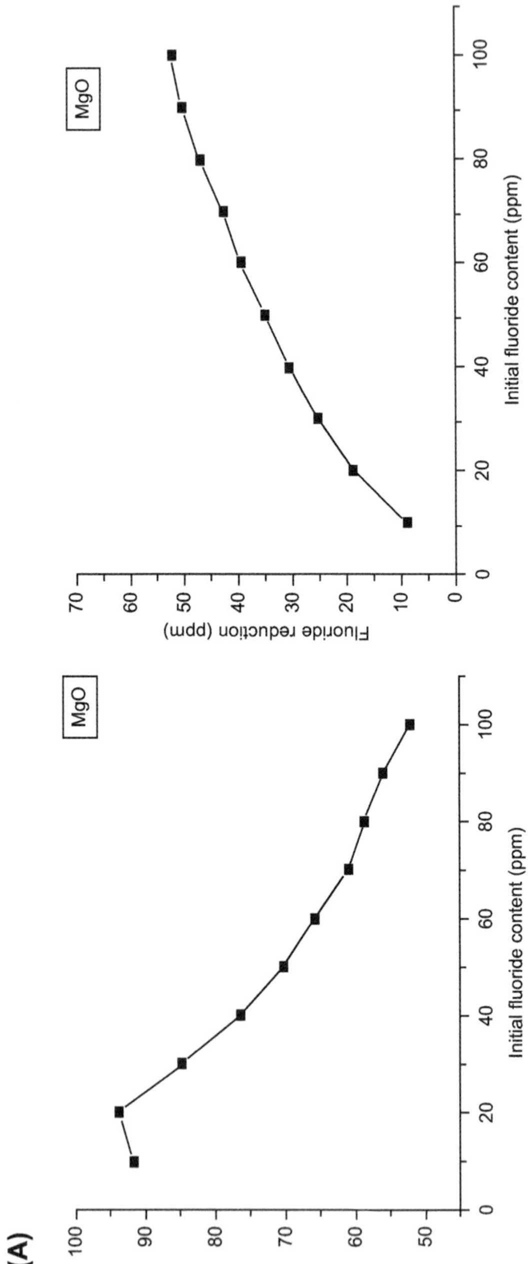

FIG. 2.10 (A) Effect of initial fluoride concentration on n-MgO. (B) Effect of initial fluoride concentration on nanocomposite.

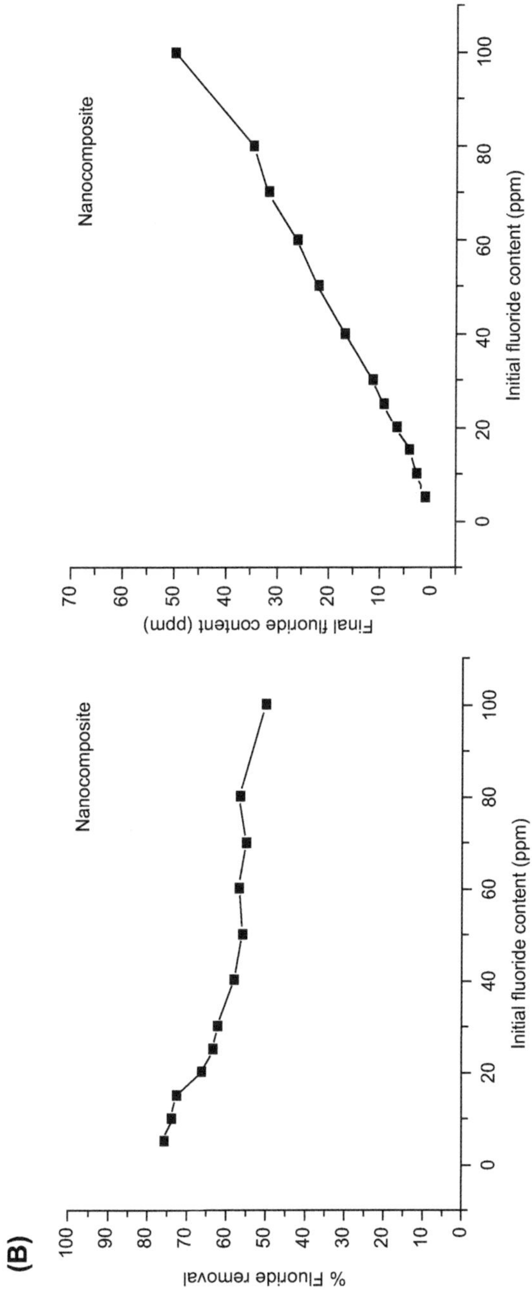

FIG. 2.10 —cont'd

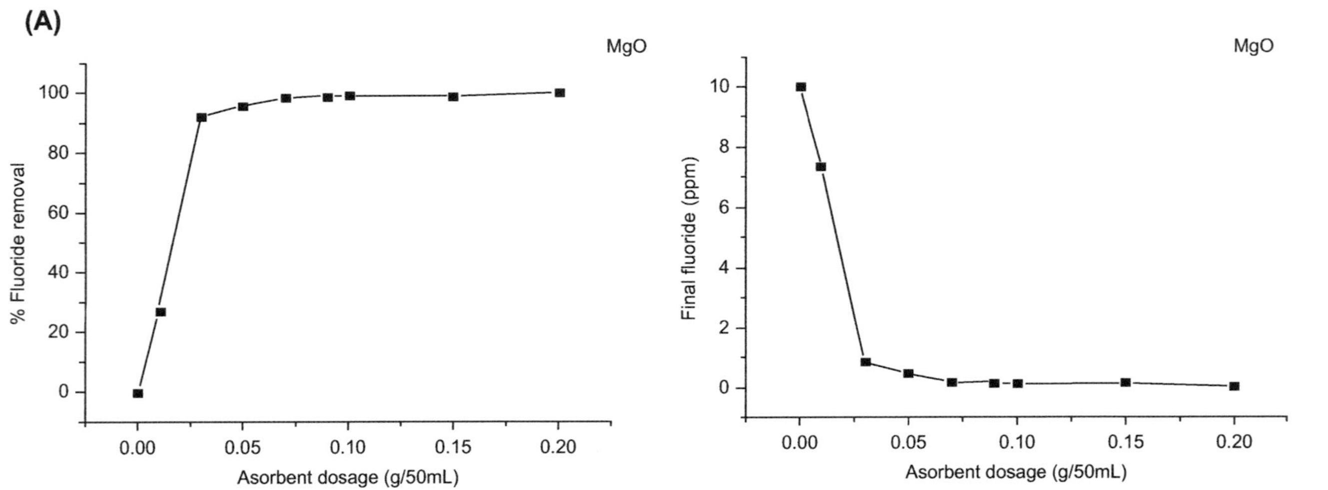

FIG. 2.11 (A) Effect of adsorbent dosage on fluoride removal using n-MgO. (B) Effect of adsorbent dosage for fluoride removal using nanocomposite.

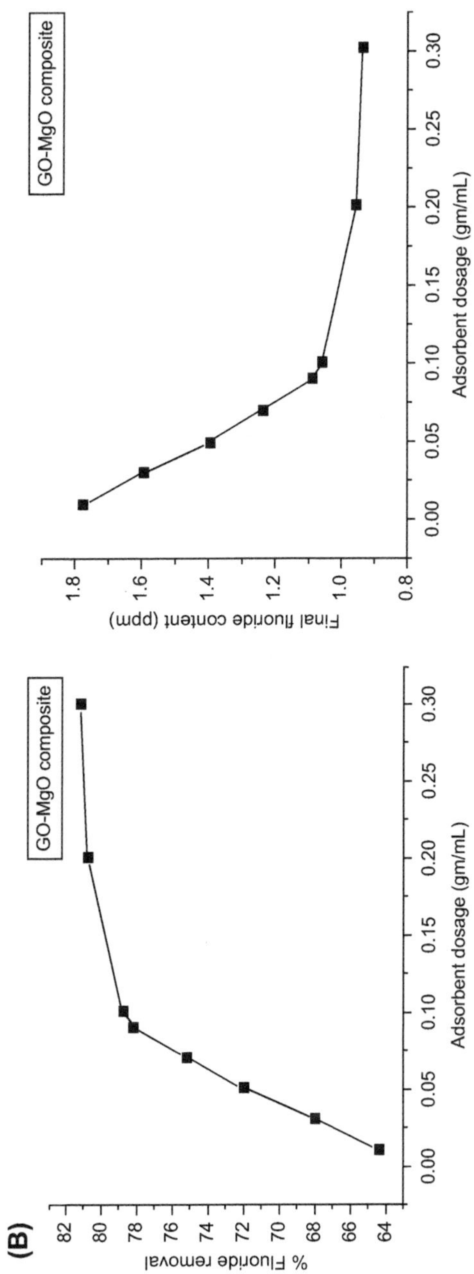

FIG. 2.11 —cont'd

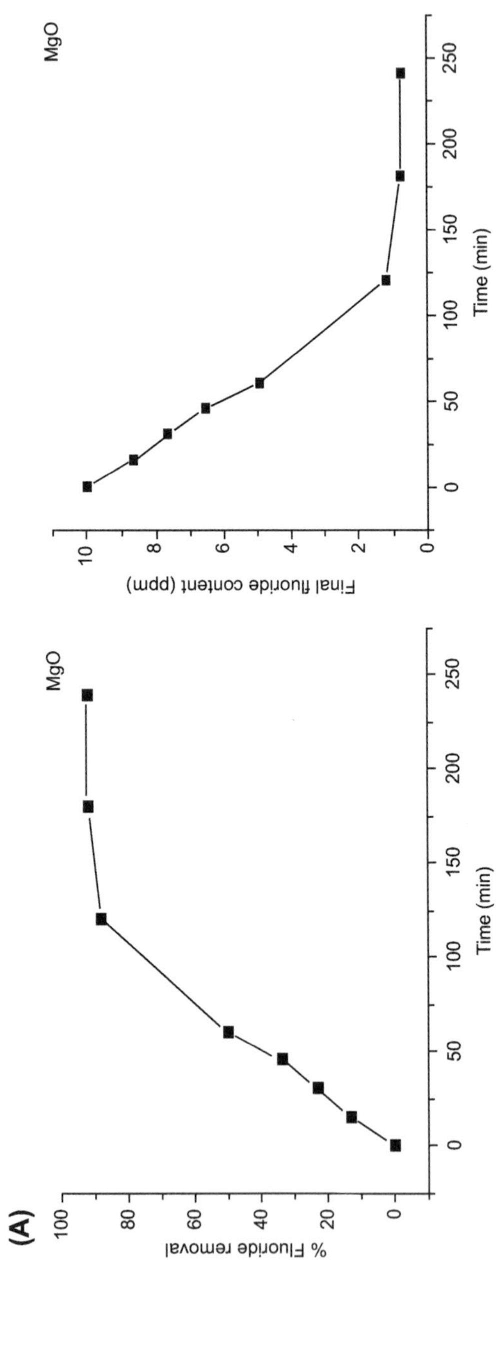

FIG. 2.12 (A) Effect of contact time for the adsorption of fluoride on n-MgO. (B) Effect of contact time for the adsorption of fluoride on nanocomposite.

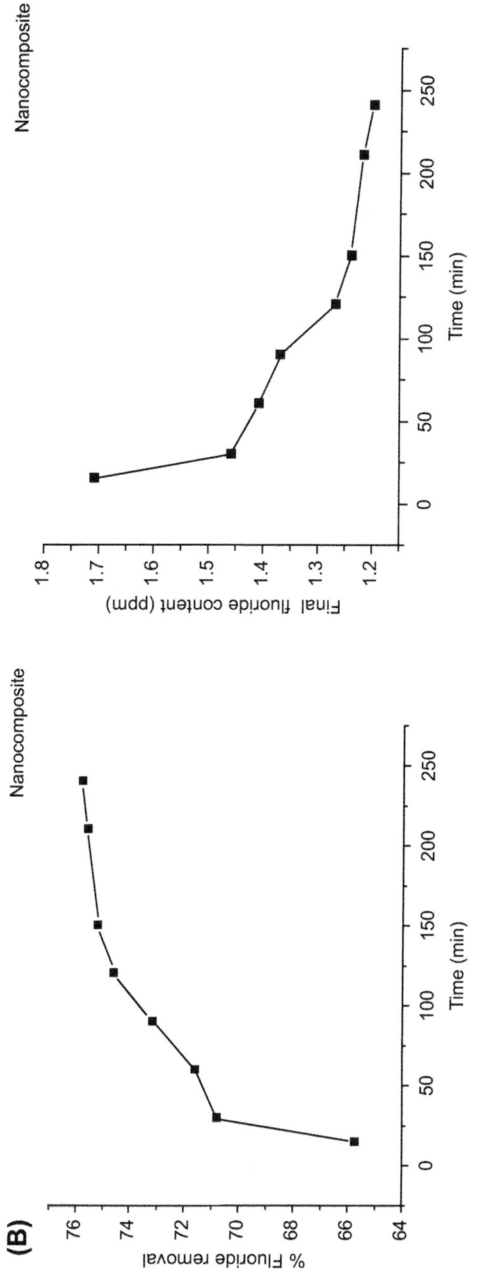

FIG. 2.12 —cont'd

In the recent years, application of nanoadsorbents has been proven to be potentially efficient for the removal of fluoride ions from ground and wastewater. This was due to the fact that small amount of nanoadsorbent is enough to treat large amount of fluoride-contaminated wastewater and bring it to acceptable limit. It makes this technique economically feasible and affordable as compared with the other techniques available for fluoride removal. Most of the earlier studies are limited to batch mode of fluoride treatment from wastewater. This technique has a great potential to be utilized in continuous mode and to be applied for large-scale removal of fluoride from ground and wastewater. Hence, the batch mode adsorption conditions should be transferred to column operations so that nanoadsorption technology may be utilized for the welfare of human beings. Hence, this chapter suggests further research on nanoadsorbents for fluoride removal from wastewater on a large scale and may provide an economically viable solution and can be extremely helpful for the society.

5. Conclusions

This chapter reveals the fact that nanotechnology is one of the efficient methods for tackling problems related to fluoride removal or water purification. Out of all reported nanoadsorbents, n-MgO and n-MgO—coated GO can be the most viable option as adsorbent for the adsorption of fluoride from drinking water which may be due to their unique two-dimensional structure, higher binding capacity, and also higher specific surface area. n-Mgo and nanocomposites were found to have the size of less than 10 and 90 nm, respectively. The BET surface area for n-MgO and nanocomposite was obtained as 20 and 42.68 m^2/g, respectively. The rate of adsorption was also found to be rapid for both adsorbents. This chapter concluded that developed nanoadsorbents were the most viable alternative for the removal of fluoride from wastewater. The batch mode adsorption conditions can be transferred to column operations so that nanoadsorption technology is utilized for the welfare of human beings by carrying out more experiments and designing the setup.

References

Ali, I., 2012. New generation adsorbents for water treatment. Chem. Rev. 112 (10), 5073—5091.

Asokan Pappua, S.R.A., Saxenaa, M., 2018. Solid wastes generation in India and their recycling potential in building materials Asokan. ISIJ Int. 58 (8), 1519—1523.

Ayoob, S., Gupta, A.K., 2007. Sorptive response profile of an adsorbent in the defluoridation of drinking water. Chem. Eng. J. 133 (1—3), 273—281.

Bhatnagar, A., Kumar, E., Sillanpää, M., 2011. Fluoride removal from water by adsorption-A review. Chem. Eng. J. 171 (3), 811—840.

Dahi, E., Mtalo, F., Njau, B., Bregnhj, H., 1996. Defluoridation using the nalgonda technique in Tanzania. In: Reaching the Unreached - Challenges for the 21st Century: Proceedings of the 22nd WEDC Conference, pp. 266—268.

Devi, R.R., Umlong, I.M., Raul, P.K., Das, B., Banerjee, S., Singh, L., 2014. Defluoridation of water using nano-magnesium oxide. J. Exp. Nanosci. 9 (5), 512–524.

Gunatilake, S.K., 2015. Methods of Removing Heavy Metals from Industrial Wastewater, vol. 1, pp. 12–18 no. 1.

Hussain, J., Sharma, K.C., Hussain, A., 2004. Fluoride in drinking water in Rajasthan and its Ill Ef- fects on human health. J. Tissue Res. 4 (2), 263–273.

Kuang, L., Liu, Y., Fu, D., Zhao, Y., 2017. FeOOH-graphene oxide nanocomposites for fluoride removal from water: acetate mediated nano FeOOH growth and adsorption mechanism. J. Colloid Interface Sci. 490, 259–269.

Li, Y., Wang, S., Cao, A., Zhao, D., Zhang, X., Xu, C., Luan, Z., Ruan, D., Liang, J., Wu, D., Wei, B., 2001. Adsorption of fluoride from water by amorphous alumina supported on carbon nanotubes. Chem. Phys. Lett. 350, 412–416.

Maliyekkal, S.M., Shukla, S., Philip, L., Nambi, I.M., 2008. Enhanced fluoride removal from drinking water by magnesia-amended activated alumina granules. Chem. Eng. J. 140 (1–3), 183–192.

Minju, N., Venkat Swaroop, K., Haribabu, K., Sivasubramanian, V., Senthil Kumar, P., 2015. Removal of fluoride from aqueous media by magnesium oxide-coated nanoparticles. Desalination Water Treat. 53 (11), 2905–2914.

Park, S., Ruoff, R.S., 2009. Chemical methods for the production of graphenes. Nat. Nanotechnol. 4 (4), 217–224.

Sánchez, A., Recillas, S., Font, X., Casals, E., González, E., Puntes, V., 2011. Ecotoxicity of, and remediation with, engineered inorganic nanoparticles in the environment. Trac. Trends Anal. Chem. 30 (3), 507–516.

Sharma, Y.C., Srivastava, V., Singh, V.K., Kaul, S.N., Weng, C.H., 2009. Nano-adsorbents for the removal of metallic pollutants from water and wastewater. Environ. Technol. 30 (6), 583–609.

Sridhar, V., Gangaraju, D., Chun, H., Park, H., 2013. Blister packing of copper hydroxide and Titania nanoparticles on graphene and its recycling. Appl. Mater. Interf. 5, 12323–12328.

Tamrakar, S., Verma, R., Sar, S.K., Verma, C., 2019. Cost effective natural adsorbents for the removal of fluoride: a green approach. Rasayan J. Chem. 12 (2), 455–463.

Teutli-Sequeira, A., Solache-Ríos, M., Martínez-Miranda, V., Linares-Hernández, I., 2015. Behavior of fluoride removal by aluminum modified zeolitic Tuff and Hematite in column systems and the thermodynamic parameters of the process. Water Air Soil Pollut. 226 (8).

Verma, N., Kaur, G., 2016. Trends on Biosensing Systems for Heavy Metal Detection, vol. 74. Elsevier Ltd.

Wang, S.G., Ma, Y., Shi, Y.J., Gong, W.X., 2009. Defluoridation performance and mechanism of nano-scale aluminum oxide hydroxide in aqueous solution. Chem. Technol. Biotechnol. 84, 1043–1050.

Zhang, J., Chen, N., Tang, Z., Yu, Y., Hu, Q., Feng, C., 2015. A study of the mechanism of fluoride adsorption from aqueous solutions onto Fe-impregnated chitosan. Phys. Chem. Chem. Phys. 17 (18), 12041–12050.

Zhang, Y., Zhu, C., Liu, F., Yuan, Y., Wu, H., Li, A., 2019. Effects of ionic strength on removal of toxic pollutants from aqueous media with multifarious adsorbents: a review. Sci. Total Environ. 646, 265–279.

Zhu, Y., Murali, S., Cai, W., Li, X., Suk, J.W., Potts, J.R., Ruoff, R.S., 2010. Graphene and graphene oxide: synthesis, properties, and applications. Adv. Mater. 22 (35), 3906–3924.

Chapter 3

Batik became two sides of blade for the sustainable development in Indonesia

N. Kusumawati[1], E. Rahmadyanti[2], M.M. Sianita[1]
[1]*Analytical Chemistry Laboratory, Mathematics and Natural Science Faculty, Universitas Negeri Surabaya, Surabaya, Indonesia;* [2]*Water and Irrigation Laboratory, Engineering Faculty, Universitas Negeri Surabaya, Surabaya, Indonesia*

1. The glimpse of batik

Batik is a familiar thing for the Indonesian people today. Batik is one of the unique archipelago heritages. Its uniqueness is shown with various kinds of motives that have their meaning. According to Asti M. and Ambar B. Arini (Musman and Ambar, 2011) based on their etymology and terminology, batik is a series of words "mbat" and "tik." "Mbat" in Javanese can be interpreted as blowing or throwing many times, while "tik." comes from the word dot. So, batik means throwing dots repeatedly on the fabric. There are also who said that the word batik comes from the word "amba" which means wide fabric and "titik." This means that batik is the dots drawn on wide fabric media to produce beautiful motifs. In the Indonesian Big Dictionary, batik has the meaning of the pictorial fabric that was specifically made by writing down or spreading the dyes resist on the fabric, then it is processed in a certain way.

Based on the manufacturing technique, batik can be divided into three types, namely handwriting, stamping, and painting (Lisbijanto, 2013). Handwriting batik is generally produced manually using a tool called a "canting" to coat the dyes resist (beeswax) on a batik motif. The batik motifs produced will never be the same. The high level of patience in working on the complexity of motifs causes the price of handwriting batik to be more expensive. Because of its exclusivity, this type of batik is generally worn by kings, royal officials, and nobles as a symbol of luxury.

In contrast to handwriting batik, stamping batik is produced by utilizing the stamp from copper material. In this case, the stamp replaces the canting's role

Green Chemistry and Water Remediation: Research and Applications.
https://doi.org/10.1016/B978-0-12-817742-6.00003-7

in superimposing dyes resist on the fabric. The lower level of complexity of the work makes this type of batik production need a shorter time, but at the same time, this ease makes the stamping batik has a lower artistic value. Furthermore, manufacturing that can be done in bulk makes the price of stamping batik much cheaper than handwriting.

Similar to the type of handwriting batik, painting batik is done manually using the help of a brush in superimposing dyes resist and applying color to certain motifs. The production of painting batik generally does not follow the standards of the applicable batik characteristics in a specific region and is more directed to the idea of a painter. As handwriting batik, this type of batik is generally expensive as a consequence of limited production with exclusive ideas.

2. The batik-producing region in Indonesia

Almost every region in Indonesia has batik with their unique motifs. Behind each of these distinctive motifs is stored a different history. Three regions are the biggest batik producers in Indonesia, namely Pekalongan, Yogyakarta, and Solo.

2.1 Pekalongan batik

Following the geographical conditions of Pekalongan on the north coast of West Java, the growth of batik in this area began during the Islamic era of the XV–XVI centuries. The development orientation of coastal batik art is influenced by the culture of the palace because batik begins within the palace walls. At present, Pekalongan is the largest coastal batik-producing region with a variety of colors and unique peculiarities and character motifs that are different from other regions producing batik. The background color on this batik does not follow the standard of other regions which generally only uses two colors. Even though it has similarities with Yogya or Solo batik, the Pekalongan Batik motif is more free and attractive because it is modified with many colors. Pekalongan Batik motifs include Jlamprang, Jawa Hokokai, pagi sore, Jawa Baru, terang bulan, tiga negeri, sogan Pekalongan, Tribusana, Petani, and dan Osdekan (Pratiwi, 2013; CV Raman Edukasitama, 2009).

2.2 Yogyakarta batik

Yogyakarta Province is one area that also produces traditional batik fabric with a variety of motifs. Yogyakarta Palace has become one of the institutions that is actively producing batik. Since the reign of Hamengkubuwono I, batik has is become one of the activities carried out by people in the palace environment, especially women. This activity is passed down from generation to generation. The courtiers who mastered batik skills then taught it to the women in the neighborhood as one of the must-have skills (Zuraida, 2010).

Batik in the Yogyakarta Palace became one of the pinnacles of the Indonesian batik art splendor. According to Kusrianto (2013), Yogyakarta batik is also known as Vorstenlenden batik, which means royal territories, such as Yogyakarta and Surakarta Palace. Geographically, batik outside the Vorstenlenden Region is called coastal batik. Because it came from one source, namely the Mataram Kingdom, the batik palace of Surakarta and Yogyakarta had many similarities in motives, although in subsequent developments began to emerge several differences. Yogyakarta batik has a characteristic of the wider white fields as well as the application of the brown and black color dominance and also geometric motifs that are made much larger than the Surakarta batik (El-Kautsar, 2018). Yogyakarta batik also has a variety of motifs that are very numerous and beautiful, ranging from ceplok, kawung, parang (parang barong and parang rusak), lereng, nitik, to isen motifs.

2.3 Solo batik

Beginning with the Pajang Kingdom more than four centuries ago, Kyai Ageng Henis was the first figure to introduce batik in the village of Laweyan, which at that time was part of the Pajang kingdom. The Solo batik characteristics, especially the classic or traditional ones, are easy to recognize because they are different from the others, especially in terms of color. Solo batik has a more static color, which symbolizes the land color such as brown, beige, white, and blue. Specifically, Solo batik is dominated by brown color (sogan) as a symbol of fertile clay soil. Furthermore, the combination of small geometric shapes located in the dyeing is another characteristic of Solo batik.

The Solo batik motifs, variety that is influenced by symbolic meanings of Hindu culture are the next characteristics of Solo batik. Various distinctive motifs include Parang Kusumo, Sidoasih, Truntum, Kawung, and Sekar Jagat (Robi'ah) (Fig. 3.1).

FIGURE 3.1 (A) Pekalongan; (B) Yogyakarta; (C) Solo batik characteristic motifs.

3. Positive and negative impacts of the batik industry in Indonesia

The batik industry has contributed to Indonesia's gross domestic product an average of 104.638 trillion rupiahs in 2002–2006 and was able to absorb an average of 5,400,000 workers/year. In the same year, productivity of 19,500,000 rupiah/worker/year was produced. These worker's productivity is higher than the national productivity of workers/year (18 million rupiahs). In 2006, Indonesia's batik industry exported 81.5 trillion rupiahs. This value is 9.13% of total national exports (Kompas, 2008). This condition ranks Indonesia 43rd in the creative economy index published by the Economic Forum (World Economic Forum, 2007; Gatut and Aryanto, 2010).

More specifically, Central Java Batik exports to the United States reached 29.3 million USD in 2007. This figure shows an increase of 20.24% compared with exports of the previous year (24.4 million USD). The export value reflects 36.46% of Indonesia's total batik exports in 2007 (Surakarta, 2009). A total of 48,300 batik business units with 792,300 workers were recorded this year. The value of batik production this year has reached 2.9 trillion rupiahs. However, various factors ranging from rising prices of raw materials to the lack of batik workers from time to time until the Government's policy has caused an unstable rate of this industry development in the period 2007–2009.

UNESCO's recognition of batik, as one of Indonesia's original cultural heritages in 2009 (Kusumawati et al., 2012; Mukimin et al., 2018), has increased the popularity of batik so that it leads to the stable development of the industry going forward even today. This recognition is the background of the growth in the number of small and medium industries and the value of Indonesia's batik exports significantly from year to year. Data from the Industry Ministry of Indonesia Republic further strengthen this fact, which has reported an increase in the number of batik business units by 14.7% in 2011–2015, from 41,623 units to 47,755 units. The same increase was also detected in the number of workers who could be absorbed by the creative industries during 2011–2015. 14.7% of growth has shown an upward trend in employment in this industry (Table 3.1).

However, this positive contribution was followed by several negative impacts on the environment because wastewater was produced. Like the home industry in general, the batik industry also produces several impacts that are classified into positive and negative impacts. The concomitant consumption of water at each stage of production and the absence of waste treatment installations in the batik home industry cause an intensive volume of wastewater generated by this industry. Batik liquid waste generally comes from the process of dyeing, dye removal from fabric fibers, and rinsing, which are mostly discharged to the nearest body of water without proper treatment. Some other nonbiodegradable pollutants, including iron (II) sulfate ($FeSO_4$), alum ($K_2SO_4.Al_2(SO_4)_3.24H_2O$), calcium oxide (CaO), and wax (beeswax), are also

TABLE 3.1 Batik business development in Indonesia in 2011–2015 (Pujiastuti, 2015).

Year	Business unit	Production value (trillion rupiahs)	Raw material (trillion rupiahs)	Export value (trillion rupiahs)
2011	41,623	4.137	1.994	43.961
2012	43,704	4.344	2.094	46.159
2013	45,015	4.474	2.157	47.543
2014	46,365	4.608	2.221	48.970
2015	47,755	4.746	2.288	50.439

found contained in this wastewater. The presence of these pollutants further increases the potential for environmental pollution, especially in aquatic environments. This condition will eventually lead to some environmental and health problems (Mukimin et al., 2018; Birgani et al., 2016).

Sustisna et al. (2016) reported that in the majority of textiles/batik synthetic dyes, such as black, red, and yellow gold remazole, only 5% were able to bind to fibers and the rest (95%) would be disposed of as waste. While the complex molecular structure makes most textile dyes have high stability, it tends to be nonbiodegradable in nature. In subsequent developments, increasing pH, chemical oxygen demand (COD), total suspended solid (TSS), and concentrated color appearance will appear along with the higher concentration in nature (Mukimin et al., 2018; Minke and Rott, 1994; Stylidi, 2003; Kitture et al., 2011).

A number of cases of water pollution related to the presence of the batik industry have been found in Indonesia, where according to the Water Resources Management Office, the four rivers in Pekalongan, which is one of the largest batik-producing regions in Indonesia, have been contaminated so that they can no longer be used according to community needs. Water from these four rivers can no longer even be used to water rice fields or fishing ponds. This records that at least 5 million liters of waste are produced every day by the batik industry throughout Pekalongan City.

To overcome this problem, Pekalongan has built a total of four communal Wastewater Treatment Plants (WWTPs), namely in the areas of Beard, Kauman, Pringlangu, and Banyurip. Not only that, as many as 80 small WWTPs were built through the Community Empowerment Institution budget spread in four subdistricts. However, of the total waste produced by the batik industry, all WWTPs in Pekalongan City can only accommodate 45%, while the rest is just thrown into the river. This condition has become the background of the Pekalongan City Environment Agency to encourage batik artisans to be able to make communal WWTPs in their respective business environments. Furthermore, the results of the observation by the Environment Agency of Pekalongan City showed that the waste dumped into the river was not only from Pekalongan City but also from the surrounding area. This is evidenced by the detection of similar pollution in the river border area (Susanto, 2019).

Similar cases of pollution are also found in the Cirebon. As reported by Sulaksono et al. (2015), Unesco's acknowledgment that batik was a native Indonesian culture in 2009 had caused a significant increase in the number of batik SMEs in Cirebon to 530 SMEs with total labor of 4408 people. Of the 530 SMEs, as many as 302 SMEs are concentrated in the Trusmi Wetan and Kulon, Kali Tengah, and Wotgali areas with a total production capacity of 454,625 m/year. With this production capacity, Sulaksono et al. (2015) reported the total pollutant load of wastewater in the Trusmi batik cluster which reached 5.9–39.5 tons/year for biological oxygen demand (BOD_5), 112–426 tons/year for COD, and 4.88–16.3 tons/year for TSS. The poor condition of the waters in Cirebon is strengthened by the news of the national newspaper, which states that rivers in the Cirebon region suffered up to 90% damage and the majority of the river was heavily polluted. As a percentage, 80% of damage is caused by human behavior, while the remaining 20% is caused by natural factors. Of the 54 km coast, there are 18 rivers with 6 of them already heavily polluted. The rivers are the Ciberes, Cipager, Cimanis, Jamblang, Kumpul Kuista, and Soka Rivers. However, IKM batik is not the only cause of pollution in this region (Hanan, 2016).

As the two biggest batik-producing regions in Indonesia, water pollution in the Yogyakarta and Solo regions is unavoidable. As reported by the Solo City Environment Agency, environmental pollution due to batik industry waste is increasingly worrying. The batik industry caused the condition of the Bengawan Solo tributary which flows in the middle of the Solo city to be polluted by heavy metals. River water pollution in the Solo due to batik industry wastewater was found to have exceeded the quality standard threshold. Two rivers, namely Premulung and Jenes, are even included in the category of heavy pollution. This assessment is based on the analysis of water quality parameter results, such as BOD and COD levels, heavy metal content, color, odor, and taste.

Referring to the results of the analysis, Kali Pepe, Kali Anyar, and Gajahputih are included in the Class III classification category of river water pollution. This means that even though the water is polluted with sewage, it can still be used for agriculture and fisheries. Meanwhile, two rivers polluted by heavy metals, the Premulung and Jenes Rivers, have been classified as Class IV, which are not even suitable for agricultural or fisheries irrigation. This is because the chemical content caused by batik wastewater is too high. To overcome this problem, a variety of efforts continue to be promoted by the City Government, including building a communal WWTP (IPAL). However, the existence of communal WWTPs is also still very minimal, as many as two units, each located in Laweyan and Sondakan. The existence of the two WWTPs is unable to accommodate the waste from all batik artisans. Laweyan Communal WWTP, for example, is only able to accommodate nine business units from around 25 MSMEs in the region. Likewise with the communal WWTP in Sondakan, from about 10 batiks SMEs in the region, only 6 of which can be managed by the WWTP (Ermawati, 2017; Balai Besar Kerajinan dan Batik, 2010).

In line with pollution in Solo, the recognition of batik as Indonesia's original cultural heritage by Unesco also triggered an increase in batik production in Yogyakarta which led to the emergence of several environmental problems in the region. Detected environmental problems are dominated by large amounts of wastewater generated from the batik production process. This condition is estimated to be related to the high water consumption in each stage of batik production on average, which reaches 25–50 m^3/m of batik fabric. This fact is reinforced by data from the Ministry of Industry of the Indonesia Republic in 2017 which calculates the need for clean water in the total batik production in Indonesia per year which reaches 500,000,000 m/year. With this total production, the batik industry in Indonesia needs 25,000,000 m^3/year of clean water. This value is equivalent to the provision of clean water needs for 2500 households (Balai Besar Kerajinan dan Batik, 2010). Of this volume, almost 85% will become batik wastewater with a thick color and a pungent odor. Besides, batik wastewater produced generally has characteristics of high temperature, acidity (pH), BOD, COD, and TSS (Rohasliney and Subki, 2011). High temperatures trigger a decrease in the content of dissolved oxygen in the water so that it has the potential to cause the death of organisms and a high content of organic waste will increase the conversion of nitrogen into nitrate compounds which cause a foul odor (Sastrawijaya, 1991; Indrayani, 2018).

Four rivers in Yogyakarta, namely Winongo, Code, Manunggal, and Gajah Wong, are listed in the critical category. The four rivers that flow from the west to the east of Yogyakarta City are included in the heavily polluted category. The pollution content in the four reaches 200–300 mg/L, far exceeding the established standard, which is 50 mg/L. The most severe pollution points are in the area around dams, riverbanks, and bridges. Disposing of solid or wastewater into rivers without prior treatment makes river water conditions in Yogya increasingly worrisome. This fact is supported by the results of the KLHK survey which shows that as many as 73% of rivers in Indonesia, including Yogya, are in the heavily polluted category (Susanto, 2019).

3.1 Management of the negative impact

The textile industry consumes large amounts of water at each stage of the process. This condition makes this industry one of the largest industrial wastewater producers in the world (Anjaneyulu et al., 2005; Nunez et al.,

2019). Textile industry wastewater is characterized by some characteristics, such as high content of organic and inorganic pollutants, heavy metals, and dyes (Nunez et al., 2019; Verma et al., 2012). In this regard, it is very important to find a treatment that allows the reuse of treated water so that it can reduce further water consumption in the textile industry.

Batik as part of the textile industry also triggers the same negative impact. Many physical, chemical, as well as biological treatment has been attempted and evaluated to overcome this problem. Physical processes such as coagulation, electrocoagulation, membrane filtration, and adsorption, as well as chemicals such as ozonation and photocatalysis are often applied for the separation of dye material from wastewater (Dutta et al., 2018 May; Khemila et al., 2018; Melo et al., 2018; Moawed and El-Shahat, 2016; Nidheesh et al., 2018; Moawed et al., 2018). In some cases, biological processes are also applied for textile dye removal. The bacteria, archaea, fungi, and algae/plant metabolic ability is detected capable of oxidizing or reducing the organic as well as inorganic compounds (Bhatia et al., 2017; Khandare and Govindwar, 2015; Paździor et al., 2019).

Lee et al. (2015) recommend the use of aluminum sulfate–aloe vera (ALAV) and magnesium sulfate–aloe vera (MGAV) hybrid coagulants for methylene blue (MB) treatment. The results showed that ALAV is capable of producing dye removal percentages of 50%–55% at pH 6, while MGAV at 60%–70% at pH 12.5. Azzaz et al. (2018) reported a combination of cationic dye adsorption treatment using sawdust and anodic oxidation to increase the potential for the dyes and water reuse by the waste-producing industry. The results show the acquisition of MB with a purity level of up to 92%. The high level of purity of the dye produced by desorption from sawdust material increases the potential for reuse of MB wastewater. Shen et al. 2019 in their publications also reported on the use of coagulation for the treatment of textile industry wastewater. Through two stages of the process, 88% COD, 86% PVA, and 99% dye removal have been produced in this process. In line with Lee at al. (2015) and Shen et al. (2019), Kuppusamy et al. (2017) have published the MB, acridine orange (AO), and malachite green (MG) separation using oak acorn peel coagulants. The results show a successful separation of 82% for MB, 86% for AO, and 66% for MG.

Zodi et al. (2013) reported their efforts in separating Direct Red 81 dyes in the textile industry wastewater using continuous flow electrocoagulation. The results show the efficiency of the separation of the DR 81 dye reaching 90.2%, COD reduction of 76.1%, and turbidity reduction of 99.90%. In line with Zodi et al. (2013); Khemila et al. (2018) published the Acid Red 336 textile dye wastewater treatment obtained from the Tindal Textile Industry located in M'Silla town (south-eastern of Algeria) using photovoltaic electrocoagulation. The results show a successful reduction in turbidity up to 99% and dye separation up to 95%. Nunez et al. (2019) in their publications reported the effectiveness of electrocoagulation in removing pollutants in the textile industry wastewater. The results show the success of dye separation, which in this case is a reactive dye (Optilan MF and Novacron NF), which reaches 86%, turbidity 82%, and COD 59%. To study the potential of reusing water from sewage treatment, Nunez et al. (2019) have used it in dyeing wool fibers, and the results show no significant effect on the color quality of wool produced.

Membrane technology is a further alternative that has also been successfully used to remove dyes in textile wastewater and has been proven to produce cost savings obtained from the reuse of recycled water in industrial processes (Lucas et al., 2007; Alventosa-deLara et al., 2012). Alventosa-deLara et al. (2012) have reported the separation of Reactive Black 5 (RB5) using ultrafiltration (UF) membranes from ceramic. Utilization of INSIDE CéRAM (Tami Industries) multichannel tubular ceramic membrane has resulted in a successful separation of 95.2%. In line with this, Li et al. (2019) have attempted to remove cationic, anionic, and neutral dyes using UiO-66-NH_2/graphene oxide (GO) composite membrane. The experimental results for mixed dye filtration show that membranes can effectively remove MB, RB, and CR in binary and ternary systems through secondary filtering. Only on MB/CR binary systems, MB and CR can be removed simultaneously through one-time filtering. Khumalo et al. (2019) reported the removal of Congo Red dyes using UF nano-enabled PSf/PVA/ZnO membranes. The results show the success of the Congo Red removal of 53.5%. This result will decrease along with the repetition of the separation process.

The adsorption process is also one of the promising technologies in this field, through the transfer of the dye phase to the adsorbent (Moawed et al., 2018; Yang et al., 2018; Yang and Hong, 2018). This procedure also allows regeneration, recovery, and recycling of adsorbent material, offering relatively more affordable costs and shorter processing times compared with other treatment methods (Moawed et al., 2018; Li et al., 2018; Lou et al., 2018; Zheng et al., 2018). Various dyeing wastewater treatment methods, including physical, chemical, and biological techniques, have been optimized for dye removal (Gupta and Suhas, 2009). However, complex chemical structures have made them have high resistance to degradation. Dyes are generally stable against erobic digestion, light, heat, as well as oxidizing agents (Crini, 2006). These weaknesses have made the adsorption technology widely used to treat dyeing liquid waste, mainly because of its economical and highly efficient cost (Qadeer, 2007). The conventional adsorbent that is most often used because of its amazing absorption performance is activated carbon (Gupta and Suhas, 2009). However, high prices and complicated regeneration processes have ultimately limited its application (Deniz and Kepekci, 2016; Silva et al., 2016). Besides, this material is ineffective and has a low sensitivity to dispersed dyes (Crini, 2006). To reduce costs and increase adsorption efficiency, many alternative materials have been engineered to outperform ancient activated carbon capabilities. In recent decades, some new clay/zeolite and their composites, biosorbents, agricultural solid waste, industrial by-products, and composite-based adsorbents as well as various other materials have been developed to remove dyes. Chaari et al. (2019) reported the use of adsorbents from natural clays with high smectite contents from the Jebel Romana (South Tunisia) and Khledia (North Tunisia) regions for anionic (Acid Brown 75, AB75) and cationic (Basic Yellow 28, BY28) dyes separation. The maximum adsorption capacity based on the Langmuir model is 76.92 mg/g for BY28 and 8.33 mg/g for AB75. Bentahar et al. (2019) have also studied the removal of base dyes using Moroccan clays. Based on the Langmuir model, the maximum monolayer adsorption obtained for MB dyes was 166.7 mg/g and 18.7 mg/g, respectively, by Rhassoul clay and Red clay.

Furthermore, adsorption technology using biosorbent also showed no less good results. The use of microbial biomass as a biosorbent is considered beneficial because of the low contamination of hazardous wastes and the ease of regeneration and reuse (Balai Besar Kerajinan dan Batik, 2010). Juchen et al. (2018) published their efforts to biosorption of reactive blue BF-5G using malt baggase, a by-product of the brewing process. The Langmuir Isotherm model shows a maximum biosorption capacity of 42.58 mg/g of 14 g/L malt baggase at pH 3. While Tonato et al. (2019) reported biosorption of procion red H-E7B (PR H-E7B) dyes from aqueous solution using residual biomass of *Nigrospora* sp. obtained from oil extraction using supercritical technology. Based on the Freundlich isotherm model, the maximum capacity for biosorption obtained at 298 K is 188.79 mg/g.

Synthetic dyes in textile wastewater have caused many severe biological problems. This waste is very colorful with varying degrees of COD. Simple biological and chemical treatments are inadequate for decolorization and reduction of organic matter (Koch et al., 2002; Swaminathan et al., 2005). High barriers to biochemical oxidation make chemical oxidation necessary for the toxic complex organic compounds treatment. In this case, ozone (O3) becomes a versatile and environmentally friendly oxidizing agent. Wang et al. (2003) published the results of his research on the removal of C.I Remazol Black 5 dyes with Advanced Oxidation Process (AOP) uses ozone. The results showed that after 360 min of ozonation, there was a decrease in dye levels up to 100%, COD_t/COD_0 44%, and TOC/TOC_0 25%. However, at the same time, an increase in BOD_5/COD_t was generated by 13.8%. This phenomenon is predicted to be the result of an increase in metabolic activity due to the abundance of more easily degraded compounds that appear to be less toxic. In line with Wang et al. (2003), Swaninatham et al. (2005) reported the decomposition of intermediate dyes ((H-acid) 1-amino-8-naphthol-3,6 disulfonic acid) using ozonation. Application of pH 11.5 in 30 min of the process with 4.33 mg/L of ozone can produce 98.7% dye and 87.8% COD removal. This procedure has produced acetic and oxalate acid as the main products of oxidation. At almost the same time, a literature reports that AOP uses UV-TiO_2 as one of the most effective dye removal methods because of its ability to produce hydroxide groups from the TiO_2 photocatalytic activity and its flexibility under various experimental conditions. This process is reported to be able to produce the complete mineralization of organic compounds into carbon dioxide (CO_2) (Al-Mamun et al., 2019). Nguyen et al. (2018) reported the degradation of MB and methyl orange (MO) dyes by photocatalysis with palladium-doped TiO_2. The results show 100% degradation for both, after 80 and 120 min of processing, respectively. In line with this, Sane et al. (2018) reported a similar finding, in which efforts were made to remove some reactive dyes, including Reactive Green 19 (RG19), Reactive Orange 84 (RO84), Reactive Violet 1 (RV1), and Reactive Yellow 81 (RY81), with the photocatalytic degradation method using CeO_2. The results show that 95% of removal has been obtained for RG19, RO84, and RY81. A higher percentage of removal was generated for RV1, which is more than 99.9%. Meanwhile, a dye removal by UV and simulated solar photocatalysis using TiO_2/ZnO/rGO has been published by Hieu et al. (2020). TZR synergism shows higher photocatalytic activity compared with each of its constituents. After 120 min of UV radiation, 99.6% degradation and 59.0% mineralization was produced for MB dyes. Comparable removal efficiency is also shown for rhodamine-B (RB) and MO. After 180 min of UV-C exposure, 99.2% degradation and 48.0% mineralization for RB and 99.4% degradation and 44.2% mineralization for MO were produced.

Compared with physical and chemical methods, biological methods present advantages in being cheaper and environmentally friendly. The catalytic activity of dye decolorization was found to be produced by various erobic and anaerobic bacteria as well as fungi. Dos Santos et al. (2007) reported the dye decolorization acceleration by adding mediation compounds and/or changing the process conditions (Dos Santos et al., 2007; Turgay et al., 2011). Processes in microbial systems generally result from the actions of many different microorganisms with high levels of interaction with each other (Turgay et al., 2011; Vandecasteele, 2003). Consortia generally have better resistance to changes than pure culture (Turgay et al., 2011; Forgasc et al., 2004; Watanabe and Baker, 2000). The consortium can also present a higher chance of complete mineralization (Turgay et al., 2011; Pearce et al., 2003). Because of its superior properties, including its ability to produce electrons that can be used to cut azo bonds, anaerobic microorganisms are more widely used to decolorize azo dyes in textile wastewater (Turgay et al., 2011; Delee et al., 1998; Laurenco et al., 2001).

In addition to bacteria, the metabolism of archaea, fungi, algae, or specific plants is often used to trigger oxidation or decrease the content of the organic and inorganic compounds (Bhatia et al., 2017; Khandare and Govindwar, 2015; Pazdzior et al., 2018). Optimization of processes in nature is ensured by providing appropriate operational conditions, such as the availability of oxygen concentration, pH, and access to the nutrients needed (Khandare and Govindwar, 2015; Pazdzior et al., 2018; Punzi et al., 2015).

System applications including anaerobic and erobic processes have been extensively studied to investigate the completeness of the most widely used dyes (azo dyes) biodegradation. Rasool and Lee (2016) studied the textile wastewater treatment using anaerobic biological sulfate reduction process. The results show that the presence of lactate and ethanol as cosubstrate has produced the highest efficiency of contaminant removal, which has detected a decrease in COD of 3000 mg/L, the color of 98.23%, and sulfate of 98.42%. In line with this, Shoukat et al. (2019) have reported treatment using a combined hybrid anaerobic–aerobic treatment and the results show that the proposed system effectively produces a removal efficiency of 99.5% for COD, 99.3% for Total Kjeldahl Nitrogen, and 78.4% for dyes.

Despite having high removal efficiency, physical and chemical treatment applications are limited by the need for a large number of chemical reagents in the process and posttreatment regeneration (Merzuok et al., 2010). In line with this, biological treatments are relatively less desirable due to the complexity of the operational conditions required, such as the need for large contact spaces, high sensitivity to wastewater complexity, and long processing times (Nguluben et al., 2017).

The presence of various wastewater treatment technologies in the textile industry, especially dyes, has shown to all of us how almost all human beings on this earth are more focused on processing rather than prevention. This is unfortunate considering that as the owner of the third-largest area of tropical rainforest in the world, Indonesia has a variety of high abundance biological resources that can be utilized as potential alternative materials that can replace the role of synthetic dyes that are toxic and not environmentally friendly in dyeing textiles, especially batik. Through the application of natural dyeing in the batik production process, it will not only be able to suppress the presence of toxic and hazardous materials in textile wastewater but also be able to present an improved quality of health of its users, especially because of the high content of antioxidant and antimicrobial compounds in natural dyes.

3.2 Natural dyeing as the most effective prevention solution

Natural dyes have been used for thousands of years. Although it began to be abandoned shortly after synthetic dyes were discovered, now natural dyes have begun to regain their place. Serious problems, such as visible residues in wastewater (Glower, 1998; Ayed et al., 2011; Solís et al., 2012; Li et al., 2015; Frose et al., 2018) and the formation of the toxic amines (Frose et al., 2018; Yamjala et al., 2016; Sarıkaya et al., 2012) by synthetic dyes, have triggered the re-emergence of the natural dyes used. Under these conditions, natural dyes offer an attractive alternative. However, in practice, the resistance of these types of dyes to various environmental conditions, such as washing, drying with sunlight, exposure to heat during ironing, and perspiration, are the main challenges that must be answered immediately. The study results show that discoloration due to exposure to sunlight or UV light, in particular, has shown the worst fastness of yellow coloring, followed by red, indigo, and finally brown (Taylor, 1986). In general, these results indicate that the "darker" color is predicted to be more stable to light.

Increased environmental awareness has supported the development of environmentally friendly textile products and processes (Haws et al., 2014; Silva et al., 2020). Many textile processes and products are reevaluated due to the resulting social and environmental damage, and hence various alternative production and consumption continue to be explored by many researchers (Silva et al., 2020; Niinimaki and Hassi, 2011). Because it triggers fewer allergies and is less toxic and produces biodegradable wastewater, natural dyes are a more environmentally friendly alternative. Synthetic dyes, especially azo dyes that release carcinogenic aromatic amines, present a risk of allergies and cancer for users (Silva et al., 2020; Shahid et al., 2013; Brookstein, 2009). This is the background of restrictions in some countries (Silva et al., 2020; European Parliament and Council of the European Union, 2002). The potential of Indonesia's large biological resources for textile natural dyes makes it important to obtain extractions that allow their use in a variety of dye media. This has become one of the competitive strategies that distinguishes it from other textile companies in meeting the market segments of textile products that are environmentally friendly and user-friendly (Silva et al., 2020; Niinimaki and Hassi, 2011; Jena et al., 2015). This condition has led to increased research on natural dyes in recent years.

Natural dyes have been around for hundreds of years. For example, indigo has been known for more than 5000 years and catechu has been used as a chocolate dye in India for more than 2000 years (Wickens, 1983; Zerin et al., 2019). Not limited to the two materials, a large number of different sources of natural dyes, such as flowering plants, fungi, mosses, insects, shellfish, and various minerals, are also available in abundance around us.

Bhuiyan et al. (2017) reported its success in cotton fibers dyeing using henna (*Lawsonia inermis* L). Henna is also known as lawsone or hennotannic acid (see Fig. 3.2A1). The red-orange pigment is the main element of this material (Rehman et al., 2012). Based on industry classification, lawsone is Natural Orange 6 and CI 75,480 known as 2-hydroxy-1,4-naphthoquinone. This material belongs to the group of substantive dyes for protein fibers and other textile fibers with the appearance of the orange shades on the substrate (Guljarani et al., 1992). However, as published by Bhuiyan et al. (2017), the application of mordant gives a significant influence on the color shades produced by Lawsone, where when dyeing is done without mordanting, light shades of orange are produced, while when accompanied by mordanting using iron, copper, alumina, and rare earth, respectively, black-gray, gray-green, orange, and orange colors have been produced.

Ford et al. (2018) in his publication reported dyeing textiles using dyer's madder root (*Rubia tinctorum* L.). This dye is known as C. I. Natural Red 8 in

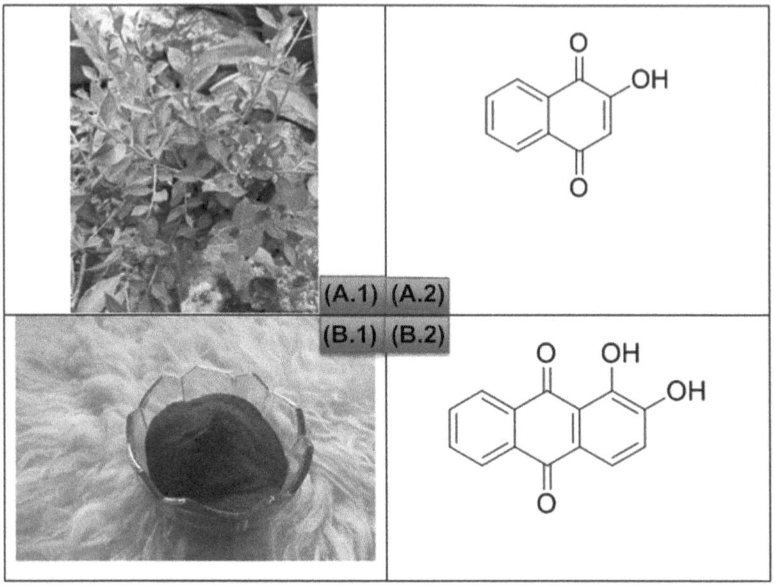

FIGURE 3.2 (A1) *Lawsonia inermis* L; (B1) dye's madder; and the molecular structure of the (A2) red-orange pigment (2-hydroxy-1,4-naphthoquinone); and (B2) alizarin, one of the anthraquinone compounds in dye's madder root.

the Color Index. This material has been used as a red dye for centuries. More than 35 anthraquinoid compounds can be extracted from the old roots of this plant (Ford et al., 2018; Derksen et al., 2002; Wijnsma and Verpoorte, 1986). Fig. 3.2B2 shows the molecular structure of Alizarin, the anthraquinone compounds contained in this material.

On the same track, Adeel et al. (2017) showed the dyeing results using marigolds flower accompanied by the application of γ-radiation. This flower contains lutein, a natural carotenoid dye, along with flavonoids and vitamin C (Vankar and Shanker, 2009; Prabhu and Bhute, 2012). Increased strength of the yellow-orange color is obtained through the use of 7% tannic acid and 5% Cu, respectively, as pre- and post-mordant. Furthermore, the optimal operational conditions have resulted in 3–4 to 4 fastness to light, 4 to 4–5 for wet washing as a wet and dry rub.

In 2011, Samanta and Konar tested the use of catechu from *Areca catechu* heartwood as a natural textile dye (Samanta and Konar, 2011). The application of Catechu in textile dyeing produces a dark brown, golden brown, and yellowish-brown color. Besides having excellent compatibility with other dyes, Catechu has high fastness. The better Catechu-fiber reaction can be obtained by enzymes pretreatment and immersed fiber exposure in the air before rinsing. The results show a better and faster dyes absorption, and better dyeing results in cotton fabrics compared with metal mordants pretreatment (Vankar and Shanker, 2008; Vankar and Shukla, 2019).

In textile dyeing, Rehman et al. (2013) reported a textile dyeing using a natural dye extract of flavonoids from irradiated onion shell powder. The maximum brown color strength and acceptable fastness are obtained from dyeing in irradiated textile fibers at 60 °C and by using 10% and 6% alum, respectively, as pre- and post-mordant. In line with this, Siddiqui et al. (2017) also reported the natural dyeing of cotton and silk fabrics using an onion (*Allium cepa*) shell (Fig. 3.3).

Orellana bixa seeds are also reported to have potential as natural textile dyes. Selvi et al. (2013) reported the success of natural dyeing and finishing on the leather using this dye. A yellowish-red appearance with good fastness (scale 4/5 on the gray scale) against wet and dry rubbing. Fairly good color fastness (scale 3 on the gray scale) to light is also obtained from this dyeing. In line with this, Sinha et al. (2013) have published optimization of the extraction of yellowish-red dyeing from Bixa Orellana seeds by microwave-assisted extraction. Bixa Orellana and the annatto molecular structure that dominate the color of this commodity are shown in Fig. 3.4B1,2.

FIGURE 3.3 (A1) Marigold flower; (B1) Catechu; and the molecular structure of (A2) lutein; (B2) tannin.

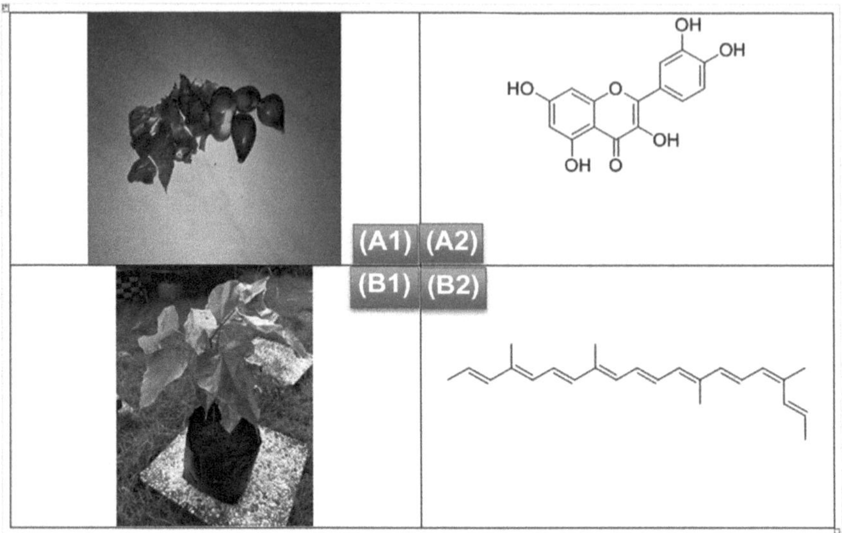

FIGURE 3.4 (A1) Onion shell; (B1) Orellana's bixa; and the molecular structure of (A2) quercetin; (B2) annatto.

Bougainvillea is the next commodity for synthetic dyes substitute. Kumar et al. (2013) reported the utilization of a red-purple dye extract from the flower bracts of *Bougainvillea glabra* (Fig. 3.5 A1). Betalain pigments (see Fig. 3.5 A2), especially betacyanin, are responsible for the bract color appearance (Kumar et al., 2013; Moreno et al., 2008; Piattelli, 1981). Betacyanin is a water-soluble red-purple pigment (Kumar et al., 2013; Moreno et al., 2008) (see Fig. 3.5 A3), with a color that is not pH-dependent and has higher stability than anthocyanin (Kumar et al., 2013; Tanaka et al., 2008). In acid and base media, dark magenta and bright yellow will be produced, respectively. Wool dyeing is generally optimized using lanthanum and chromium mordants (Vankar and Shukla, 2019; Mathur and Bhandari, 2001). The use of lanthanum and chromium combination is expected to reduce the amount of chromium in wastewater. Although it can minimize damage to wool fibers, the use of chromium mordant has an opposite effect to the concept of natural dyeing.

The ability of purple cabbage or *Brassica oleracea* var. Capitata F. Rubra in natural dyeing has also been long known. Following the change in chromophore produced by each pH, in acidic media, this pigment forms a reddish color, whereas in alkaline media it will form a blue color (see Fig. 3.5 B2). The multicolor appearance in textiles was evaluated by Senthilkumar et al. (2002), Vankar and Shukla (2019). Ticha et al. (2016) became the next researcher that studies this dyeing. Ticha et al. (2016) tried to increase the dyeability of this commodity in modified cotton fibers using ultrasonic energy. In this case, cationization pretreatment has been proven to be able to improve the better

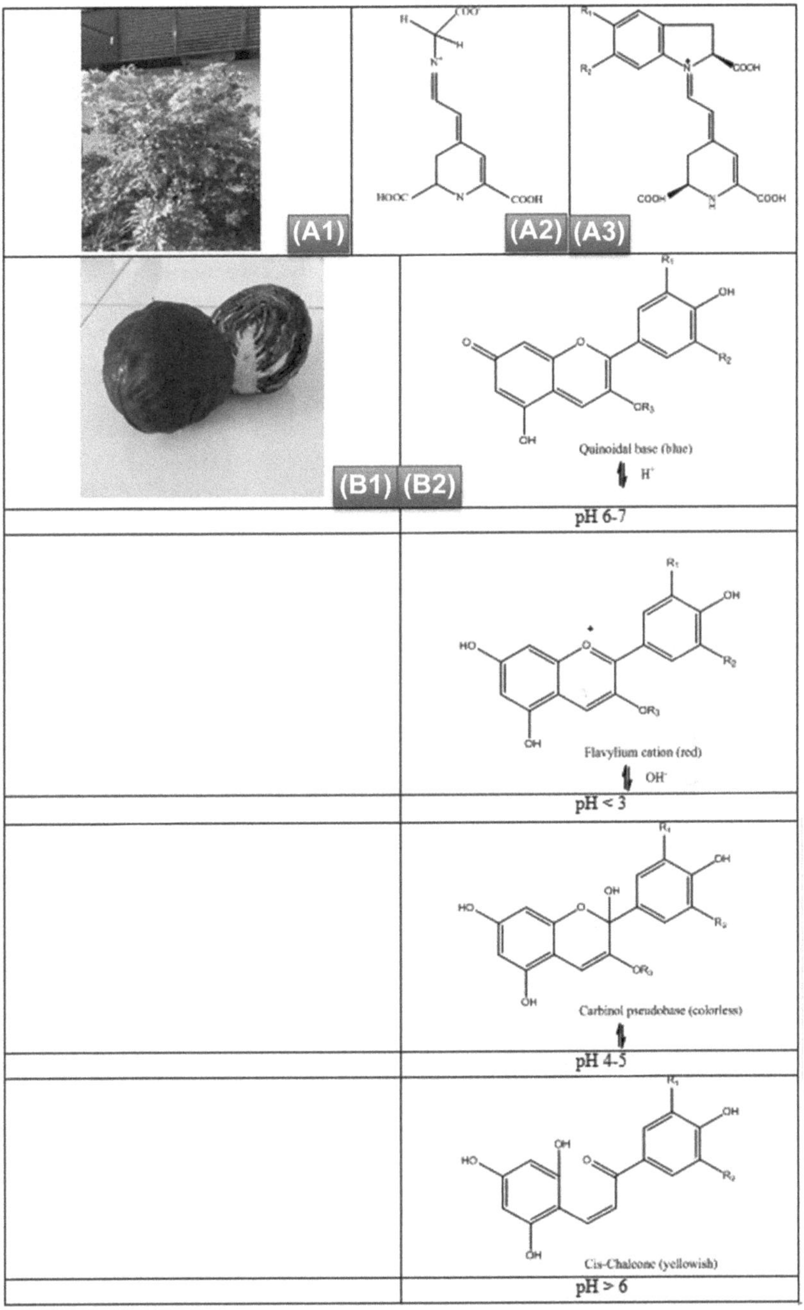

FIGURE 3.5 (A1) Bougainvillea; (B1) Purple cabbage; and the molecular structure of (A2) betalain; (A3) betacyanin; (B2) anthocyanin in a different pH.

quality and fastness in cotton dyeing using red cabbage. The specific color of purple cabbage is produced by the anthocyanin pigment. Silk and wool can also be dyed using purple cabbage. The different mordants application has triggered the appearance of the different shades. The stannic produces better color strength than alum and copper mordant in silk. Slightly different, the stannous chloride mordant shows much better color strength than stannic and potassium dichromate in wool (Vankar and Shukla, 2019; Vankar and Shanker, 2006).

Not inferior to the others, indigo is one of the earliest natural dyes found and most often used in textiles (Fig. 3.6 A). Indigo is a water-insoluble blue pigment from the leaves of the Indigo tinctoria (David and Richardson, 2016), with a molecular structure as shown in Fig. 3.6 B, which is known for its excellent brightness and color stability on textiles. Blyskal (2016) has evaluated its application to wool and its effect on interactions with microorganisms. Indigo dyeing shows the characteristics (L* 20.91; a* -0.80; b* -10.89) in wool fiber. Furthermore, better resistance to microbes was found in the fiber treated by indigo.

Tannin-rich fruits, leaves, and myrobalan skins have been widely used as triggers for the colors yellow, brown, and black in cotton and silk and as a substitute for metal mordants. Ketapang or Indian almond (*Terminalia catappa*), which is often also called Javanese almond or sea, is one of the most famous myrobalans (Fig. 3.7). Its use with aluminum mordants produces yellow and black when shared with iron mordants (Asian Textile Studies, 2016). Iron acetate fixation produces black, alizarin and alum trigger the appearance of red, while the combination of the two has produced a brown color. Slightly different, Myrobalan wood tends to trigger reddish-brown color (Asian Textile Studies, 2016). Cotton and silk dyeing using myrobalan is done by the post-mordant method with metal salts other than lead. Specifically, the result of silk dyeing then must be coated with soda ash paste.

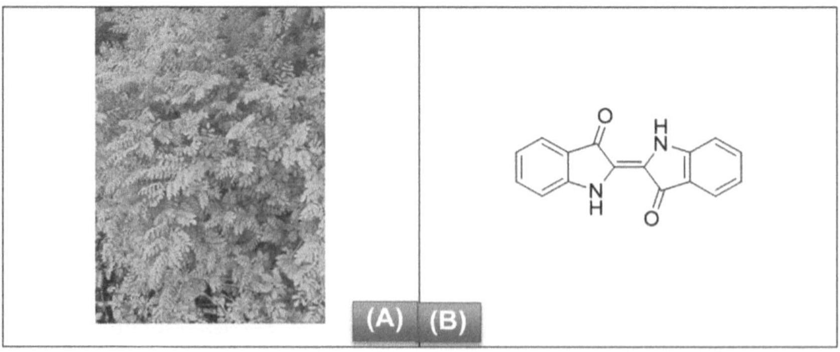

FIGURE 3.6 (A) Indigo tinctorial; and the molecular structure of (B) Indigo.

FIGURE 3.7 Myrobalan.

Water guava (*Syzygium aqueum*) is a type of biological resource that is easily found in almost every region in Indonesia (Danarto et al., 2019; Kusumawati and Samik, 2019). As with leaves in general, the tannin content in it allows its use as a natural dyeing agent. Tannins are polyphenol compounds that are soluble in polar solvents. Its use in dyeing together with metal mordants, such as iron, calcium, copper, and magnesium, triggers the appearance of dark brown or even black (Kusumawati et al., 2018a,b).

Like water guava, although it has scattered tannin content in leaves, fruit, flowers, and skin, the tannin content in the guava (*Psidium guajava*) leaves is much more. Han and Lee (2011) have evaluated the success of silk natural dyeing using *P. guajava*. The results show that the color fastness to wet washing reaches 3–4 when used with Sn mordant and 4 with Fe mordant. Fade resistance to dry washing and sunlight also shows very good results, reaching 4–5 and 3–4, respectively, on the gray scale. Higher fastness (>4) is only obtained when mordant Fe is used. Very high antibacterial activity (99.9%) was also found in fabrics dyed with this material. Kusumawati et al. (Santoso and Wijiastuti, 2018) have evaluated the color intensity and fastness produced by this material in cotton fibers. The appearance is light brown, dark brown to black, respectively, in fixation using alum, lime, and iron (II) sulfate. Furthermore, this dyeing produces a color intensity of 90.94%–92.75% with fastness to wet washing reaching 3–4 on the gray scale.

Brazilwood (*Caesalpinia sappan* L.) is also one of the biological resources that has a high potential for natural dyeing. The main compounds of

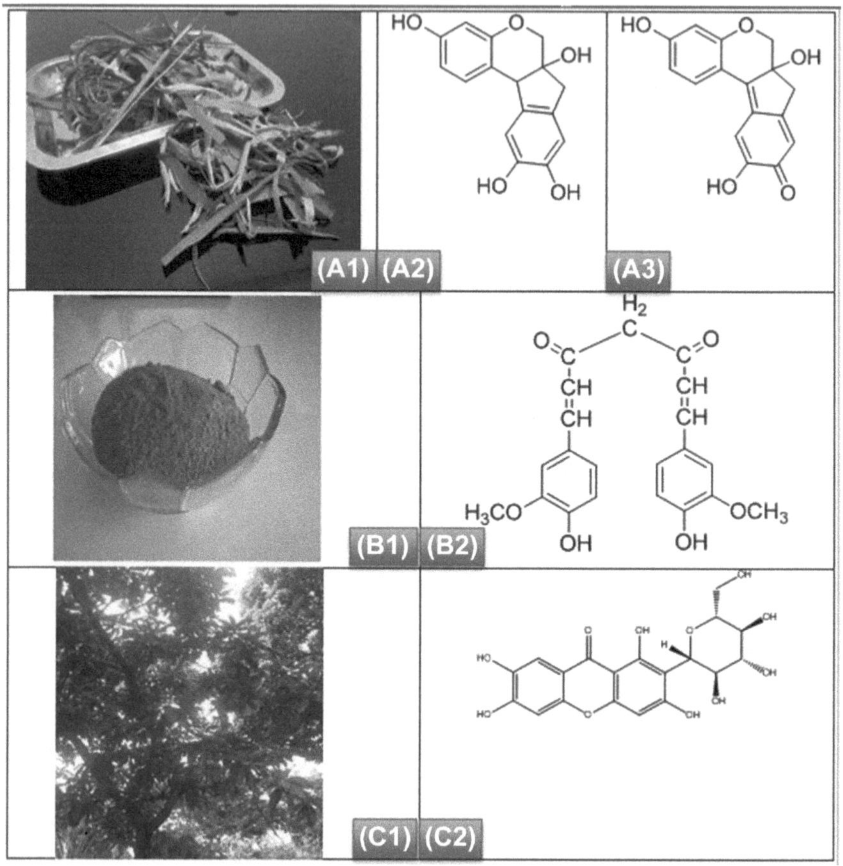

FIGURE 3.8 (A1) *Caesalpinia sappan* L.; (B1) turmeric; (C1) *Mangifera indica*; and the molecular structure of (A2) brazilin; (A3) brazilein; (B2) curcumin; (C2) mangiferin.

Brazilwood are homoisoflavonoid brazilin (Fig. 3.8A1) and brazilein (Fig. 3.8A2). Brazilein is the main chromophore in Brazilwood obtained from brazilin oxidation (Ferreira et al., 2004; Edwards et al., 2003; Nowik, 2001; Manhita et al., 2012). Manhita et al. (2012) have evaluated the effect of aging on the Brazilwood color quality, and Padfield and Landi (1966) have evaluated the fastness of this dye in wool and other fibers. Several types of mordant are used and the results show that the light fastness is still low. Zarkogianni et al. (2010) also reported cotton and wool dyeing with Brazilwood. The results show increased lightfastness through pre-mordanting applications using ferrous metals, while fastness to rubbing was found not to be improved.

Turmeric rhizomes have been widely recognized as natural dyes for cotton fibers because of their rich phenolic compounds called curcuminoids (Teli et al., 1994; Bhatti et al., 2009). More specifically, the active dyeing agent in curcuminoid is curcumin or Natural Yellow 3 (C I 75,300). The molecular structure of the curcumin is shown in Fig. 3.13B2.

Some researchers report dyeing cotton using turmeric. Bhatti et al. (2009) evaluated the effect of γ-radiation on the strength and colorfastness of *Curcuma longa* L. The yellow appearance with a color strength of more than 100% at pH 6 and increased light, wash, and rub fastness after irradiation and premordanting dyeing. Not limited to cotton dyeing, Zhou and Tang (2016) have published silk dyeing using curcumin. Specifically, Zhou and Tang (2016) studied the modification of curcumin using a reactive UV absorber and its application in silk dyeing. The yellow color arises from curcumin dyeing, while brownish-yellow appears from modified curcumin dyeing. Furthermore, curcumin dyeing has resulted in fastness on a scale of 2 to light, 4 to dry rubbing, and 3–4 to wet rubbing. While modified curcumin dyeing results in fastness on a scale of 4 to light, as well as fastness on a scale of 5, each of which is the dry and wet rub.

Mango (*Mangifera indica* L.) is a group of plants that has the potential to be used as a substitute for synthetic dyes because the leaves are one of the main sources of mangiferin, the dominant polyphenol. Furthermore, its antioxidant and antimicrobial properties will not only produce good color strength and colorfastness but also protect from free radical attacks and microorganisms (Sanchez-Sanchez et al., 2017; Kusumawati et al., 2018a,b). Baliarsingh et al. (2013) have evaluated the pre-mordanting dyeing of cotton and silk using this material. To replace the role of metal mordant which is strongly attached to the textiles natural dyeing and is considered to reduce the environmental friendliness of natural dyes, cationic (cetyltrimethylammonium bromide) and anionic surfactants (sodium dodecylbenzene sulfonate) are used. The results show a significant interaction between natural dyes with both surfactants. The use of cationic surfactants produces a higher color strength than anionics. Cotton and silk dyed with this procedure also show excellent antimicrobial activity against bacteria.

Indonesia is a tropical country with abundant natural resources, including coconut. In 2017, 2,868,686 tons of coconut were produced (Fitriyah and Ciptandi, 2018). As a side effect, at the same time 1,004,040 tons of coconut fiber waste per year were also produced. The tannin content which is quite high (3.12%) in coconut fiber makes this material potential to be used as a textile natural dye. Tannins are polyphenol compounds (quinone pigments) which trigger the appearance of brown color in the fiber (Prayitno and Nurimaniwati, 2003; Kusumawati et al., 2019; Ammayappan and Mosess, 2007; Ali et al., 2010).

Lamtoro (*Leucaena leucocephala* L.) or Chinese petai is also detected as a textiles natural color producer. The high tannin content in Lamtoro leaves triggers the appearance of yellowish-green color in the fiber (Anggrain and Hidayati, 2014). Anggraini (Anggrain and Hidayati, 2014) has studied batik dyeing using lamtoro leaves. The results show the influence of the dyeing frequency (20, 25, and 30 times) on the color evenness. However, no similar effect was found in colorfastness to washing, in which all of them produced 4 to 4–5 on the gray scale.

Cherry is a tropical fruit plant that is easily found in Indonesia. The contents of flavonoids, tannins, triterpenoids, saponins, and polyphenols make cherry leaves potential as natural dyes that also produce antioxidant and antimicrobial properties in dyed fibers. Santoso et al. (Nieuwolt, 1982) evaluated the fixer's effect on the shade, intensity, and fastness produced by this material in cotton. The results show the appearance of light brown color in alum, yellowish-brown in lime, and black in iron (II) sulfate fixation. The resulting color intensity is 79.51%–90.66% with a fastness of 3–4 on the gray scale.

Mangosteen peel is one of the plant wastes developed for the textiles natural dyeing in Indonesia (Farida and Nisa, 2015) and Malaysia (Ferreira et al., 2004), due to anthocyanin content (Chairat et al., 2007) (Fadri et al., 2015) which triggers red and tannin (Kusumawati et al., 2019) which triggers brown color. Basitah (2015) studied the use of this material for silk dyeing. The results showed the appearance of brown, red, and purple colors on silk fabrics, each after going through a pre-, simultaneous-, and post-mordanting process. Basitah also reports good fastness to light and washing. In line with this, Kusumawati et al. (2017) reported cotton dyeing using mangosteen peel waste. The results show the green appearance of iron (II) sulfate fixation, light brown from alum and dark brown from a lime. This pre-mordanting dyeing produces 90.15%–91.40% color intensity in iron (II) sulfate fixation, 61.56% –83.83% in alum, and 84.89%–92.77% in lime, with 1–10 times dyeing frequency.

4. Utilization of natural dyes in the batik production process in Indonesia

Although considered to have disadvantages compared with synthetic dyes, especially in color variations, as well as the color strength and fastness, intensive research in this field has produced many breakthroughs that can restore the position of natural dyes as potential dyes in textile production, especially batik. This is evidenced by the rampant reproduction of natural dyed batik in almost all corners of Indonesia.

4.1 Central Java

From Semarang-Central Java, "Si Putri," "Zie," and "Gemawang" batik has become a pioneer in this matter. *Cocos nucifera* fiber, *T. catappa* leaves, *Terminalia bellirica* fruit, Swietenia mahagony root bark, Brazilwood, and *Ceriops candolleana* bark become natural dyes developed by "Si Putri" batik (Affan, 2019) (Fig. 3.9). Meanwhile, the abundance of wood sawdust waste, Mangrove propagules (fruit), mahogany bark, and Indigofera leaves became

FIGURE 3.9 "Si Putri" (A); "Gemawang" (B); "Zie" Natural dyed batik.

the mainstay of "Zie" batik to produce natural dyes (Fig. 3.9 B). Specifically, there are also four natural dye batik businesses located together with "Zie" batik (Gunungpati), namely Delima, Crystal, Mangosteen, and Citra Batik. Specifically, the natural dyes batik production in Gunungpati is supported by the provision of six ha of nursery and planting land for raw material commodity of natural dyeing paste (Jateng, 2019).

The distinctive color shades of the "Gemawang" (Fig. 3.9 A) natural dye batik located in Jambu District are also obtained from Mahogany bark and Terminalia Bellirica fruits as "Si Putri" and "Zie" batik. Not only limited to both, but this batik business also colors the batik using Allium Cepa, *Cudrania javanensis* Trécul, and many other extracts. In general, the batik business in this region requires one month of dyeing to produce quality colors. This time is much longer than synthetic dyeing which is only one day (Pimay, 2019).

"Marenggo" in Sleman, "Jarum" in Klaten, "Jolawe" and "coconut fiber wagging" in Bantul, as well as "Giriloyo" batik are known as producers of natural dyeing batik in Yogyakarta (Fig. 3.10). In contrast to Semarang, the

FIGURE 3.10 "Marenggo" (A); "Jarum" (B); "Jolawe" (C); "Giriloyo" (D); "Coconut fiber wagging" (E) natural dyed batik.

raw material for dyeing in this region was obtained from *Nephelium lappaceum*, *M. indica*, *Chromolaena odorata*, and *Physalis angulata* leaves (Natalia, 2016). In line with this, to support its overseas marketing, "Jarum" batik relies on the availability of *M. indica* leaves, mahogany bark, and *C. nucifera* fiber for its batik dyeing (Perdana, 2018). Not much different, to replace synthetic dyes rich in heavy metals, mercury, and chrome, "Jolawe" batik relies on biological resources such as Mahogany roots, *C. nucifera* fibers, and Indigofera leaves (Gudegnet, 2017).

Do not want to miss to revive the natural dyes batik production, Yogyakarta Regional Government has given full support to the "Giriloyo" batik center to become a "pilot project" for other regions. Different from other batik centers in Yogyakarta, "Giriloyo" batik characterizes its production in the typical green color motif of the *M. indica* leaf and the distinctive blue color of the Indigofera leaf (Republika, 2014). Not inferior to others, "coconut fiber wagging" batik also promotes environmental friendliness through natural dyeing from wood materials, such as the brown from *Peltophorum pterocarpum*, vintage green from *T. bellirica*, red from Brazilwood, yellow from Mahogany, and brownish-red from *Tectona grandis*. Enrichment of color shades is done through fixer selection (Jogja, 2018).

Solo became the next region in Central Java that played a role in reducing water pollution. The development of an eco-friendly batik center in Laweyan-Solo is a form of support for Solo to become an eco-friendly culture city. Several attempts were made to improve the practicality of application as well as the strength and fastness of natural dyes. Thus, batik can continue to be marketed without ignoring environmental conditions (Solopos, 2012). In addition to producing batik, Laweyan batik villages also market natural dyes from *Indigofera tinctoria* and wood extracts, each of which produces blue and brown colors (Fig. 3.11 A).

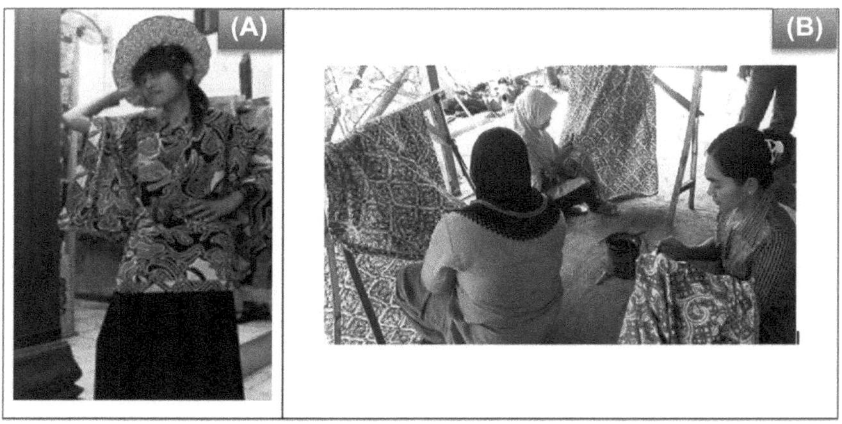

FIGURE 3.11 "Laweyan" (A); "Tawangsari" (B) natural dyed batik.

The increasing demand for natural dyes batiks lately one of which can be proven from the high demand for natural dyes produced by UNS Ecody. This is the background for the continued intensive cultivation of raw materials for natural dyes in Tawangsari Sukoharjo. The success of reviving the natural dyes batik glory has the potential to have a large impact on foreign exchange, given that more than 90% (700,000 tons) used by the textile industry so far are synthetic dyes. Moreover, the fulfillment of the need for dyes with a total value of 44 trillion rupiahs was carried out by import (Krjogja, 2018) (Fig. 3.11 B).

Natural dye batik is not only found in Semarang, Yogyakarta, and Solo but also Pekalongan. "Artho Moro," "Laba-laba," and "Pesindon" batik are three of the few natural dye batik businesses in Pekalongan. All three depend on the ability of Indigofera Tinctoria to produce blue. Furthermore, specifically for "Artho Moro" and "Laba-laba" batik, the dyeing development was carried out using *T. bellirica* seeds which present a yellow and Pelterhorum pterocarpum bark for brown, reddish, and brick red colors (Pekalongan, 2017).

The construction of the new icon "Kampung Batik Warna Alam" in Pekalongan is expected to be an innovation in the development of batik crafts in the future. The emergence of this new icon also further strengthens the dominance of Pekalongan batik. Although there is only 7.5% of the total batik business in this area, the success of the three in running the natural dye batik business is expected to motivate other batik businesses in the future (Pekalongan, 2017).

4.2 West Java

Not only Central Java but also West Java, especially in Cirebon, has also begun to produce natural dyes batik. "Ciwaringin" is one of the Cirebon Batiks which is famous for its unique motifs and dyes. This batik relied on dyeing from *Durio zibethinus* and *M. indica* peel, mahogany and *Ceiba pentandra* bark, and *C. nucifera* fiber (Syafei, 2018). "Pangandaran" is no less unique than the "Ciwaringin" batik. Utilization of Dahon rind sap with reddish to purplish colors is the main uniqueness of this batik. The lack of this natural dyeing material uses causes batik "Pengandaran" to be very distinctive (Berita Pangandaran, 2019).

4.3 East Java

Similar to Central and West Java, East Java also began producing natural dyes batik. "Seng" batik became one of the pioneers in the natural dyes batik production in the region (Fig. 3.12). The batik center at Sengguruh-Kepanjen utilizes the leaves of *Annona muricata*, *M. indica* and *N. lappaceum*, *C. nucifera* fiber, and mahogany bark for dyeing (Hapsari, 2017). To enrich the shades of his batik, Seng batik began to innovate by dyeing using coffee waste (Erwin, 2019).

"Majapahit" batik is an iconic product of Mojokerto which is also environmentally friendly (Fig. 3.12 A). This batik is dominated by brown color,

FIGURE 3.12 "Seng" (A); "Majapahit" (B); "Kembang Turi" (C) natural dyed fabric.

which is obtained from *Theobroma cacao* L. seed husk waste. This waste is mostly produced from the cocoa plantation industry in the region. A better intensity brown color is generally obtained from blends with other natural dyes, such as *T. bellirica* fruit skin extract, while the combination with *C. javanensis* Trécul and *C. candolleana* bark, as well as mahogany wood powder, is usually used to improve colorfastness (Budianto, 2019). More rapid development is shown by the "Kembang Turi" batik in Candimulyo-Jombang (Fig. 3.12 B). This batik is produced using *M. indica* leaf and mahogany bark extract. Consistency in maintaining product quality, including through dyeing for 2–30 days, makes this batik business have loyal customers to Singapore (Desa Kreatif, 2018). In different locations, the "Peri Kecil" natural dye batik produced in Bangkalan-Madura is no less popular. To get the distinctive color, "Peri Kecil" batik was dyed using trees and fruit bark as well as the roots of various plants (Redaksi Pesona, 2019).

Not limited to the level of SMEs, from academics, the author became one of the activists in the batik's natural dyes development. Fifteen natural dyes have been evaluated for their character and performance on cotton fibers. The intended ingredients include *M. indica*, *S. aqueum*, *P. guajava*, *Muntingia calabura*, *L. inermis*, young *T. grandis*, and *T. catappa* leaves, *Garcinia mangostana* and Citrus X Sinensis peel, *C. longa*, *C. sappan* bark, Phylantus

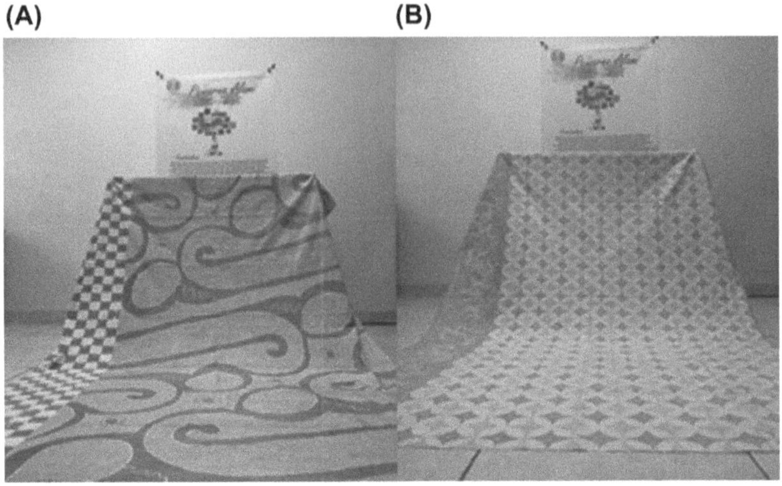

FIGURE 3.13 (A) The color shades produced from natural dyeing, and (B) natural dyeing batik product produced by the author.

reticulatus poir fruit, Rosa flowers, Pandanus, and Indigofera. Color enrichment has also been done through double and triple component dyeing. While using the right mordant and fixer will increase colorfastness. So far, the authors have evaluated the application of pre-, simultaneous-, and post-mordanting dyeing methods with fixation using alum, lime, and iron (II) sulfate. Fig. 3.13 presents various dyeing products and natural dyeing batik produced in collaboration with the batik business in Pacitan, Lamongan, and Sumenep-East Java.

Although some natural dye batik businesses have been found in several regions in Indonesia, especially in Java, compared with the synthetic dye batik business, the number of businesses is still much smaller. Therefore, to further enhance the existence of environmentally friendly batik businesses and divert people's interest from synthetic dye batik, many efforts must be made, such as the policies enactment from the Indonesia Government to require the use of natural dye batik by civil servants in several official activities, initiating establishment of a natural dye batik business in more areas, facilitating the cultivation of natural dye commodities and the management of agricultural wastes that have the potential as natural dyes, also facilitating the marketing of natural written batik products, both at home and abroad.

References

Adeel, S., Gulzar, T., Azeem, M., Saeed, M., Hanif, I., Iqbal, N., January 1, 2017. Appraisal of marigold flower based lutein as natural colourant for textile dyeing under the influence of gamma radiations. Radiat. Phys. Chem. 130, 35–39.

Affan, F.M., 2019. Batik pewarna alam Si Putri dipamerkan dalam ajang privat fashion show di hotel Alaska Semarang [Internet]. Tribun Jateng, Semarang [updated 2019 Nov 24; cited 2019 Oct 15]. Available from: https://jateng.tribunnews.com/2018/03/09/batik-pewarna-alam-si-putri-dipamerkan-dalam-ajang-privat-fashion-show-di-hotel-alaska-semarang.

Al-Mamun, M.R., Kader, S., Islam, M.S., Khan, M.Z.H., October 2019. Photocatalytic activity improvement and application of UV-TiO_2 photocatalysis in textile wastewater treatment: a review. J Environ Chem Eng 7 (5), 103248.

Ali, S., Saleemm, H., Hussain, T., September 2010. Effect of Tannin Acid and Metallic Mordant on the dyeing Properties of natural dye extracted from Acacia nilotica bark. Asian J. Chem. 22 (9), 7065–7069.

Aliya. Batik pekalongan. Jakarta CV Raman Edukasitama, 2009.

Alventosa-deLara, E., Barredo-Damas, S., Alcaina-Miranda, M.J., Iborra-Clar, M.J., March 30, 2012. Ultrafiltration technology with a Ceramic Membrane for reactive dye removal: optimization of membrane performance. J. Hazard Mater. 209, 492–500.

Ammayappan, A., Mosess, J., August 2007. An overview on application of tanning in textile processing. Man Made Textiles in India 50 (8), 293–297.

Anggrain, N.T., Hidayati, L., May 2014. Pengaruh frekuensi celupan terhadap hasil jadi pewarnaan batik dengan daun Lamtoro pada kain katun. Jurnal Tata Busana 3 (2), 34–38.

Anjaneyulu, Y., Chary, N.S., Raj, D.S.S., November 2005. Decolourization of industrial effluents available methods and emerging technologies-a review. Rev. Environ. Sci. Biotechnol. 4 (4), 245–273.

[place unknown] Asian Textile Studies [Internet], 2016. David and Sue Richardson. Natural Brown Dyes; [date unknown] [cited 2019 sep 30]; [about 4 screens]. Available from: http://www.asiantextilestudies.com/brown.html.

Ayed, L., Mahdhi, A., Cheref, A., Bakhrouf, A., July 1, 2011. Decolorization and degradation of Azo dye Methyl Red by an isolated Sphingomonaspaucimobilis: biotoxicity and metabolites characterization. Desalination 274 (1–3), 272–277.

Azzaz, A.A., Jellali, S., Akrout, H., Assadi, A.A., Bousselmi, L., November 10, 2018. Dynamic investigations on cationic dye desorption from chemically modified lignocellulosic material using a low-cost eluent: dye recovery and anodic oxidation efficiencies of the desorbed solutions. J. Clean. Prod. 201, 28–38.

Balai Besar Kerajinan dan Batik, 2010. Pengolahan Limbah Industri Batik. Balai Besar Kerajinan dan Batik, Yogyakarta.

Baliarsingh, S., Jena, J., Das, T., Das, N.B., October 2013. Role of cationic and anionic surfactants in textile dyeing with natural dyes extracted from waste plant materials and their potential antimicrobial properties. Ind. Crop. Prod. 50, 618–624.

Basitah, T., August 1, 2015. Extraction, Characterization, and application of natural dyes from the fresh rind of index color 5 mangosteens (*Garcinia mangostana* L.). Int. J. Chem. Mol. Nucl. Mater. Metallurg. Eng. 9 (7), 871–874.

Batik wastewater treatment using TiO_2 Nanoparticles coated on the surface of Plastic Sheet [Internet]. In: Sutisna, W.E., Rokhmat, M., Rahman, D.Y., Murniati, R., Khairurrijal, Abdullah, M. (Eds.), September 10, 2016. Engineering Physics International Conference. Elsevier Ltd, Bandung ID. London (UK). Available from: https://doi.org/10.1016/j.proeng.2017.03.015 [cited 2019 Aug 9].

Bentahar, Y., Draoui, K., Hurel, C., Ajouyed, O., Khairoun, S., Marmier, N., June 2019. Physico-chemical characterization and valorization of swelling and non swelling Moroccan clays in basic dye removal from aqueous solutions. J. Afr. Earth Sci. 154, 80–88.

Berita Pangandaran [Internet], 2019. Harapanrakyat.com, Pangandaran. Unik, batik asli pangandaran ini gunakan pewarna alami dari getah buah dahon; 2019 03 Oct [cited 2019 Oct 25]; [about 1 screen]. Available from: https://www.harapanrakyat.com/2019/10/batik-asli-pangandaran-ini-gunakan-pewarna-alami/.

Bhatia, D., Sharma, N.R., Singh, J., Kanwar, R.S., October 2017. Biological methods for textile dye removal from wastewater: a review [cited 2019 Aug 19] Crit. Rev. Environ. Sci. Technol. 0 (0), 1–41. https://doi.org/10.1080/10643389.2017.1393263. Available from: https://www.researchgate.net/publication/320545333_Biological_methods_for_textile_dye_removal_from_wastewater_A_Review.

Bhatia, D., Sharma, N.R., Singh, J., Kanwar, R.S., October 20, 2017. Biological methods for textile dye removal from wastewater: a review. Crit. Rev. Environ. Sci. Technol. 47 (19), 1836–1876.

Bhatti, I.A., Adeel, S., Jamal, M.A., Safdar, M., Abbas, M., December 7, 2009. Influence of Gamma radiation on the colour strength and fastness properties of fabric using Turmeric (*Curcuma longa* L.) as natural dye. Radiat. Phys. Chem. 79 (5), 622–625.

Bhuiyan, M.A.R., Islam, A., Ali, A., Islam, M.N., November 20, 2017. Color and chemical constitution of natural dye henna (*Lawsonia inermis* L) and its application in the coloration of textiles. J. Clean. Prod. 167, 14–22.

Birgani, P.M., Ranjbar, N., Abdullah, R.C., Wong, K.T., Lee, G., Ibrahim, S., et al., 2016. An efficient and economical treatment for batik textile wastewater containing high levels of Silicate and organic pollutants using a sequential process of acidification, Magnesium Oxide, and palm shell-based activated Carbon application [cited 2019 Aug 6] J. Environ [Internet] 184 (2), 229–239. https://doi.org/10.1016/j.jenvman.2016.09.066. Available from:

Blyskal, B., December 2016. Indigo dyeing and microorganism-polymer interaction. J. Cult. Herit. 22, 974–983.

Brookstein, D.S., July 2009. Factors associated with textile pattern dermatitis caused by contact allergy to dyes, finishes, foams, and preservatives. Dermatol. Clin. 27 (3), 309–322.

Budianto, E.E., 2019. Emak-emak di Mojokerto kembangkan batik ramah lingkungan, seperti apa? [Internet]. Mojokerto: detikNews [updated 2019 31 Jul; cited 2019 25 Oct]. Available from: https://news.detik.com/berita-jawa-timur/d-4646051/emak-emak-di-mojokerto-kembangkan-batik-ramah-lingkungan-seperti-apa.

Chaari, I., Fakhfakh, E., Medhioub, M., Jamoussi, F., March 2019. Comparative study on adsorption of cationic and anionic dyes by smectite rich natural clays. J. Mol. Struct. 1179, 672–677.

Chairat, M., Bremner, J.B., Chantrapromma, K., December 2007. Dyeing of cotton and silk yarn with the extracted dye from the fruit hull of Mangosteen, Garcinia mangostana Linn. J. Fib. Poly. 8 (6), 613–619.

Crini, G., June 2006. Non-conventional low-cost adsorbents for dye removal: a review. Bioresour. Technol. 97 (9), 1061–1085.

Danarto, S.A., Budiharta, S., Fauziah, August 2019. Tree species preference and rehabilitation perspective by local community: case study in Bondowoso, East Java, Indonesia. Asian J. Forest. 3 (2), 54–63.

David, Richardson, S., 2016. Asian Textile Studies[Internet] [place unknown]. David [updated 2016 jan 24; cited 2019 sep 27]. Available from: http://www.asiantextilestudies.com/indigo.html.

Delee, W., O'Neill, C., Hawkes, F.R., Pinheiro, H.M., August 19, 1998. Anaerobic treatment of textile effluents. J. Chem. Technol. Biotechnol. 73 (4), 323–335.

Deniz, F., Kepekci, R.A., July 2016. Dye biosorption onto pistachio by-product: a green environmental engineering approach. J. Mol. Liq. 219, 194–200.

Derksen, G.C.H., Niederlander, H.A.G., van Beek, T., November 29, 2002. Analysis of anthraquinones in Rubia tinctorum L. by liquid chromatography coupled with diode-array UV and mass spectrometric detection. J. Chromatogr. A 978 (1–2), 119–127.

Desa Kreatif [Internet], 2018. Derapdesa, Jombang. Batik warna alam jombang, diminati hingga luar negeri; 2018 22 Des [cited 2019 Nov 5]; [about 1 screen]. Available from: https://derapdesa.id/batik-warna-alam-jombang-diminati-hingga-luar-negeri/.

Dos Santos, A.B., Cerventes, F.J., Van Lier, J.B., September 2007. Review paper on current technologies for decolourisation of textile wastewaters: perspectives for anaerobic biotechnology. Bioresour. Technol. 98 (12), 2369–2385.

Dutta, A.K., Ghorai, U.K., Chattopadhyay, K.K., Banerjee, D., May 2018. Title of article. Removal of textile dyes by carbon nanotubes: a comparison between adsorption and UV assisted photocatalysis. Physica E Low Dimens Syst Nanostruct 99, 6–15.

Edwards, H.G., de Oliviera, L.F., Nesbitt, M., November 25, 2003. Fourier-transform Raman characterization of Brazilwood trees and substitutes. Analyst 128 (1), 82–87.

El-Kautsar, C.R., 2018. Peran Kesultanan Yogyakarta dalam perkembangan Batik Klasik di Yogyakarta [thesis on the Internet]. Fakultas Bahasa dan Seni Universitas Negeri Yogyakarta, Yogyakarta [cited 2019 Aug 5]. Available from: https://www.google.com/url?sa=t&rct=j&q=&esrc=s&source=web&cd=1&cad=rja&uact=8&ved=2ahUKEwiqt52to_rlAhUszjgGHVfYCmAQFjAAegQIAxAC&url=https%3A%2F%2Fcore.ac.uk%2Fdownload%2Fpdf%2F154913893.pdf&usg=AOvVaw0QfhICEhjOOMe6JUxI_NXz.

Ermawati, R., 2017. Limbah industri batik cemari sungai-sungai di Solo [Internet]. Solopos.com, Solo ID [cited 2019 Aug 15]. Available from: https://www.solopos.com/limbah-industri-batik-cemari-sungai-sungai-di-solo-856479.

Erwin, M., 2019. Griya Batik Seng Akan Berinovasi Produksi Batik Dengan Pewarna Berbahan Dasar Limbah Kopi [Internet]. Malang: Suryamalang.com [updated 2019 23 Jul; cited 2019 25 Oct]. Available from: https://suryamalang.tribunnews.com/2019/07/23/griya-batik-seng-akan-berinovasi-produksi-batik-dengan-pewarna-berbahan-dasar-limbah-kopi.

European Parliament and Council of the European Union, 2002. Directive of 19 July 2002 amending for the nineteenth time Council Directive 76/769/EEC relating to restrictions on the marketing and use of certain dangerous substances and preparations (azocolourants), 2002/61/EC Off. J. Eur. Communities - Legislation 243, 15–18.

Fadri, R.A., Salvia, Putri, S.K., Evawati, N.R., Hasniyati, R., 2015. Secure/safe, healthy, intact and halal of Mangosteen small medium enterprise. Mar [cited 2019 Oct 11] Int. J. Adv. Sci. Eng. Inf. Technol. 5 (3), 242–246. Available from: https://www.researchgate.net/publication/289500909_SecureSafe_Healthy_Intact_and_Halal_of_Mangosteen_Small_Medium_Enterprise.

Farida, R., Nisa, F.C., April 2015. Ekstraksi Antosianin limbah kulit manggis metode microwave assisted extraction (lama ekstraksi dan rasio bahan: pelarut). Jurnal Pangan dan Agroindustri 3 (2), 326–373.

Ferreira, E.S.B., Hulme, A.N., McNab, H., Quye, A., July 5, 2004. The natural constituents of historical textile dyes. Chem. Soc. Rev. 33 (6), 329–336.

Fitriyah, H., Ciptandi, F., December 2018. Pengolahan limbah sabut kelapa tua sebagai pewarna alam pada produk fesyen. E-Proc. Art Design 5 (3), 2534–2552.

Ford, L., Rayner, C.M., Blackburn, R.S., July 2018. Degradation of Lucidin: new insights into the fate of this natural pigment present in dyer's madder (*Rubia tinctorum* L.) during the extraction of textile artefacts. Dyes Pigments 154, 290–295.

Forgasc, E., Cserhati, T., Oros, G., Welander, U., September 2004. Removal of synthetic dyes from wastewaters: a review. Environ. Int. 30 (1), 953–971.

Frose, A., Schmidtke, K., Sukmann, T., Junger, I.J., Ehrmann, A., 2018. Application of natural dyes on diverse textile materials. Optik 181, 215–219.

Gatut, B., Aryanto, V., January 2010. Batik industry of Indonesia: the rise, fall and prospects. Studies in Business and Economics 5 (3), 156–170 [cited 2019 Aug 5].

Glower, B., January 1998. Doing what comes naturally in the dyehouse. J. Soc. Dye. Colour. 114 (1), 4–7.

Gudegnet, 2017. Batik Jolawe (Batik Dengan Pewarna Alami) [Internet]. Gdegnet, Yogyakarta [updated 2017 May 2; cited 2019 Oct 16]. Available from: https://gudeg.net/direktori/7614/batik-jolawe-batik-dengan-pewarna-alami.html.

Guljarani, M., Gupta, D., Aggarwal, V., Jain, M., 1992. Some studies on natural yellow dyes, Part III: quinones: henna, dolu. Indian Textil. J. 102 (3), 76–83.

Gupta, V.K., Suhas, June 2009. Application of low-cost adsorbents for dye removal – a review. J. Environ. Manag. 90 (8), 2313–2342.

Han, M.-R., Lee, J.-S., October 31, 2011. Natural dyeing of fabrics with Guava (*Psidium guajava* L.) leaf extract - characteristics of Guava (*Psidium guajava* L.) leaf extract and dyeability with silk fabrics. J. Korean Soc. Cloth. Ind. 13 (5), 778–789.

Hanan, S., 2016. 90 persen sungai di Cirebon tercemar [Internet]. Pikiran Rakyat, Jawa Barat ID [cited 2019 Aug 15]. Available from: https://www.pikiran-rakyat.com/jawa-barat/2016/07/30/90-persen-sungai-di-cirebon-tercemar-376120.

Hapsari, D.A.A., 2017. Batik seng di Kabupaten Malang gunakan bahan pewarna alami [Internet]. Timesindonesia, Jakarta [updated 2017 2 Oct; cited 2019 25 Oct]. Available from: https://www.timesindonesia.co.id/read/news/157872/batik-seng-di-kabupaten-malang-gunakan-bahan-pewarna-alami.

Haws, K.L., Winterich, K.P., Naylor, R.W., July 2014. Seeing the world through GREEN-tinted glasses: green consumption values and responses to environmentally friendly products. J. Consum. Psychol. 24 (3), 336–354.

Hieu, N.C., Lien, T.M., Van TTT, Juang, R.S., February 2020. Enhanced removal of various dyes from aqueous solutions by UV and simulated solar photocatalysis over TiO_2/ZnO/rGO composites. Separat. Purificat. Technol. 232, 1–14.

Indrayani, L., November 2018. Pengolahan limbah cair industri batik sebagai salah satu percontohan IPAL batik di Yogyakarta. Ecotrophic 12 (2), 173–184.

Jateng, H., 2019. Melestarikan Batik Pewarna Alam Khas Kampong Malon [Internet]. Humas Jateng, Semarang [updated 2019 Jan 11; cited 2019 Oct 15]. Available from: https://humas.jatengprov.go.id/detail_berita_gubernur?id=2054.

Jena, B., Das, B.P., Khandual, A., Sahu, S., Behera, L., February 14-15, 2015. Ecofriendly processing of textiles [Internet]. In: 4th International Conference on Materials Processing and Characterization. Elsevier Ltd., Andhra Pradesh IN. London (UK) [cited 2019 Sep 17]. Available from: https://www.sciencedirect.com/science/article/pii/S2214785315002655.

Jogja, T., 2018. Batik Kibasan Sabut Kelapa Dari Bantul Pakai Bahan Pewarna Alami [Internet]. Tribun Jogja, Yogyakarta [updated 2018 Dec 14; cited 2019 Oct 24]. Available from: https://jogja.tribunnews.com/2018/12/14/batik-kibasan-sabut-kelapa-dari-bantul-pakai-bahan-pewarna-alami.

Juchen, P.T., Piffer, H.H., Veit, M.T., Goncalves, G.D.C., Palacio, S.M., Zanette, J.C., December 2018. Biosorption of reactive blue BF-5G dye by malt bagasse: kinetic and equilibrium studies. J. Environ. Chem. Eng. 6 (6), 7111–7118.

Khandare, R.V., Govindwar, S.P., December 2015. Phytoremediation of textile dyes and effluents: current scenario and future prospects [cited 2019 Aug 19] Biotechnol. Adv. 33 (8), 1697–1714. https://doi.org/10.1016/j.biotechadv.2015.09.003. Available from: www.sciencedirect.com/science/article/abs/pii/S0734975015300367.

Khandare, R.V., Govindwar, S.P., December 2015. Phytoremediation of textile dyes and effluents: current scenario and future prospects. Biotechnol. Adv. 33 (8), 1697–1714.

Khemila, B., Merzouk, B., Chouder, A., Zidelkhir, R., Lapicque, F., March 2018. Removal of a textile dye using photovoltaic electrocoagulation. Sustain Chem Pharm 7, 27–35.

Khumalo, N.P., Vilakati, G.D., Mhlanga, S.D., Kuvarega, A.T., Mamba, B.B., Li, J., et al., October 2019. Dual-functional ultrafiltration nano-enabled PSf/PVA membrane for the removal of Congo red dye. J. Water Proc. Eng. 31, 1–8.

Kitture, R., Koppikar, S.J., Kaul-Ghanekar, R., Kale, S.N., January 2011. Catalyst efficiency, photostability and reusability study of ZnO nanoparticles in visible light for dye. J. Phys. Chem. Solid. 72 (1), 60–66 (degradation).

Koch, M., Yediler, A., Lienert, D., Insel, G., Kettrup, A., January 2002. Ozonation of hydrolyzed Azo dye reactive yellow 84 (CI). Chemosphere 46 (1), 109–113.

Kompas, 2008. Industri Kreatif di Indonesia 12 September.

Krjogja, 2018. Dengan Pewarna Alami Batik Makin Keren, Nih Buktinya [Internet]. Solo: Krjogja [updated 2018 Sept 1; cited 2019 Oct 24]. Available from: https://krjogja.com/web/news/read/76305/Dengan_Pewarna_Alami_Batik_Makin_Keren_Nih_Buktinya.

Kumar, S.N., Ritesh, S.K., Sharmila, G., Muthukumaran, C., July 2013. Extraction optimization and characterization of water soluble red purple pigment from floral bracts of Bougainvillea glabra [cited 2019 Sep 25] Arab. J. Chem. 10 (Suppl. 2), 145–150. Available from: https://www.sciencedirect.com/science/article/pii/S1878535213002475.

Kuppusamy, S., Venkateswarlu, K., Thavamani, P., Lee, Y.B., Naidu, R., Megharaj, M., April 2017. Quercus robur a corn peel as a novel coagulating adsorbent for cationic dye removal from aquatic ecosystems. Ecol. Eng. 101, 3–8.

Kusrianto, A., 2013. Batik Filososi, Motif & Kegunaan. Penerbit Adi, Yogyakarta.

Kusumawati, N., Samik, K.A., 2019. Quality improvement of ABBS hand writing batik production through standardization of natural dyeing using water guava and mango leaves. Adv. Eng. Res. 1, 29–32.

Kusumawati, N., Wijiastuti, A., Rahmadyanti, E., May 2012. Operating conditions optimization on Indonesian batik; dyes wastewater treatment by Fenton oxidation and separation using ultrafiltration membrane [cited 2019 Aug 6] J. Environ Sci Eng [Internet] 1 (5A), 672. Available from: http://connection.ebscohost.com/c/articles/85651204/operating-conditionsoptimization-indonesian-batik-dyes-wastewater-treatment-by-fenton-oxidationseparation-using-ultrafiltration-membrane.

Kusumawati, N., Santoso, A.B., Sianita, M.M., Muslim, S., June 2017. Extraction, characterization, and application of natural dyes from the Fresh Mangosteen (*Garcinia mangostana* L.) peel. Int. J. Adv. Sci. Eng. Inf. Technol. 7 (3), 878–884.

Kusumawati, N., Mufida, L., Annisa, R.A., 2018a. Pewarna Alami Untuk Tekstil. Unesa University Press, Surabaya ID.

Kusumawati, N., Samik, Santoso, A.B., Wijiastuti, A., 2018b. Development of textile natural dyeing using hybrid dyes from Mango Leaves Turmeric. In: Seminar Nasional Kimia; 2018 Sep 22; State University of Surabaya ID. UNESA, Surabaya (ID), pp. 50–55.

Kusumawati, N., Samik, Muslim, S., 2019. Exploration and standarization of coconut fiber waste utilization in batik dyeing process. Adv. Eng. Res. (in press in IOP series).

Laurenco, N.D., Novais, J.M., Pinheiro, H.M., January 15, 2001. Effect of some operational parameters on textile dye biodegradation in a sequential batch reactor. J. Biotechnol. 89 (2–3), 163–174.

Primary treatment of dye wastewater using Aloe Vera-aided aluminium and Magnesium hybrid coagulants. In: Lee, K.E., Hanafiah, M.M., Halim, A.A., Mahmud, M.H. (Eds.), August 19-20, 2015. International Conference on Environmental Forensics. Elsevier B.V., Putrajaya MY. Amsterdam (NL) https://doi.org/10.1016/j.proenv.2015.10.010 [cited 2019 Aug 26]. Available from:

Li, H.-X., Xu, B., Tang, L., Zhang, J.H., Mao, Z.H., 2015. Reductive decolorization of indigo carmine dye with *Bacillus* p. MZS10. Int. Biodeterior. Biodegrad. 103, 30–37.

Li, C., Lou, T., Yan, X., Long, Y., Wang, X., January 2018. Fabrication of pure chitosan nanofibrous membranes as effective absorbent for dye removal. Int. J. Biol. Macromol. 106, 768–774.

Li, J., Gong, J.-L., Zeng, G.-M., Zhang, P., Song, B., Cao, W.-C., et al., December 2019. The performance of UiO-66-NH_2/Graphene Oxide (GO) composite membrane for removal of differently charged mixed dyes. Chemosphere 237, 1–13.

Lisbijanto, H., 2013. Batik. Edisi Pertama. Graha Ilmu, Yogyakarta.

Lou, T., Cui, G., Xun, J., Wang, X., Zhang, J., January 2018. Synthesis of a terpolymer based on Chitosan and Lignin as an effective flocculant for dye removal. Colloids Surf. A Physicochem. Eng. Asp. 537, 149–154.

Lucas, M.S., Dias, A.A., Sampaio, A., Amaral, C., Peres, J.A., March 2007. Degradation of a textile reactive Azo dye by a combined chemical-biological process: fenton's reagent-yeast. Water Res. 41 (5), 1103–1109.

Manhita, A., Santos, V., Vargas, H., Candeias, A., Ferreira, T., Dias, C.B., October 30, 2012. Ageing of Brazilwood dye in wool – a chromatographic and spectrometric study. J. Cult. Herit. 14 (6), 471–479.

Mathur, J.P., Bhandari, C.S., December 2001. Physico-chemical study of dyed wool: Part I-Bougainvillaea as wool colourant with mixed mordant. Indian J. Fibre Text. Res. 26 (4), 432–437.

Melo, R.P.D., Neto, E.L.B., Nunes, S.K.S., Dantas, T.N.C., Neto, A.A.D., January 2018. Removal of Reactive Blue 14 dye using micellar solubilization followed by ionic flocculation of surfactants. Separ. Purif. Technol. 191, 161–166.

Merzuok, B., Madani, K., Sekki, A., January 15, 2010. Using electrocoagulation-electroflotation technology to treat synthetic solution and textile wastewater, two case studies. Desalination 250 (2), 573–577.

Minke, R., Rott, U., 1994. Anaerobic Treatment of Split FLow Wastewater and Concentrates from the Textile Processing Industry [cited 2019 Aug 9] Water Sci. Technol. vol. 40 (1), 169–176. https://doi.org/10.1016/S0273-1223(99)00377-7. Available from:

Moawed, E.A., El-Shahat, M.F., January 2016. Equilibrium, kinetic and thermodynamic studies of the removal of Triphenyl Methane dyes from wastewater using Iodopolyurethane powder. J. Taibah. Univ. Sci. 10 (1), 46–55.

Moawed, E.A., Wahba, A.E., Gabr, R.A., December 2018. Synthesis and application of LGB/St/Al_2O_3 biocomposite for sensitive detection and efficient removal of brilliant green dye from wastewater. J. Environ. Chem. Eng. 6 (6), 7225–7232.

Moreno, D.A., García-Viguera, C., Gil, J.I., Gil-Izquierdo, A., July 1, 2008. Betalains in the era of global agri-food science, technology and nutritional health. Phytochemistry Rev. 7 (2), 261–280.

Mukimin, A., Vistanty, H., Zen, N., Purwanto, A., Wicaksono, K.A., 2018. Performance of bio-equalization-electrocatalytic integrated method for pollutants removal of hand-drawn batik wastewater. J. Water Proc. Eng. 21, 77–83.

Musman, A., Ambar, A.B., 2011. Warisan Adiluhung Nusantara. ANDI, Yogyakarta.

Natalia, M.D., 2016. Perempuan Inspiif Penggiat Muda Batik Pewarna Alam [Internet]. Solopos, Yogyakarta [updated 2016 Sept 12; cited 2019 Oct 16]. Available from: https://m.solopos.com/perempuan-inspiratif-penggiat-muda-batik-pewarna-alam-752407/4.

Nguluben, T., Gumbo, J.R., Masindi, V., Maity, A., April 15, 2017. An update on synthetic dyes adsorption onto clay based minerals: a state-of-art review. J. Environ. Manag. 191, 35–57.

Nguyen, C.H., Fu, C.C., Juang, R.S., November 2018. Degradation of Methylene Blue and Methyl Orange by Palladium-doped TiO_2 photocatalysis for water reuse: efficiency and degradation pathways. J. Clean. Prod. 202, 413–427.

Nidheesh, P.V., Zhou, M., Oturan, M.A., April 2018. An overview on the removal of synthetic dyes from water by electrochemical advanced oxidation processes. Chemosphere 197, 210–227.

Nieuwolt, S., 1982. Introduction: the importance of climate in agricultural planning. In: Climate and Agricultural Planning in Peninsular Malaysia. Unit Percetakan MARDI, Serdang ID.

Niinimaki, K., Hassi, L., November 2011. Emerging design strategies in sustainable production and consumption of textiles and clothing. J. Clean. Prod. 19 (16), 1876–1883.

Nowik, W., 2001. The possibility of differentiation and identification of red and blue "soluble" dyewoods. Dyes Hist Archaeol 16, 129–144.

Nunez, J., Yeber, M., Cisternas, N., Thibaut, R., Medina, P., Carrasco, C., June 5, 2019. Application of electrocoagulation for the efficient pollutants removal to reuse the treated wastewater in the dyeing process of the textile industry. J. Hazard Mater. 371, 705–711.

Padfield, T., Landi, S., September 7, 1966. The light-fastness of the natural dyes. Stud. Conserv. 11 (4), 181–196.

Pazdzior, K., Bilinska, L., Ledakowicz, December 10, 2018. A review of the existing and emerging technologies in the combination of AOPs and biological processes in industrial textile wastewater treatment. Chem. Eng. J. 376, 120597.

Paździor, K., Bilińska, L., Ledakowicz, S., November 15, 2019. A review of the existing and emerging technologies in the combination of AOPs and biological processes in industrial textile wastewater treatment. Chem. Eng. J. 376, 1–94.

Pearce, C.I., Lloyd, J.R., Guthrie, J.T., September 2003. The removal of colour from textile wastewater using whole bacterial cells: a review. Dyes Pigments 58 (3), 179–196.

Pekalongan, 2017. Perajin Batik Dilatih Produksi Batik Warna Alam [Internet]. Pekalongan: Pekalongan [updated 2017 Sept 12; cited 2019 Oct 24]. Available from: https://weblama.pekalongankota.go.id/berita/perajin-batik-dilatih-produksi-batik-warna-alam.

Perdana, 2018. Geliat Produksi Batik Pewarna Alam Desa Jarum [Internet]. Solo: Jawa Pos [updated 2018 Sep 27; cited 2019 Oct 16]. Available from: https://radarsolo.jawapos.com/read/2018/09/27/96869/geliat-produksi-batik-pewarna-alam-desa-jarum.

Piattelli, M., 1981. The betalains: structure, biosynthesis, and chemical taxonomy. Biochem. Plants 7, 557–575.

Pimay, I.A., 2019. Batik Gemawang Khas Kabupaten Semarang Yuk Belajar Membatik [Internet]. Kompas, Semarang [updated 2019 May 3; cited 2019 Oct 16]. Available from: https://travel.kompas.com/read/2019/05/03/092100227/batik-gemawang-khas-kabupaten-semarang-yuk-belajar-membatik-?page=all.

Prabhu, K.H., Bhute, A.S., 2012. Plant based natural dyes and mordants: a Review. J. Nat. Prod. Plant Resour. 2 (6), 649–664.

Pratiwi, E., 2013. Perkembangan Batik Pekalongan Tahun 1950-1970 [thesis on the Internet]. Universitas Negeri Semarang, Semarang(ID) [cited 2019 Aug 2]. Available from: https://lib.unnes.ac.id/19819/.

Prayitno, E.K., Nurimaniwati, 2003. Proses Ekstraksi Bahan Pewarna Alam Dari Limbah Kayu Mahoni. Puslitbang Teknologi Maju BATAN, Yogyakarta, pp. 208–213.

Pujiastuti, L., 2015. Diakui Dunia, Ekspor Batik RI Meningkat Setiap Tahun [Internet]. Detik Finance, Jakarta ID [cited 2019 Aug 6]. Available from: https://finance.detik.com/berita-ekonomi-bisnis/d-3034083/diakui-dunia-ekspor-batik-ri-meningkat-setiap-tahun.

Punzi, M., Nilsson, F., Anbalagan, A., Svensson, B.M., Jönsson, K., Mattiasson, B., et al., July 2015. Combined anaerobic-ozonation process for treatment of textile wastewater: removal of acute toxicity and mutagenicity. J. Hazard Mater. 292, 52–60.

Qadeer, R., February 2007. Adsorption behavior of Ruthenium ions on activated charcoal from Nirtic Acid medium. Colloids Surf., A 293 (1–3), 217–223.

Rasool, K., Lee, D.S., April 2016. Effect of ZnO nanoparticles on biodegradation and biotransformation of co-substrate and sulphonated azo dye in anaerobic biological sulfate reduction processes. Int. Biodeterior. Biodegrad. 109, 150–156.

Redaksi Pesona [Internet], 2019. Pesona Indonesia, Jombang. Mengintip pembuatan batik pewarna alam di Bangkalan; 2019 22 Oct [cited 2019 Nov 5]; [about 1 screen]. Available from: https://pesona.travel/front/search?
q=Mengintip+Pembuatan+Batik+Pewarna+Alam+di+Bangkalan.

Rehman, F-u, Adeel, S., Qaiser, S., Bhatti, L.A., Shahid, M., Zuber, M., November 2012. Dyeing behavior of gamma irrardiated cotton fabric using lawson dye extracted from henna leaves (*Lawsonia inermis*). Radiat. Phys. 81 (11), 1752–1756.

Rehman, F.U., Adeel, S., Shahid, M., Bhatti, I.A., Nasir, F., Akhtar, N., et al., November 1, 2013. Dyeing of γ-irradiated cotton with natural flavonoid dye extracted from irradiated onion shells (*Allium cepa*) powder. Radiat. Phys. Chem. 92, 71–75.

Republika, 2014. Batik Giriloyo, Batik Khas Dengan Warna Alami [Internet]. Republika, Yogyakarta [updated 2014 Nov 28; cited 2019 Oct 16]. Available from: https://www.republika.co.id/berita/gaya-hidup/trend/14/11/28/nfraju-batik-giriloyo-batik-khas-dengan-warna-alami.

Robi'ah B., Tiga Industri Batik Terbesar di Indonesia [Internet]. Place unknown, Publisher unknown. Date unknown [cited 2019 Aug 5]. Available from: https://www.academia.edu/9751237/Tiga_Industri_Batik_Terbesar_di_Indonesia.

Rohasliney, H., Subki, N.S., December 2011. A preliminary study on batik effluent in kelantan state: a water quality perspectiv. In: International Conference on Chemical, Biological, and Environment Science [Internet] (Bangkok).

Samanta, A.K., Konar, A., November 14, 2011. Dyeing of textiles with natural dyes. Natural Dyes 3, 30–56.

Sanchez-Sanchez, J., Fernandez-Ponce, M.T., Casas, L., Mantell, C., de la Ossa, M., October 2017. Impregnation of mango leaf extract into a polyester textile using supercritical carbon dioxide. J. Supercrit. Fluids 128, 208–217.

Sane, P.K., Tambat, S., Sontakke, S., Nemade, P., August 2018. Visible light removal of reactive dyes using CeO_2 synthesized by precipitation. J. Environ. Chem. Eng. 6 (4), 4476–4489.

Santoso, A.B., Wijiastuti, A., 2018. National Strategic Research Report 2018: Standarisasi Preparasi Dan Prosedur Pewarnaan Batik Menggunakan Pewarna Alam Sebagai Wujud

Penguatan Dan Pengembangan Umkm Batik Andalan Koridor Ekonomi (Ke) Jawa. Universitas Negeri Surabaya, Surabaya ID. Retrieved from Research and community service institutions.

Sarıkaya, R., Selvi, M., Erkoç, F., August 2012. Evaluation of potential genotoxicity of five food dyes using the somatic mutation and recombination test. Chemosphere 88 (8), 974−979.

Sastrawijaya, A.T., 1991. Pencemaran Lingkungan. Surabaya. Penerbit Rineka Cipta.

Selvi, A.T., Aravindhan, R., Madhan, B., Rao, J.R., May 1, 2013. Studies on the application of natural dye extract from Bixa orellana seeds for dyeing and finishing of leather. Ind. Crop. Prod. 43, 84−86.

Senthilkumar, P., Umasankar, P., Sujatha, B., January 2002. Ultrasonic dyeing of cotton fabric with cabbage leaves. Man Made Text. India 45 (2), 71−74.

Shahid, M., Ul-Islam, S., Mohammad, F., August 15, 2013. Recent advancements in natural dye applications: a review. J. Clean. Prod. 53, 310−331.

Shen, C., Pan, Y., Wu, D., Liu, Y., Ma, C., Li, F., et al., October 15, 2019. A crosslinking-induced precipitation process for the simultaneous removal of poly(vinyl alcohol) and reactive dye: the importance of covalent bond forming and magnesium coagulation. Chem. Eng. J. 374, 904−913.

Shoukat, R., Khan, S.J., Jamal, Y., June 2019. Hybrid anaerobic-aerobic biological treatment for real textile wastewater. J. Water Proc. Eng. 29, 1−8.

Siddiqui, I., Siddiqui, K., Khan, N., 2017. Red onion peel (*Allium cepa*) pigment a good dye source for dyeing of cotton and silk. Man Made Text. India 45 (4), 130−132.

Silva, T.L., Ronix, A., Pezoti, O., Souza, L.S., Leandro, P.K.T., Bedin, K.C., dkk, November 2016. Mesoporous activated carbon from industrial laundry sewage sludge: adsorption studies of reactive dye Remazol brilliant blue- Rev. Chem. Eng. J. 303, 467−476.

Silva, P.M.D.S., Fiaschitello, T.R., de Queiroz, R.S., Freeman, H.S., da Costa, S.A., Leo, P., et al., February 2020. Natural dye from Croton urucurana Baill. bark: extraction, physicochemical characterization, textile dyeing and color fastness properties. Dyes Pigments 173, 1−14.

Sinha, K., Chowdhury, S., Saha, P.D., Datta, S., January 1, 2013. Modeling of microwave-assisted extraction of natural dye from seeds of Bixa orellana (Annatto) using response surface methodology (RSM) and artificial neural network (ANN). Ind. Crop. Prod. 41, 165−171.

Solís, M., Solís, A., Pérez, H.I., Manjarrez, N., Flores, M., December 2012. Microbial decolouration of Azo dyes: a review. Process Biochem. 47 (12), 1723−1748.

Solopos, 2012. Batik Pewarna Alam: Eksotis Dan Ramah Linkungan [Internet]. Solo: Solopos [updated 2012 Jan 19; cited 2019 Oct 24]. Available from: https://www.solopos.com/batik-pewarna-alam-eksotis-dan-ramah-lingkungan-156965.

Stylidi, M., 2003 Feb [cited 2019 Aug 9]. Pathways of solar light-induced photocatalytic degradation of Azo Dyes in aqueous TiO_2 suspensions. Appl. Catal. B [Internet] 40 (4), 271−286. https://doi.org/10.1016/S0926-3373(02)00163-7. Available from:

Sulaksono, A., Effendi, H., Kurniawan, B., July 2015. Title of article. Kajian beban pencemaran limbah cair industri kecil menengah (IKM) batik klister Trusmi Kabupaten Cirebon. Jurnal Pengelolaan Sumberdaya Alam dan Lingkungan 5 (1), 17−24.

Surakarta, D., 2009. Data Industri Kreatif Batik di Surakarta. Tidak dipublikasikan.

Susanto, B., 2019. DLH Kota Pekalongan Sadari Baru 4 Persen Limbah Batik Terolah IPAL, Sisanya Terbuang Ke Sungai [Internet]. Tribun Jateng, Pekalongan ID [cited 2019 Aug 15]. Available from: https://jateng.tribunnews.com/2019/04/09/dlh-kota-pekalongan-sadari-baru-45-persen-limbah-batik-terolah-ipal-sisanya-terbuang-ke-sungai.

Swaminathan, K., Pachhade, K., Sandhya, S., December 2005. Decomposition of a dye intermediate, (H-acid) 1 amino-8-naphthol-3,6 disulfonic acid in aqueous solution by ozonation. Desalination 186 (1–3), 155–164.

Syafei, F.R., 2018. Batik Ciwaringin Cirebon Andalkan Pewarna Alami Agar Lebih Menawan [Internet]. Ayobandung.com, Cirebon [updated 2018 22 Oktober; cited 2019 25 Oktober]. Available from: https://www.ayobandung.com/read/2018/10/22/39489/batik-ciwaringin-cirebon-andalkan-pewarna-alami-agar-lebih-menawan.

Tanaka, Y., Sasaki, N., Ohmiya, A., May 2008. Biosynthesis of plant pigments: anthocyanins, betalains and carotenoids. Plant J. 54 (4), 733–749.

Taylor, G.W., June 1986. Natural dyes in textile applications. Rev. Prog. Coloration 16 (1), 53–61.

Teli, M.D., Nayak, A.N., Nawathe, V.B., Adivarekar, R.V., 1994. Dyeing of cotton with turmeric. J. Text. Assoc. 1, 249–255.

Ticha, M.B., Haddar, W., Meksi, N., Guesmi, A., Mhenni, June 14, 2016. Improving dyeability of modified cotton fabrics by the natural aqueous extract from Red Cabbage using ultrasonic energy. Carbohydr. Polym. 154, 287–295.

Tonato, D., Drumm, F.C., Grassi, P., Georgin, J., Gerhardt, A.E., Dotto, G.L., dkk, October 2019. Residual biomass of Nigrospora sp. from process of the microbial oil extraction for the biosorption of procion red H-E7B dye. J. Water Proc. Eng. 31, 1–6.

Turgay, O., Ersoz, G., Atalay, S., Forss, J., Welander, U., May 2011. The treatment of Azo dyes found in textile industry wastewater by anaerobic biological method and chemical oxidation. Separ. Purif. Technol. 79 (1), 26–33.

Vandecasteele, F.P., 2003. Constructing efficient microbial consortia using a genetic algorithm [dikutip 2 September 2019]. In: Biological Applications for Genetic Evolutionary Computation (Bio-GEC) [Internet]; 2003 Apr 5-7 April. Springer Berlin Heidelberg, Berlin Heidelberg, pp. 69–71. Available from: https://link.springer.com/chapter/10.1007%2F978-3-540-24653-4_13.

Vankar, P.S., Shanker, R., August 2006. Dyeing cotton, silk and wool with *Brassica oleracea* or purple cabbage. Asian Dyers 3 (4), 75–77. +79-80.

Vankar, P.S., Shanker, R., September 30, 2008. Ecofriendly ultrasonic natural dyeing of cotton fabric with enzyme pretreatments. Desalination 230 (1–3), 62–69.

Vankar, P.S., Shanker, R., June 2009. Eco-friendly pretreatment of silk fabric for dyeing with Delonix regia extract. Color. Technol. 125 (3), 155–160.

Vankar, P.S., Shukla, D., 2019. Newer Natural Dyes for Various Textiles [Internet]. Chapter 1, Sources, chemistry and applications; [cited 2019 Sep 23]. Woodhead, India, pp. 1–69. https://doi.org/10.1016/B978-0-08-102686-1.00001-9. Available from:

Verma, A.K., Dash, R.R., Bhunia, P., January 2012. A review on chemical coagulation/flocculation technologies for removal of colour from textile wastewaters. J. Environ. Manag. 93 (1), 154–168.

Wang, C., Yediler, A., Lienert, D., Wang, Z., Kettrup, A., August 2003. Ozonation of an Azo dye C.I. Remazol Black 5 and toxicological assessment of its oxidation products. Chemosphere 52 (7), 1225–1232.

Watanabe, K., Baker, P., January 1, 2000. Environmentally relevant microorganisms. J. Biosci. BioEng. 89 (1), 1–11.

Wickens, H., 1983. Natural Dyes for Spinners and Weavers. BT Batsford, London, p. 96p.

Wijnsma, R., Verpoorte, R., 1986. Anthraquinones in the rubiaceae. Fortschr Chem Org Naturst/ Prog Chem Org Nat Prod. 49, 79–149.

World Economic Forum, 2007. Economic Creativity Index Ranking [Internet]. Seputar Indonesia [cited 2019 Aug 5]. Available from: www.seputarindonesia.com.

Yamjala, K., Nainar, M.S., Ramisetti, N.R., February 1, 2016. Methods for the analysis of Azo dyes employed in food industry – a review. Food Chem. 192, 813–824.

Yang, J.X., Hong, G.B., February 2018. Adsorption behavior of modified glossogyne tenuifolia leaves as a potential biosorbent for the removal of dyes. J. Mol. Liq. 525, 289–295.

Yang, R., Li, D., Li, A., Yang, H., January 2018. Adsorption properties and mechanisms of palygorskite for removal of various ionic dyes from water. Appl. Clay Sci. 151, 20–28.

Zarkogianni, M., Mikropoulou, E., Varella, E., Tsatsaroni, E., August 27, 2010. Colour and fastness of natural dyes: revival of traditional dyeing techniques. Color. Technol. 127 (1), 18–27.

Zerin, I., Farzana, N., Sayem, A.S.M., Anang, D.M., Haider, J., 2019. Potentials of natural dyes for textile applications [cited 2019 Sep 17]. In: Reference Module in Materials Science and Materials Engineering [Internet], vol. 8, pp. 38–40. https://doi.org/10.1016/B978-0-12-803581-8.11668-6. Available from: http://e-space.mmu.ac.uk/623886/, 2.

Zheng, X., Li, X., Li, J., Wang, L., Tang, K., February 2018. Efficient removal of anionic dye (Congo red) by Dialdehyde microfibrillated cellulose/chitosan composite film with significantly improved stability in dye solution. Int. J. Biol. Macromol. 107, 283–289.

Zhou, Y., Tang, R.-C., November 2016. Modification of Curcumin with a reactive UV absorber and its dyeing and functional properties for silk. Dyes Pigments 134, 203–211.

Zodi, S., Merzouk, B., Potier, O., Lapicque, F., Leclerc, J.-P., April 19, 2013. Direct red 81 dye removal by a continuous flow electrocoagulation/flotation reactor. Separ. Purif. Technol. 108, 215–222.

Zuraida, H., 2010. Sejarah Batik Tradisional Imogiri 1935-1942 [thesis on the Internet]. Fakultas Sastra Universitas Sanata Dharma Yogyakarta, Yogyakarta(ID) [cited 2019 Aug 2]. Available from: https://www.google.com/url?sa=t&rct=j&q=&esrc=s&source=web&cd=1&ved=2ahUKEwj_25KjjPjlAhUxzjgGHfHDC3UQFjAAegQIAhAE&url=https%3A%2F%2Frepository.usd.ac.id%2F27376%2F2%2F034314007_Full%255B1%255D.pdf&usg=AOvVaw3xNkgXt0nnmy0TR2qMS1K9.

Chapter 4

Biosorption of dyes

Jindrayani Nyoo Putro[a], Yi-Hsu Ju[b,c], Felycia Edi Soetaredjo[d], Shella Permatasari Santoso[d], Suryadi Ismadji[d]
[a]*Department of Chemical Engineering, National Taiwan University of Science and Technology, Taipei City, Taiwan;* [b]*Graduate Institute of Applied Science and Technology, National Taiwan University of Science and Technology, Taipei City, Taiwan;* [c]*Taiwan Building Technology Center, National Taiwan University of Science and Technology, Taipei City, Taiwan;* [d]*Department of Chemical Engineering, Widya Mandala Surabaya Catholic University, Surabaya, East Java, Indonesia*

1. Introduction

Contamination of the aquatic environment by dyestuffs causes severe problems for humans and aquatic biota. Environmental problems caused by the pollution of dyestuffs become more complicated because most of the dyes discharged into the environment are synthetic dyes. Mostly, these synthetic dyes are designed to be resistant to degradation by microorganisms. Once the synthetic dyes release into the environment, it will resist biological degradation for a long time. Therefore, a treatment process should be conducted before the wastewater containing dyes releases to the water environment. Currently, several processes for the treatment of the wastewater containing dyes are available. In terms of effectiveness, the adsorption process is the most suitable method for treating wastewater containing dyes, especially at low concentrations. The success of treating wastewater containing dyes using the adsorption process is highly dependent on the adsorption capacity of the adsorbent used.

The use of a variety of environmentally friendly and inexpensive adsorbents to remove dyes from water or wastewater has been studied by thousands of researchers in the last few decades. Several current review articles discussed the removal of dyes from water or wastewater and are also available in the scientific journal publications (Shamsollahi and Partovinia, 2019; Zhou et al., 2019; Afroze and Sen, 2018; Anastopoulos et al., 2017; Bhatnagar et al., 2010, 2015; Anastopoulos; Kyzas, 2014). These environmentally friendly and inexpensive adsorbents include agricultural wastes and its modified forms, seeds and leaves of various plants, industrial by-products, different kinds of microorganisms (living or dry ones), and natural polymer composites.

Green Chemistry and Water Remediation: Research and Applications.
https://doi.org/10.1016/B978-0-12-817742-6.00004-9

The studies of dyes removal using biosorbents have been conducted for more than three decades; however, the biosorption study still very attractive for the scientists indicated by a large number of publications in this area in recent years. Newly modified biosorbents including multifunctional and composite biosorbents have been prepared and used for the color removal from aqueous solutions. Therefore, reviews about the current progress of dyes biosorption are still required. This chapter summarizes and discusses the up-to-date progress of biosorption of dyes using newly modified biosorbents. Various aspects of dyes biosorption are given and comprehensively explained.

2. Classification of synthetic dyes

Synthetic dyes are developed to fulfill the human needs of modern and biologically and physically resistant coloring agents. Usually, synthetic dyes are classified according to their application, chemical structure, ionic forms in the solution, as well as its color (Zhou et al., 2019). Classification of dyes based on their chemical structure possibly is the most appropriate way since it offers many advantages (Gregory, 1990). Based on this method of classification, dyes can be categorized as follow:

- Azo dyes: CI disperse yellow 16, CI solvent yellow 14, disperse blue, reactive brown, CI acid black 1, CI direct green 26, etc.
- Anthraquinone dyes: quanizarin, CI disperse red 15, CI disperse violet 1, disperse torquise, etc.
- Benzodifurano dyes: disperse red.
- Polycyclic aromatic carbonyl dyes (anthraquinonoid vat dyes): the blue, indanthrone, CI vat black, the green, the brown, etc.
- Indigoid dyes: tyrian purple.
- Polymethine and related dyes: cyanine, carbocyanine, azacarbocyanines, etc.
- Styryl dyes: malononitrile, CI disperse yellow 31, dicyanovinyl dyes, tetracyanoethylene, etc.
- Di and tri-aryl carbonium and related dyes: diarylcarbonium dyes, malachite green, CI basic blue 9, etc.
- Phthalocyanines dyes: CI direct blue 86, hemin, etc.
- Quinophthalone dyes: CI solvent yellow 33, CI disperse yellow 54, etc.
- Sulfur dyes: CI sulfur black 1.
- Nitro and nitroso dyes: CI acid yellow 1, CI 10,305, CI disperse yellow 14, CI acid green 1, etc.
- Miscellaneous dyes: stilbene dyes, formazan dyes.

Details about properties, synthesis, and structures of synthetic dyes can be seen elsewhere (Gregory, 1990).

3. Agricultural and plant wastes as biosorbents

A variety of agricultural and plant wastes have been employed as cheap adsorbents for the removal of various dyes from aqueous solutions as listed in Table 4.1. Some of the biomass has potential application for the industrial-scale wastewater treatment process, while others possibly will end up only in the laboratory-scale application. The low adsorption capacity of agricultural and plant wastes materials is the main obstacle of using these alternative adsorbents for the industrial-scale treatment of wastewater containing dyes.

TABLE 4.1 Adsorption of various dyes using agricultural and plant wastes.

Agricultural waste	Dyes	Adsorption capacity, mg/g	Reference
Azolla filiculoides	Basic orange	833 at pH 7 and T = 30 °C	Tan et al. (2010)
Brazil nutshells	Methylene blue (MB) and indigo carmine (IC)	7.81 (MB), and 1.09 (IC)	Brito et al. (2010)
Chia seeds (*Salvia hispanica*)	Reactive yellow B2R	70.95 at pH 2 and T = 30°	Da Silva and Pietrobelli (2019)
Corn silk	Reactive blue 19 (RB19) and reactive red 218 (RR218)	71.6 (RB19) and 63.3 (RR218) at pH 2.0 and T = 25 °C	Degermenci et al. (2019)
Elaeagnus angustifolia L. fruits	Methylene blue (MB), and Indigo Carmine (IC)	344.8 (MB) and 9.7 (IC) at room temperature	Oymak and Eruygur (2019)
Glossogyne tenuifolia leaves	Congo red (CR) and malachite green (MG)	13.4 (CR), and 210.9 (MG) at T = 40 °C	Yang and Hong (2018)
Mentha pulegium	Direct Red 80 (DR80) and Acid Black 26 (AB26)	52.3 (DR80) and 46.3 (AB26) at pH 2.5 and T = 25 °C	Mahmoodi et al. (2011)
Moringa oleifera seed	Reactive red 120	413.32 at T = 50 °C and pH 1.0	Celekli et al. (2019)
Musa acuminate peel	Emerald green	10.75 at T = 25 °C	Rehman et al. (2019)
Olive stone	Alizarin Red S (ARS), and methylene blue (MB)	16.10 (ARS), and 13.2 (MB)	Albadarin and Mangwandy (2015)

Continued

TABLE 4.1 Adsorption of various dyes using agricultural and plant wastes.—cont'd

Agricultural waste	Dyes	Adsorption capacity, mg/g	Reference
Orange bagasse	Reactive blue 5G	28.9 at T = 40 °C	Fiorentin et al. (2010)
Palm date stones	Basic violet 3 (BV3), and Basic red 2 (BR2)	97.8 (BR2) and 117.3 (BV3) at pH 7.7 and T = 35 °C	Wakkel et al. (2019)
Palm kernel fiber	Methyl violet	114.2 at T = 66 °C	Ofomaja et al. (2011)
Parthenium hysterophorus L	Safranine	89.3 at pH 6	Shrivastava (2010)
Paulownia tomentosa stud leaf powder	Acid Orange 52	10.5 at T = 25 °C	Deniz and Saygideger (2010)
Pinus sylvestris L	Reactive red 195	6.69 at 20 °C, and 7.38 at 50 °C	Aksakal and Ucun (2010)
Pine tree leaves	Basic red 46	71.94 at T = 45 °C	Deniz and Karaman (2011)
Princess tree leaves	Basic red 46	43.10 at T = 45 °C	Deniz and Saygideger (2011)
Rice husk	Direct red 31 (DR31) and Direct orange 26 (DO26)	57.88 (DR31) at pH 2, and 36.14 (DO26) at pH 3, adsorption temperature was 30 °C	Safa and Bhatti (2011)
Sapindus mukorossi dead leaves	CI Reactive red 241	8.18 at T = 30 °C	Javed et al. (2019)
Sesame hull	Methylene blue	359 at T = 30 °C	Feng et al. (2011)
Solanum tuberosum peel	Emerald green	2.61 at T = 25 °C	Rehman et al. (2019)
Soybean hulls	BF−4B reactive red	19 at pH 1 and T = 45 °C	Módenes et al. (2019)
Walnut shell	Methylene blue	80.4 at T = 45 °C	Liu et al. (2019)
Water hyacinth leaves	Amaranth dye	70 at pH = 1.5 and T = 50 °C	Guerrero-Coronilla et al. (2015)

Agricultural and plant wastes biomass are categorized as lignocellulosic materials which are mainly composed of lignins, hemicellulose, and cellulose. Lignins are composed of benzene derivatives such as coumaryl, sinapyl, and coniferyl alcohols, and it has very complex and variable structures. Hemicelluloses are sugar polymers that consist of C_5 and C_6 sugar monomers. The polymerization degree in hemicelluloses is lower than celluloses. While celluloses are polysaccharides which consist of a linear chain of β-1,4-linked D-glucose monomers. The composition of lignins, hemicellulose, and celluloses in the lignocellulosic materials strongly depends on the type of biomass, geographic condition, and also season. The chemical composition variation of agricultural and plant wastes is another big challenge of using these lignocellulosic materials as the industrial biosorbents.

Low Brunauere-Emmere-Teller (BET) surface area as well as small pore volume are the characteristics of lignocellulosic materials. With this poor pore structure, natural lignocellulosic materials usually possess small adsorption capacity toward a particular type of dyes (Aksakal and Ucun, 2010; Brito et al., 2010; Deniz and Saygideger, 2010, 2011; Fiorentin et al., 2010; Albadarin and Mangwandy, 2015; Yang and Hong, 2018; Javed et al., 2019; Módenes et al., 2019; Rehman et al., 2019). Typical nitrogen sorption isotherm of agricultural waste material (cassava peel) is given in Fig. 4.1. However, in some cases, the excellent adsorption capacity was observed for some natural lignocellulosic materials (Tan et al., 2010; Celekli et al., 2019;

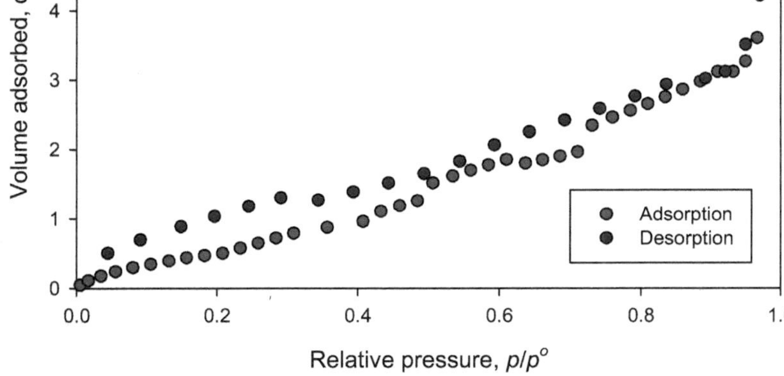

FIG. 4.1 Sorption isotherms of nitrogen onto cassava peel waste.

Oymak and Eruygur, 2019). The natural lignocellulosic materials have high adsorption capacity possible due to the excess presence of polar functional group-containing compounds (Lesmana et al., 2009). In the adsorption of dyes using lignocellulosic materials, particular surface functional groups of adsorbent play an essential role. These functional groups are likely responsible for binding with dyes molecules during the adsorption process.

Several polar and nonpolar functional groups which usually exist in the agricultural waste materials are as follow: O − H stretch (alcohols or phenols), N − H stretch (1° and 2° amines), C − H stretch (methylenes), C = O stretch (carboxylic acids, saturated aliphatic esters, carboxyls), C = C stretch (alkynes), and C − H bend (aromatic compounds). The existence of surface functional groups on the agricultural waste usually is obtained using a Fourier Transform Infra-Red (FTIR) method. In the case of the adsorption of reactive red 120 on *Moringa oleifera* seed, the carbonyl, amines, and amide groups play essential roles in the binding of the dye (Çelekli et al., 2019).

4. Thermal modification of agricultural wastes

To increase the adsorption capacity of some agricultural waste toward dyes molecules, the modification at high temperatures is one of the best options. During the thermal treatment, the lignocellulosic materials will be decomposed into some low molecular weight compounds (gases and tar) and leaving solid carbonaceous material known as biochar or the black product. The resulting biochar possesses a porous structure; however, in many cases, the direct use of this solid material as the adsorbent often gives unsatisfactory results. Activation using some chemicals (known as chemical activation) or oxidizing gases (physical activation) usually is conducted to increase the adsorption capacity of the biochar/activated carbon (Table 4.2).

Thermal modification of lignocellulosic materials (physical or chemical) produces porous materials known as activated carbons or activated biochars. Thermal modification using chemicals as the activating agent offers several advantages than the physical one (Sudaryanto et al., 2006), such as

- Lower temperature, usually between 400 and 600 °C.
- The single-step process, the activation, and the carbonization process take place in one process.
- The chemical activation process produces a higher yield of solid than physical activation.

The main drawback of the modification of lignocellulosic materials using chemical agents is often, it produces hazardous wastewater (depends on the chemical agents used), which creates another environmental problem. The physical activation is known as an environmentally friendly process since it only utilizes steam or carbon dioxide as the activating agent.

TABLE 4.2 Thermal process of some agricultural and plant wastes.

Agricultural waste	Thermal process	Application	Reference
Almond shell	Microwave-assisted process	Methylene blue adsorption	Du et al. (2016)
Coconut shell	Thermochemical activation process using $ZnCl_2$ at 700 °C	Adsorption of methylene blue, malachite green, and methyl orange	Gupta and Khatri (2019)
Corncob	Steam activation at 892 °C for 40 min	Methylene blue adsorption	Yu et al. (2019)
Corn stigmata	Hydrothermal carbonization at 180 °C followed by CO_2 activation at 900 °C	Methylene blue adsorption	Mbarki et al. (2019)
Foxtail palm	Vacuum pyrolysis at 400 −700 °C and activated using H_2SO_4, H_3PO_4, and KOH	Methylene blue adsorption	Dos Santos et al. (2019)
Mangosteen peel	$ZnCl_2$ activation at 600 °C	Methylene blue adsorption	Nasrullah et al. (2019)
Olive tree pruning and coffee husk	Chemical activation at 500 °C (K_2CO_3 and KOH), and physical activation at 910 °C (steam)	Methylene blue adsorption	Mamani et al. (2019)
Peanut shell	Carbonized at 800 °C in the presence of $FeCl_3$ in CO_2 environment	Malachite green adsorption	Guo et al. (2018)
Pineapple crown leaf	Carbonization at 500 °C followed by chemical activation (KOH) using microwave irradiation	Methyl violet adsorption	Astuti et al. (2019)
Pinecone	Pyrolized at 500 °C and activated using $ZnCl_2$	Alizarin red S adsorption	Bhomic et al. (2018)
Pinecone	Microwave-induced $ZnCl_2$ activation	Methylene blue adsorption	Ozhan et al. (2014)
Prickly pear seeds and cactus	Activated using H_3PO_4 at the temperature of 450 °C	Methylene blue adsorption	Ouhammou et al. (2019)
Shiitake mushroom	Activated using K_2CO_3 and carbonized at 800 °C	Methylene blue adsorption	Sun et al. (2019)

Continued

TABLE 4.2 Thermal process of some agricultural and plant wastes.—cont'd

Agricultural waste	Thermal process	Application	Reference
sunflower seed husk	$ZnCl_2$ was used as a chemical activating agent, and the carbonization was conducted at 500 °C using microwave irradiation as the heat source	Methylene blue adsorption	Baytar et al. (2018)
Walnut shell	Activated with phosphoric acid and carbonized at 450 °C	Malachite green adsorption	Hajialigol and Masoum (2019)
Water fern	Carbonized at 300 °C, and activated using NaOH and KOH and followed by activation at 600 °C under argon atmosphere for 3 h	Acid orange II adsorption	Emrooz et al. (2019)

In the thermal activation process, several steps of processes occurred during the heating process that leads to the pore creation. The first step of the methods is the release of free moisture content and bound water from the structure of lignocellulosic material. The evaporation of both free and bound water occurs at a temperature of around 50–200 °C. In the subsequent step of the processes, which occur at a temperature range of 200–300 °C, the hemicellulose will disappear due to the decomposition. At higher temperatures (300–360 °C), the cellulose decomposes into the smaller molecular weight of organic compounds, and the last part of organic substance in the lignocellulosic material, lignin, will decompose at a temperature of 360–500 °C. The degradation or decomposition of high molecular weight compounds into the smaller ones of the structure of lignocellulosic material will create space within the internal structure of the material. The creation of space in the internal structure of materials will increase the active surface area of the materials and later will enhance its adsorption performance.

Chemical and physical activation processes were employed by Mamani et al. (2019) to prepare the activated carbons from the coffee husk and olive tree pruning. The impregnation of the biomasses was conducted at 130 °C using potassium hydroxide (KOH) and potassium carbonate (K_2CO_3) as chemical activating agents, and the activation was performed at 800 °C. For physical activation, the carbonization was conducted at 500 °C and the gasification at 910 °C using superheated steam as the activating agent.

The chemical activation produced the carbon with a surface area over 2500 m^2/g, while for the physical process created carbon with BET surface area around 1500 m^2/g. The resulting carbons could remove more than 99% of methylene blue from aqueous solution.

5. Combination of agricultural wastes and other lignocellulosic materials with other materials to form composites

To enhance the adsorption capacity of the activated carbon derived from the agricultural waste, the combination with other materials was usually performed to produce new composite materials which have high adsorption performance or capacity. Naushad et al. (2019) developed arginine-modified–activated carbon (AGDPA@AC) for removal of methylene blue from aqueous solution. The maximum uptake of methylene blue by the AGDPA@AC was achieved at pH 8 and temperature of 25 °C (219.9 mg/g). The time required to reach the equilibrium condition was 120 min.

The combination of macroalgae *Saccharina japonica* (which is also known as kelp) with montmorillonite to form the composite of biochar–bentonite was studied by Sewu et al. (2019). The composite was utilized as the adsorbent for the removal of an anionic dye (Congo Red) and cationic dye (crystal violet). The incorporation of bentonite into biochar increased its mesopores structure, and it enhanced the adsorption performance of the composite material toward the Congo Red and crystal violet (at 30 °C).

6. Algae as biosorbents

Many research groups around the world have conducted studies on the removal of dyes from water or wastewater using microalgae as well as macroalgae for many years. Their adsorption capability and its abundant availability as diverse groups of plant-like organisms which can be found easily as either in multicellular or unicellular forms make these organisms very attractive as the alternative low-cost biosorbents (Table 4.3).

Sargassum dentifolium and Ulva fasciata (macroalgae species) have been tested for their adsorption performance to remove methylene blue from aqueous solution by Moghazy et al. (2019). There is high removal efficiency for both macroalgae due to the presence of various surface functional groups that possess high adsorption affinity toward methylene blue molecule. Different adsorption mechanisms between the surface functional groups of macroalgae with methylene blue cation occurred during the adsorption process, such as ionic interaction, coordination, ion-dipole bond, and van der Waals interaction (Moghazy et al., 2019).

Brown macroalgae *Stoechospermum marginatum* was utilized to recover Acid Blue 25, Acid Orange 7, and Acid Black 1 (acid dyes) from the aqueous

TABLE 4.3 Adsorption of dyes using algae as biosorbents.

Algae	Dye	Parameters	Adsorption capacity, mg/g	Reference
Chlorella pyrenoidosa	Direct Red-31	pH: 3–14; contact time: 0–180 min; initial dye concentration: 10–50 mg/L	30.53	Sinha et al. (2016)
	Methylene blue	pH: 6; temperature: 28 °C; time: 24 h; initial concentration of dye: 100 mg/L	101.75	Lebron et al. (2018)
	Rhodamine B	pH: 2–10; temperatures: 25, 35, 45 °C; contact time: 5–240 min; initial dye concentration: 100 mg/L	63.14	Da Rosa et al. (2018)
	Methylene blue	Modified with H_3PO_4 and $ZnCl_2$; pH: 6; temperature: 28 °C; adsorbent dose: 2 g/L	212	Lebron et al. (2019)
Chlorella vulgaris	Supranol Red 3BW, Lanaset Red 2 GA, and Levafix Navy Blue EBNA	pH: 3.85–11.40; temperature: 35–58 °C	256.4 (Supranol Red 3BW), 345 (Lanaset Red 2 GA), and 188.7 (Levafix Navy Blue EBNA)	Lim et al. (2010)
Enteromorpha prolifera	Direct Fast Scarlet 4BS	pH: 0.5–10; temperature: 30–50 °C; time: 5–540 min	318.87	Sun et al. (2019)
Fucus vesiculosus	Methylene blue	Modified with H_3PO_4 and $ZnCl_2$; pH: 6; temperature: 28 °C; adsorbent dose: 2 g/L	1162.9	Lebron et al. (2019)
Laminaria japonica	Methylene blue	pH: 2–10; temperature: 25 °C; time: 0–60 min; adsorbent dose: 0.2–2.0 g/L	549.45	Shao et al. (2017)

Sargassum dentifolium	Methylene blue	pH: 3–9; temperature: 20–50 °C; biosorbent dose: 0.25–3 g/L; contact time: 0–90 min; initial concentration of dye: 20–80 mg/L	66.6	Moghazy et al. (2019)
Spirulina maxima	Methylene blue	pH: 6; temperature: 28 °C; time: 24 h; initial concentration of dye: 100 mg/L	145.34	Lebron et al. (2018)
	Methylene blue	Modified with H_3PO_4 and $ZnCl_2$; pH: 6; temperature: 28 °C; adsorbent dose: 2 g/L	343.66	Lebron et al. (2019)
Stoechospermum marginatum	Acid Blue 25, Acid Orange 7, and Acid Black 1	pH: 2–10; temperature 27 °C; initial dye concentration: 10–50 mg/L	22.2 (Acid Blue 25); 6.73 (Acid Orange 7), and 6.57 (Acid Black 1)	Daneshvar et al. (2012)
Ulothrix sp.	Methylene blue	pH: 1.3–10.4; temperature: 20–60 °C; biosorbent dose: 0.5–5 g/L; contact time: 0.5–12 h.	86.1	Dogar et al. (2010)
Ulva fasciata	Methylene blue	pH: 3–9; temperature: 20–50 °C; biosorbent dose: 0.25–3 g/L; contact time: 0–90 min; initial concentration of dye: 20–450 mg/L	244	Moghazy et al. (2019)

solution by Daneshvar et al. (2012). Various operation parameters (pH, temperature, initial concentration of acid dyes, the particle size of biosorbent, and contact time) influence the uptake of acid dyes by *Stoechospermum marginatum*. The maximum adsorption capacities of the brown macroalgae were 6.57.6.73, and 22.2 mg/g for Acid Black 1, Acid Orange 7, and Acid Blue 25. The low adsorption capacity of brown macroalgae *Stoechospermum marginatum* toward acid dyes is due to the lack of internal micropore and mesopore structures in brown macroalgae. The adsorption of acid dyes solely occurred in the macropore structure and several surface functional groups available on the surface of the macroalgae. The FTIR analysis reveals that the amine and hydroxyl groups are mainly responsible for the uptake of the acid dyes.

Ulothrix species is a filamentous green alga which is widely found in both marine and freshwater. This species of algae had been studied for its possibility as an alternative adsorbent for methylene blue removal from aqueous solution (Dogar et al., 2010). The adsorption capability of the Ulothrix species toward methylene blue was strongly influenced by the pH of the solution and temperature of the system. The adsorption capacity of this filamentous green algae increased with the increased of the pH of the solution. The experimental adsorption results indicate that the temperature had a negative effect on the amount of methylene blue adsorbed by this biosorbent. The amount of methylene blue adsorbed by the algae decreased with increasing temperature. This phenomenon suggests that physical adsorption is the primary mechanism that controls the adsorption of methylene blue on Ulothrix species.

7. Fungus/fungi as biosorbents

Fungus/fungi have been utilized as the alternative low-cost biosorbents for the removal of dyes from the water environment by a large number of research groups. The capability of fungal biomass in decolorizing of large numbers of dyes has been discussed by Sen et al. (2016). The combination of enzyme degradation and adsorption is the primary mechanism in the decolorization of dyes by fungal biomass (Sen et al., 2016). The ability of the fungus to degrade complex synthetic dyes is mostly due to the availability of extracellular ligninolytic enzymes in the fungus itself. Several of the studies on the utilization of the fungus or fungi to decolorize dyes from the solutions are listed in Table 4.4.

The modification of fungus with chemicals is usually conducted to increase its adsorption capacity. The modification of *Aspergillus versicolor* using CTAB was studied by Huang et al. (2016) to increase its adsorption capability toward Reactive Black 5. The highest removal efficiency of Reactive Black 5 (>98%) by surfactant-modified *A. versicolor* (1.5% CTAB) was achieved at pH 2.0 and contact time of 420 min. The high removal efficiency of surfactant-modified *A. versicolor* is due to the increased active sites of surface functional groups as well as the increased BET surface area. At an acidic pH

TABLE 4.4 Adsorption of dyes using fungus/fungi as biosorbents.

Fungus/fungi	Dye	Parameters	Adsorption capacity, mg/g	Reference
Agaricus bisporus	Reactive Red 2	The fungal was modified by CTAB (cetyl trimethyl ammonium bromide). pH: 2.0–8.0, temperature: 25, 35, 45 °C; contact time: 5–90 min.	141.53	Akar and Divriklioglu (2010)
Aspergillus lentulus FJ172995	Acid Navy Blue, Orange HF, Fast Red A, Acid Sulfone Blue, and Acid Magenta	pH: 4.0–10.0; initial dye concentration: 100–900 mg/L; temperature: 30, 35, 40, 45 °C	212.92	Kaushik and Malik. (2010)
Aspergillus versicolor	Reactive Black 5	The fungal was modified by CTAB; pH: 2.0–8.0; temperature: 30 °C; adsorption time: 0–24 h	227.27	Huang et al. (2016)
Ceriporia lacerata P2	Crystal Violet	pH: 2.0–10.0; initial concentration: 10–500 mg/L; temperature: 20 °C	239.25	Lin et al. (2011)
Coriolopsis sp.	Crystal Violet, Methyl Violet, Cotton Blue, Malachite Green	Initial dye concentration: 50, 100, and 200 mg/L; temperature: 30 °C	n.a	Chen and Ting (2015a)
Cunninghamella elegans	Acid Blue 62, Acid Red 266 and Acid Yellow 49	Contact time: 0–125 min; temperature: 30 °C; pH: 3.0–11.0	1035	Russo et al. (2010)

Continued

TABLE 4.4 Adsorption of dyes using fungus/fungi as biosorbents.—cont'd

Fungus/fungi	Dye	Parameters	Adsorption capacity, mg/g	Reference
Lenzites elegans WDP2	Crystal Violet, Malachite Green, Fuschsin Basic, and Brilliant Green	pH: 7.0; temperature: 30 °C; time: 0–72 h.	n.a	Pandey et al. (2018)
Mucor circinelloides	Congo Red	Temperature: 25, 50, 75 °C; initial concentration of dye: 150, 300, 600, 800, 1000 mg/L	169.49	Azin and Moghimi (2018)
Penicillum restrictum	Reactive Yellow 145	pH: 1–8; temperature: 20, 30, and 40 °C	116.5	Caner et al. (2011)
Rhizopus arrhizus	Methylene blue	Modified with SDS (sodium dodecylsulfate) surfactant. pH: 2–12; SDS concentration: 0–20 mM; temperature: 25 °C; initial concentration of MB: 25–1100 mg/L	1666.6	Aksu et al. (2010)

(2.0), the protonation of several functional groups such as carboxyl, hydroxyl, amide, and amine groups in the surface of surfactant-modified *A. versicolor* occurred. The surface of biosorbent became positively charged, and electrostatic interaction between the surface of biosorbent and negatively charged (SO_3^-) ions of Reactive Black 5 occurred. This phenomenon significantly enhanced the uptake of the anionic dye from the aqueous solution (Huang et al., 2016).

8. Bacteria as biosorbents

As microscale organisms, bacteria also have been widely explored for its potential application as low-cost adsorbents for decolorization of wastewater containing dyes. Several recent studies on the utilization of bacteria as biosorbents for dyes removal from an aqueous environment are summarized in Table 4.5.

The activity and adaptability of bacteria in the wastewater treatment system strongly influence the effectiveness of bacteria in the decolorization process (Meerbergen et al., 2018). Bacteria possess different decolorization mechanisms compared with other conventional biosorbents such as fungus, algae, or other lignocellulosic biomass. In the conventional biosorbents, the removal of color or decolorization process is mainly due to the adsorption process, so basically, the decolorization process using conventional biosorbents will be effective if the concentration of dye in the solution or wastewater is quite low. In the decolorization process using bacteria as the biosorbent, the main decolorization mechanism is degradation rather than adsorption; therefore, this process is likely very effective for wastewater containing a high concentration of dyes (Meerbergen et al., 2018).

9. Equilibrium studies

Various adsorption isotherm equations which were initially developed for the gas-phase adsorption are employed to represent the adsorption equilibria data of various dyes by different kinds of biosorbents. Those equations are either two-parameter models (Langmuir, Freundlich, Dubinin–Radushkevich (DR), Temkin, Flory–Huggins, BET) or three-parameter models (Sips, Toth, and Redlich–Paterson). Among them, Langmuir and Freundlich are the most widely used isotherm equations to represent the adsorption equilibria data due to their simplicity.

Langmuir equation has the form as follows:

$$q_e = q_{\max}\frac{K_L C_e}{1 + K_L C_e} \tag{4.1}$$

Parameters $q_{\max}$ and K_L represent the adsorption capacity of biosorbent toward specific dye pollutants, and K_L is the adsorption affinity. In most cases,

TABLE 4.5 Adsorption of dyes using bacteria as biosorbents.

Bacteria	Dyes	Notes	Reference
Acinetobacter (ST16.16/164)	Reactive Orange 16 and Reactive Green 19	The bacteria were isolated from activated sludge of two wastewater treatment plants located in Flanders, Belgium. The influence of carbon sources, pH, temperature, dye concentration, and concentration of salt addition on the decolorization capability of Acinetobacter was explored. The incubation was conducted for 3 days. The decolorization efficiency of Acinetobacter toward both of azo dyes was > 80%.	Meerbergen et al. (2018)
Klebsiella (ST16.16/034)	Reactive Orange 16 and Reactive Green 19	The bacteria were isolated from activated sludge of two wastewater treatment plants located in Flanders, Belgium. The influence of carbon sources, pH, temperature, dye concentration, and concentration of salt addition on the decolorization capability of Klebsiella was explored. The incubation was conducted for 3 days. The decolorization efficiency of Klebsiella toward both of azo dyes was > 80%.	Meerbergen et al. (2018)
Penicillium simplicissimum	Methyl Violet, Crystal Violet, and Cotton Blue	The bacteria were isolated from wastewater at Monash Uni Malaysia. The decolorization process was conducted at 30 °C for 14 days with initial dyes concentration of 50, 100, and 200 mg/L. The removal efficiencies were 98% for Methyl Violet, 95% for Crystal Violet, and 82% for Cotton Blue.	Chen and Ting (2015b)
Pseudomonas aeruginosa	Reactive Red 21, Reactive Orange 16, and Reactive Blue 19	Jack fruit seed powder was employed as cosubstrate. Response Surface Methodology (RSM) was used to evaluate the influence of different process parameters on the decolorization performance of *Pseudomonas aeruginosa*. The bacteria were capable to remove 97.7 ± 0.3 % of Reactive Red 21, 98.9 ± 0.3 % of Reactive Orange 16, and 92.6 ± 0.4 % of Reactive Blue 19.	Mishra et al. (2019)

both of these parameters are temperature dependent. For physical adsorption (exothermic process), the values of parameters $q_{\max}$ and K_L decrease with the increase of the temperature of the system. While the opposite behavior is observed for chemical adsorption (endothermic process).

Freundlich adsorption isotherm has the mathematical expression as follows:

$$q_e = K_f C_e^{1/n} \tag{4.2}$$

The parameters K_f and n represent Freundlich adsorption capacity and heterogeneity of the system, respectively. Usually, the value of parameter n is between 1 and 10. The higher the value of parameter n, the more heterogeneous the adsorption system. Although it does not have the saturation capacity in high concentration and Henry's law at very low concentration, the Freundlich equation could represent the adsorption equilibria data of many systems very well, as indicated in Table 4.6. In all cases, the adsorption experiments were conducted at low to moderate dye concentrations, and at these ranges of concentrations, the drawback of the Freundlich model does not exist; therefore, this adsorption equation could describe the experimental data well (Febrianto et al., 2009).

Another adsorption isotherm model, which was initially developed for the adsorption of gas onto microporous adsorbents, DR equation, also widely employed to correlate the adsorption of dyes onto biosorbents (nonporous materials). The DR equation can be expressed in mathematic form as follows:

$$q_e = q_{\max} \exp\left(- \left(\frac{RTln(c_e/c_s)}{\alpha Eo} \right)^2 \right) \tag{4.3}$$

The parameter α is an affinity coefficient that is proportional to the molar volume of the adsorbate. The theoretical value of α can be estimated based on the refractive index of the adsorbate (Ismadji and Bhatia, 2001). Symbol C_s indicates the solubility of the solute in the solvent. The parameter E_o represents solid characteristic energy (usually compared to benzene, which is generally chosen as the reference compound). One of the unique features of the DR equation is in the characteristic curves. If the DR equation could represent the experimental data, all of the adsorption data at different temperatures when plotted as the logarithmic amount adsorbed (q_e) versus the square of potential energy ($RTln(C_e/C_s)$) should lie on the same straight line. However, in all of the studies which were employed, the DR equation failed to provide this characteristic curve, the judgment of the validity of the model to represent the experimental adsorption data is only based on the value of the R^2 (Daneshvar et al., 2012; Ari and Celik, 2013; Lebron et al., 2018; Deniz and

TABLE 4.6 Adsorption equations to represent the equilibria data of the adsorption of dye onto biosorbent.

Biosorbent	Dye	Isotherms used	Isotherm valid	References
Aspergillus niger	Acid Blue 161 and Procion Red MX-5B	Langmuir and Freundlich	Freundlich	Almeida and Corso (2019)
Aspergillus terreus	Acid Blue 161 and Procion Red MX-5B	Langmuir and Freundlich	Freundlich	Almeida and Corso (2019)
Aspergillus versicolor	Reactive black 5	Langmuir and Freundlich	Langmuir	Huang et al. (2016)
Banana trunk–activated carbon	Methylene blue	Langmuir, Freundlich, DR, Temkin, Frumkin, Harkins-Jura, and Smith	Freundlich	Danish et al. (2018)
Carpinus betulus sawdust	Basic red 46	Langmuir, Freundlich, and DR	Freundlich and Langmuir	Deniz and Yildis (2019)
Cashew nutshell–based carbons	Methylene blue	Langmuir and Freundlich	Freundlich and Langmuir	Spagnoli et al. (2017)
Ceriporia lacerata P2	Crystal violet	Langmuir, Freundlich, Koble-Corrigan, and Scatchard	Koble-Corrigan	Lin et al. (2011)
Chlorella pyrenoidosa	Methylene blue	Langmuir, Freundlich, DR, Temkin	Langmuir	Lebron et al. (2018)
	Direct Red-31	Langmuir and Freundlich	Langmuir and Freundlich	Sinha et al. (2016)
Chlorella vulgaris	Supranol Red 3BW	Langmuir and Freundlich	Langmuir and Freundlich	Lim et al. (2010)
Corn silk	Reactive Blue 19, and Reactive Red 218	Langmuir, Freundlich, DR, Temkin, Halsey, and Harkins-Jura	Freundlich and Temkin	Degermenci et al. (2019)

Date palm	Methylene blue	Langmuir, Freundlich, and DR	Freundlich	Esmaeili and Foroutan (2019)
Eucalyptus	Methylene blue	Langmuir, Freundlich, and DR	Freundlich	Esmaeili and Foroutan (2019)
Furfural lignocellulosic waste	Methyl Orange and Rhodamine B	Langmuir and Freundlich	Langmuir (Methyl orange), and Freundlich (Rhodamine B)	Chen et al. (2019)
Kefir grains	Remazol navy	Langmuir, Freundlich, and Temkin	Temkin and Langmuir	Erdoğdular and Apar (2019)
Laminaria japonica	Methylene blue	Langmuir and Sips	Langmuir	Shao et al. (2017)
Macroalga Stoechospermum marginatum	Acid Blue 25, Acid Orange 7, Acid Black 1	Langmuir, Freundlich, DR, and Temkin	Freundlich	Daneshvar et al. (2012)
Microalgae Scenedesmus	Methylene blue	Langmuir	Langmuir	Afshariani and Roosta (2019)
Mucor circinelloides	Congo red	Langmuir and Freundlich	Freundlich	Azin and Moghimi (2018)
Neonectria radicicola	Acid Orange 51, Reactive Red 75, and Direct Blue 86	Langmuir, Freundlich, Redlich–Peterson, Toth, and Temkin	Freundlich and Temkin	Ghariani et al. (2019)

Continued

TABLE 4.6 Adsorption equations to represent the equilibria data of the adsorption of dye onto biosorbent.—cont'd

Biosorbent	Dye	Isotherms used	Isotherm valid	References
Neurospora sitophila cells–Zeamays silk tissue biomass system	Reactive blue 49	Langmuir and Freundlich	Langmuir	Akar and Celik (2011)
Olive stone by-product	Alizarin red and methylene blue	Langmuir, Freundlich, Redlich–Peterson, and Temkin	Redlich–Peterson	Albadarin and Mangwandi (2015)
Pomegranate peel	Malachite green	Langmuir, Freundlich, Temkin, DR, and Redlich–Peterson	Langmuir	Gunduz and Bayrak (2017)
Prunus dulcis leaves	Acid Blue 113	Langmuir and Freundlich	Langmuir	Jain and Gogate (2019)
Pyracantha coccinea	Orange G	Langmuir, Freundlich, DR	Langmuir	Ari and Celik (2013)
Rhizopus oligosporus	Acid Blue 161 and Procion Red MX-5B	Langmuir and Freundlich	Freundlich	Almeida and Corso (2019)
Saccharomyces cerevisiae	Brilliant green and methylene blue	Langmuir and Freundlich	Freundlich	Ghaedi et al. (2013)
Sargassum dentifolium	Methylene blue	Langmuir and Freundlich	Freundlich	Moghazy et al. (2019)

Sargassum muticum	Methylene blue	Langmuir, Freundlich, DR, Temkin	DR and Temkin	El Atouani et al. (2019)
Sesame hull	Methylene blue	Langmuir, Freundlich, and Temkin	Langmuir	Feng et al. (2011)
Sesame straw	Methylene blue	Langmuir and Freundlich	Langmuir	Feng et al. (2017)
Sour lemon sawdust	Methylene blue	Langmuir, Freundlich, and DR	Freundlich	Esmaeili and Foroutan (2019)
soybean hulls	Reactive Red 195	Langmuir, Freundlich, Toth, Sips, Radke–Praustnitz, Temkin, and BET	Langmuir, BET, and Toth	Módenes et al. (2019)
Spirulina maxima	Methylene blue	Langmuir, Freundlich, DR, Temkin	Freundlich and Temkin	Lebron et al. (2018)
Swede rape straw	Methylene blue	Langmuir, Freundlich, and Temkin	Langmuir	Feng et al. (2013)
Trichoderma harzianum	Reactive black B	Langmuir, Freundlich, and Temkin	Freundlich	Karthik et al. (2019)
Ulva fasciata	Methylene blue	Langmuir and Freundlich	Freundlich	Moghazy et al. (2019)
Walnut shell	Malachite green	Langmuir, Temkin, Freundlich, and Dubinin–Kaganer–Radushkevich (DKR)	Langmuir	Hajialigol and Masoum (2019)
Walnut shell	Methyl blue and Congo Red	Langmuir	Langmuir	Liu et al. (2019)
Water hyacinth	Amaranth anionic dye	Langmuir, Freundlich, Temkin, Halsey, DR, Sips, Redlich–Peterson, Radke-Prausnitz, Toth	Langmuir, Sips, Redlich–Petersen	Guerrero-Coronilla et al. (2015)

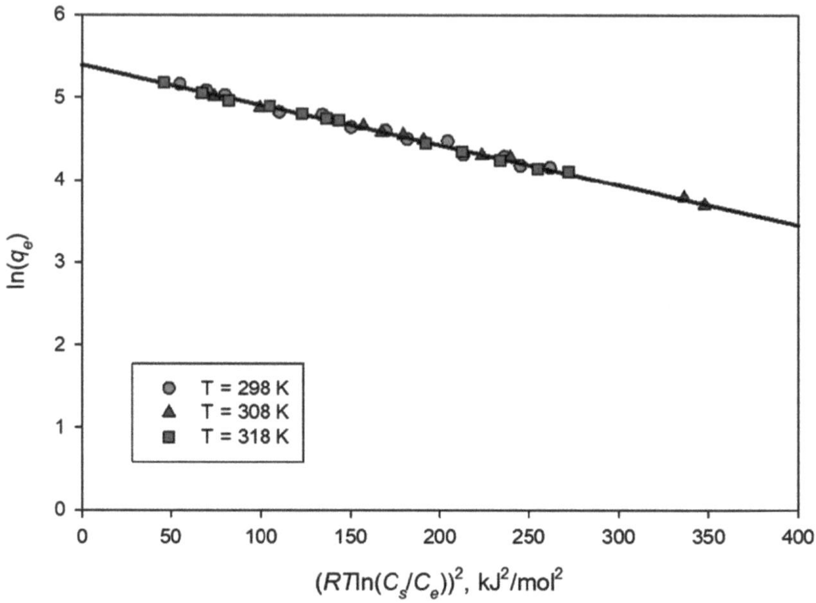

FIG. 4.2 The characteristic curve of the adsorption of methyl violet onto jackfruit peel waste activated carbon.

Yildis, 2019; Degermenci et al., 2019; El Atouani et al., 2019; Esmaeili and Foroutan, 2019). The characteristic curve of the adsorption of methyl violet onto jackfruit peel waste-activated carbon is given in Fig. 4.2.

Based on Fig. 4.2, it is evident that the DR equation can represent the equilibria data of the adsorption of methyl violet onto jackfruit peel waste–activated carbon well. As mentioned previously that the DR equation was developed for the adsorption of gas onto porous solids, therefore, this model is mostly applicable to represent the adsorption equilibria data of various substances onto porous materials such as activated carbons.

Temkin equation is also a popular model to represent the adsorption of dyes onto various biomass systems. Similar to the previous adsorption isotherm models, the Temkin model was also developed for the gas-phase adsorption (adsorption of H_2 onto Pt electrodes). The mathematic expression of Temkin equation is as follows:

$$q_e = \frac{RT}{b_T} ln(K_T C_e) \tag{4.4}$$

The parameter b_T represents the heat of adsorption, while the binding energy of the adsorption at equilibrium condition is given by symbol K_T. Temkin equation is also frequently used by many researchers to correlate the biosorption of dyes onto biosorbents such as adsorption of methylene blue on banana trunk–activated carbon (Danish et al., 2018), adsorption of methylene blue on *Chlorella pyrenoidosa* (Lebron et al., 2018), adsorption of acid dyes onto brown macroalgae *Stoechospermum marginatum* (Daneshvar et al., 2012), adsorption of Reactive Blue 19, and Reactive Red 218 on corn silk (Degermenci et al., 2019), adsorption of Remazol navy on kefir grains (Erdoğdular and Apar, 2019), adsorption of Acid Orange 51, Reactive Red 75, and Direct Blue 86 on Neonectria radicicola (Ghariani et al., 2019), etc. In several systems, this equation gave better performance than other adsorption equations in correlating the experimental adsorption data of dyes onto biosorbents (Lebron et al., 2018; Degermenci et al., 2019; El Atouani et al., 2019; Ghariani et al., 2019). In general, the adsorption of dyes onto biosorbents involvs very complex phenomena; several factors such as the pH of the solution, solubility of dyes in water, and the surface chemistry of the biosorbents play crucial roles on the uptake of dye molecules by biosorbents. Since the development of the Temkin adsorption equation is only based on straightforward assumptions, the complex phenomena of dyes adsorption by the biosorbents cannot be captured by this equation, and therefore, in most cases, this equation fails to represent the adsorption equilibria data.

Compared with the other two parameter adsorption equations, the Halsey isotherm is seldom used to represent the equilibrium data of the adsorption of dyes onto biosorbents. Halsey isotherm was derived according to the condensation of gases in a multilayer condition (the distance relatively far from the surface of the solid). The Halsey isotherm equation can be written as follows:

$$q_e = \left(\frac{K_H}{C_e}\right)^{1/n_H} \tag{4.5}$$

where K_H and n_H are the parameters of the Halsey isotherm equation. This equation fails to represent the adsorption of Amaranth anionic dye onto water hyacinth (Guerrero-Coronilla et al., 2015).

Similar to Halsey isotherm equation, the BET isotherm is also rarely employed to correlate the adsorption data of dyes onto biosorbents. The BET model for liquid-phase adsorption has the form as follows:

$$q_e = q_{\max}\frac{BC_e}{\left(C_e - C_s^*\right)[1 + (B-1)(C_e/C_s)]} \tag{4.6}$$

The development of the BET model used a similar kinetic concept of Langmuir isotherm (rate of adsorption = rate of desorption). Unlike the Langmuir equation, the restriction that one adsorption site can be occupied by one molecule of adsorbate has been omitted in the BET model. The BET model has been successfully employed to correlate the adsorption data of Reactive Red 195 on the soybean hull (Módenes et al., 2019).

In several cases, the available two-parameter adsorption equations could not represent the adsorption equilibria data well, and for this reason, three-parameter adsorption isotherms were developed. Sips isotherm is the three-parameter model which was developed to overcome the drawback of the Freundlich equation; this equation possesses saturation capacity at relatively high concentration of the adsorbate.

$$q_e = q_{\max}\frac{(K_sC_e)^n}{1+(K_sC_e)^n} \tag{4.7}$$

Similar to Freundlich, the parameter n in the Sips equation also represents the heterogeneity of the adsorption system. For the homogeneous system, the parameter n in Sips equation is close to unity, and it will become the Langmuir equation.

As a three-parameter equation, the Sips equation is more superior in representing the adsorption equilibria than other two-parameter equations. Sips equation could represent the adsorption equilibria for many systems such as water hyacinth—Amaranth anionic dye (Guerrero-Coronilla et al., 2015), acid-treated coffee husk–Malachite Green (Murthy et al., 2019), *Eugenia umbelliflora* Berg—Methylene Blue (Postai and Rodrigues, 2018), *Anethum graveolens*—Methylene Blue (Hamitouche et al., 2017), Beechwood–Direct Red (Muntean et al., 2017), etc. Since the Sips equation also does not have a correct Henry's law at very low concentration, therefore, this equation mostly fails to represent the adsorption data at low concentration.

Toth equation is another empirical three-parameter adsorption isotherm, this equation is able to represent many adsorption equilibria of heterogeneous systems.

$$q_e = q_{\max}\frac{K_TC_e}{\left(1+(K_TC_e)^n\right)^{1/n}} \tag{4.8}$$

Toth equation has a correct Henry law for very low concentration. Being an empirical three-parameter equation, Toth should be able to describe the experimental data for most complex adsorption systems. However, this equation failed to represent the adsorption equilibria of particular systems such as water hyacinth—Amaranth anionic dye (Guerrero-Coronilla et al., 2015), Neonectria radicicola—Acid Orange 51, Neonectria radicicola—Reactive Red 75, and Neonectria radicicola—Direct Blue 86 (Ghariani et al., 2019). While for the biosorption Reactive Red 195 onto soybean hull, Toth gave the best correlation compared with other available isotherm equations (Módenes et al., 2019).

Table 4.6 indicates that Redlich–Paterson isotherm is also a popular model to represent the equilibria data of the adsorption of dyes on biosorbents. The Redlich–Paterson isotherm has the form as follows:

$$q_e = \frac{K_RC_e}{1+a_RC_e^{\beta}} \tag{4.9}$$

For several adsorption systems, the Redlich–Paterson could fit the experimental data well, as indicated in Table 4.6. Eq. (4.9) combines both characteristics of Freundlich and Langmuir equations. When β is equal to 0, Eq. (4.9) reduces to Henry's law form, and if β is equal to 1, it becomes a Langmuir equation.

10. Kinetic studies

For the design of efficient and effective adsorption system, both equilibria and kinetic data are required. Many kinetic models are available to represent the adsorption kinetic of dyes onto biosorbents. The most widely used models to describe the kinetics of adsorption of dyes onto biosorbents are pseudo–first-order and pseudo–second-order equations. Both of those equations were developed according to the concept of chemical reaction that occurs on the surface of adsorbent (Plazinski et al., 2009).

In 1898, Lagergren proposed the pseudo–first-order equation. This empirical equation was developed to describe the adsorption of malonic and ocalic acids on the surface of the charcoal (Plazinski et al., 2009). The pseudo–first-order equation has the following mathematical form:

$$q_t = q_e(1 - \exp(k_1 t)) \tag{4.10}$$

where q_t is the amount of dye adsorbed by biosorbent at time t, while q_e and k_1 are the parameters of the pseudo–first order which indicate the amount adsorbed at equilibrium condition and time scaling factor. The parameters q_e and k_1 are usually obtained through the linear regression method. Since k_1 is a time scaling factor, the higher the value of this parameter, the sooner the system reaches equilibrium (shorter time required) as illustrated in Fig. 4.3.

Usually, the parameter k_1 is dependent on the pH of the solution, temperature, as well as the initial concentration of the solute (Albadarin and Mangwandi, 2015; Akar and Celik, 2011; Akar and Divriklioglu, 2010).

Pseudo–second-order equation was proposed by Blanchard et al. (1984), which has the form as follows:

$$q_t = \frac{q_e^2 k_2 t}{1 + q_e k_2 t} \tag{4.11}$$

Similar to the parameter k_1 in pseudo–first order, the parameter k_2 is also a time scaling factor. The effect of k_2 on fractional loading of dye on biosorbent is depicted in Fig. 4.4.

The higher the value of k_2, the shorter the time required by the system to reach the equilibrium condition. Similar to k_1, k_2 is also influenced by the adsorption condition such as initial concentration, pH of the solution, and temperature. In most adsorption systems, the pseudo–second order gave better

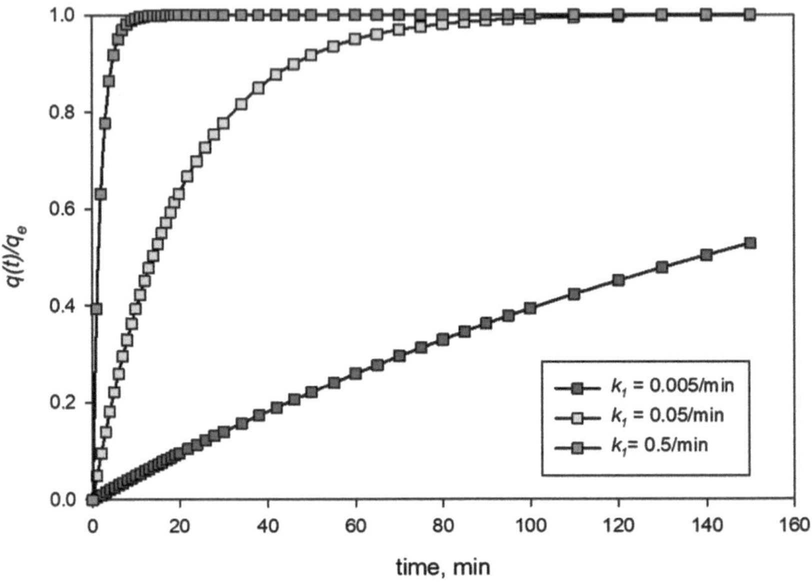

FIG. 4.3 The effect of k_1 on fractional loading of dye on biosorbent.

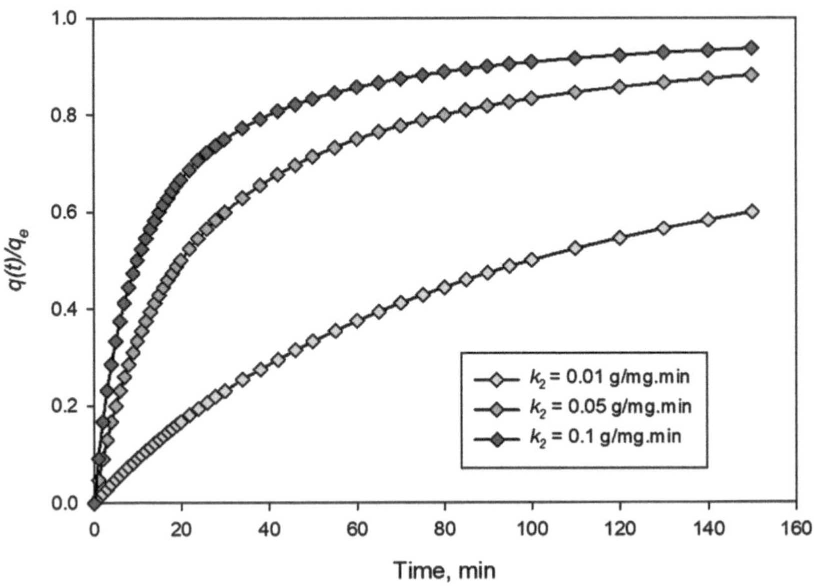

FIG. 4.4 The effect of k_2 on fractional loading of dye on biosorbent.

representation of the experimental data compared with the pseudo−first order (Da Silva et al., 2019; Liu et al., 2019; Módenes et al., 2019; Moghazy et al., 2019; Rehman et al., 2019; Lebron et al., 2018; Huang et al., 2016; Ghaedi et al., 2013). The parameter q_e in pseudo−second order equation is less sensitive from the effect of the random experimental error (Plazinski et al., 2009); therefore, this equation gives better prediction than the pseudo−first-order equation.

11. Thermodynamic of adsorption of dyes on biosorbents

The thermodynamic of adsorption is also one of the important information which is required for the design of the adsorption systems. Unfortunately, there is no single equipment that can be used to determine the thermodynamics of adsorption directly. The only option to obtain the thermodynamic of adsorption is through the interpretation of the adsorption isotherms using Van't Hoff equation

$$lnK_D = \frac{\Delta S^o}{R} - \frac{\Delta H^o}{RT} \tag{4.12}$$

where K_D is the linear sorption distribution coefficient, ΔS^o and ΔH^o are the standards of entropy and enthalpy change, respectively. In most cases, the value of K_D was obtained through a simple relation $K_D = q_e/C_e$ which is not a correct one (Da Silva et al., 2019; Degermenci et al., 2019; Deniz and Yildiz, 2019; Danish et al., 2018; Daneshvar et al., 2012; Akar and Celic, 2011; Caner et al., 2011; Akar and Divriklioglu, 2010). Through this simple relation, in one adsorption isotherm, there are many q_e and C_e, and any single relation between q_e and C_e will produce one value of K_D, and it is not clear which value of K_D should be chosen. The value of K_D should be calculated based on the method of Khan and Singh (1987). Through a straight-line plot of $\ln(q_e/c_e)$ versus C_e, the value of K_D will be obtained (see Fig. 4.5). Through Van't Hoff equation, the value of the standard of entropy and enthalpy change can be obtained.

Another valuable thermodynamic property of adsorption is Gibb's free energy change (ΔG^o) which can be calculated through the following relation:

$$\Delta G^o = -RTlnK_D \tag{4.13}$$

If the value of ΔG^o is negative, it means that the adsorption process is spontaneous and feasible in the term of thermodynamic (Daneshvar et al., 2012; Akar and Celic, 2011; Caner et al., 2011; Aksakal and Ucun, 2010; Brito et al., 2010). The positive value of ΔG^o indicates the adsorption of dyes onto the surface of biosorbents is not a spontaneous process; the thermodynamic equilibrium condition between adsorption and desorption processes could not be achieved (Degermenci et al., 2019; Guerrero-Coronilla et al., 2015).

The exothermicity and endothermicity of the adsorption process can be determined from the value of ΔH^o. If the uptake of adsorbate by adsorbent

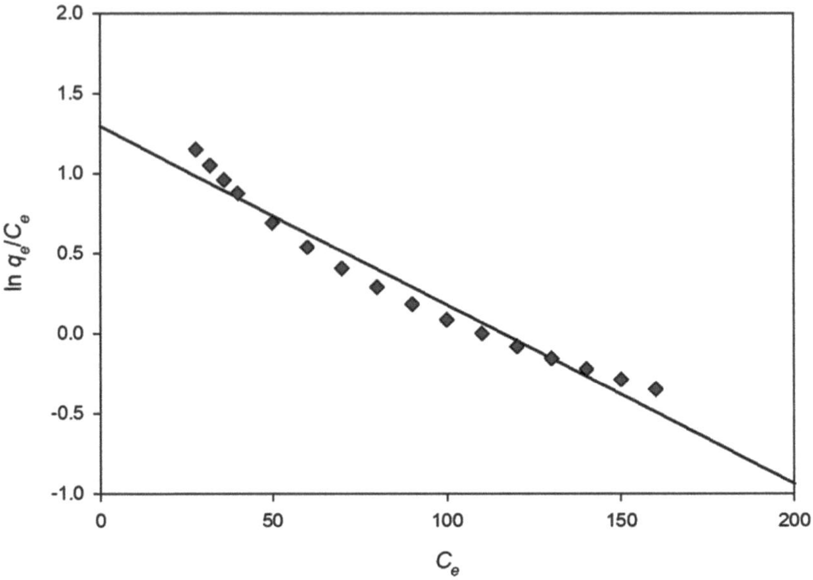

FIG. 4.5 Plot for determination of the value of K_D.

increases with temperature, the adsorption process is endothermic which is indicated by the positive value of ΔH^o (Degermenci et al., 2019; Guerrero-Coronilla et al., 2015; Daneshvar et al., 2012; Akar and Celic, 2011; Aksakal and Ucun, 2010). While the negative value of ΔH^o suggests that the adsorption process is exothermic, in this case, the temperature has a negative influence on the amount adsorbed by the adsorbent (Brito et al., 2010).

The randomness of the adsorption system is measured by ΔS^o. Positive value of ΔS^o indicates that adsorbate molecules possess a high affinity toward the active sites of the adsorbent. The adsorption system becomes more random between both interfaces of solute and adsorbent as the entropy of the system increases (Degermenci et al., 2019; Daneshvar et al., 2012; Akar and Celic, 2011; Aksakal and Ucun, 2010). When the values of ΔS^o is negative, the adsorption systems become more order (Guerrero-Coronilla et al., 2015; Brito et al., 2010).

12. Conclusions

Various low-cost biosorbents have explored its potential application as industrial adsorbents for the removal of dyes from aqueous solutions or wastewater. The performance of the biosorbents in removing dyes is strongly influenced by the pH and temperatures of the solution. Among various adsorption isotherm equations employed to correlate the experimental data, Langmuir and Freundlich could represent most of the adsorption systems very well. For kinetic models, the pseudo–second order is the best choice to represent the adsorption kinetic data. The thermodynamic of adsorption was obtained through the interpretation of the adsorption isotherms using Van't Hoff equation. From the results of the studies, it seems the utilization of these low-cost biosorbents in industrial scale is still far from reality.

References

Afroze, S., Sen, T.K., 2018. A review on heavy metal ions and dye adsorption from water by agricultural solid waste adsorbents. Water Air Soil Pollut. 229, 225.

Afshariani, F., Roosta, A., 2019. Experimental study and mathematical modeling of biosorption of methylene blue from aqueous solution in a packed bed of microalgae Scenedesmus. J. Clean. Prod. 225, 133–142.

Akar, T., Celik, S., 2011. Efficient biosorption of a reactive dye from contaminated media by Neurospora sitophila cells – zeamays silk tissue biomass system. J. Chem. Technol. Biotechnol. 86, 1332–1341.

Akar, T., Divriklioglu, M., 2010. Biosorption applications of modified fungal biomass for decolorization of Reactive Red 2 contaminated solutions: batch and dynamic flow mode studies. Bioresour. Technol. 101, 7271–7277.

Aksakal, O., Ucun, H., 2010. Equilibrium, kinetic and thermodynamic studies of the biosorption of textile dye (Reactive Red 195) onto Pinus sylvestris L. J. Hazard Mater. 181, 666–672.

Aksu, Z., Ertugrul, S., Donmez, G., 2010. Methylene Blue biosorption by Rhizopus arrhizus: effect of SDS (sodium dodecylsulfate) surfactant on biosorption properties. Chem. Eng. J. 158, 474–481.

Albadarin, A.B., Mangwandi, C., 2015. Mechanisms of alizarin red s and methylene blue biosorption onto olive stone by-product: isotherm study in single and binary systems. J. Environ. Manag. 164, 86–93.

Almeida, E.J.R., Corso, C.R., 2019. Decolorization and removal of toxicity of textile azo dyes using fungal biomass pelletized. Int. J. Environ. Sci. Technol. 16, 1319–1328.

Anastopoulos, I., Bhatnagar, A., Hameed, B.H., Ok, Y.S., Omirou, M., 2017. A review on waste-derived adsorbents from sugar industry for pollutant removal in water and wastewater. J. Mol. Liq. 240, 179–188.

Anastopoulos, I., Kyzas, G.Z., 2014. Agricultural peels for dye adsorption: a review of recent literature. J. Mol. Liq. 200, 381–389.

Ari, A.G., Celik, S., 2013. Biosorption potential of Orange G dye by modified Pyracantha coccinea: batch and dynamic flow system applications. Chem. Eng. J. 226, 263–270.

Astuti, W., Sulistyaningsih, T., Kusumastuti, E., Ratna, G.Y., Thomas, S., Kusnandi, R.Y., 2019. Thermal conversion of pineapple crown leaf waste to magnetized activated carbon for dye removal. Bioresour. Technol. 287, 121426.

Azin, E., Moghimi, H., 2018. Efficient mycosorption of anionic azo dyes by Mucor circinelloides: surface functional groups and removal mechanism study. J. Environ. Chem. Eng. 6, 4114–4123.

Baytar, O., Sahin, O., Saka, C., 2018. Sequential application of microwave and conventional heating methods for the preparation of activated carbon from biomass and its methylene blue adsorption. Appl. Therm. Eng. 138, 542–551.

Bhatnagar, A., Sillapaa, M., Witek-Krowiak, A., 2015. Agricultural waste peels as versatile biomass for water purification – a review. Chem. Eng. J. 270, 244–271.

Bhatnagar, A., Vilar, V.J.P., Botelho, C.M.S., Boanventura, R.A.R., 2010. Coconut-based biosorbents for water treatment—a review of the recent literature. Adv. Colloid Interface Sci. 160, 1–15.

Bhomic, P.C., Baruah, A.S.M., Pongener, C., Sinha, D., 2018. Activated biocarbon for adsorption of anionic dye from aqueous solution: isotherm, kinetic, thermodynamic and regeneration studies. Sust. Chem. Pharm. 10, 41–49.

Blanchard, G., Maunaye, M., Martin, G., 1984. Removal of heavy metals from waters by means of natural zeolites. Water Res. 18, 1501–1507.

Brito, S.M.D.O., Andrade, H.M.C., Soares, L.F., de Azevedo, R.P., 2010. Brazil nutshells as new biosorbent to remove methylene blue and indigo carmine from aqueous solutions. J. Hazard Mater. 174, 84–92.

Caner, M., Kiran, I., Ilhan, S., Pinarbasi, A., Iscen, C.F., 2011. Biosorption of reactive yellow 145 dye by dried penicillum restrictum: isotherm, kinetic, and thermodynamic studies. Separ. Sci. Technol. 46, 2283–2290.

Celekli, A., Al-Nuaimi, A.S., Bozkurt, H., 2019. Adsorption kinetics and isotherms of reactive red 120 on Moringa oleifera seed as an eco-friendly process. J. Mol. Struct. 1195, 168–178.

Chen, S.H., Ting, A.S.Y., 2015a. Biodecolorization and biodegradation potential of recalcitrant triphenylmethane dyes by *Coriolopsis* sp. isolated from compost. J. Environ. Manag. 150, 274–280.

Chen, S.H., Ting, A.S.Y., 2015b. Biosorption and biodegradation potential of triphenylmethane dyes by newly discovered Penicillium simplicissimum isolated from indoor wastewater sample. Int. Biodeterior. Biodegrad. 103, 1–7.

Chen, X., Li, H., Liu, W., Zhang, X., Wu, Z., Bi, S., Zhang, W., 2019. Effective removal of methyl orange and rhodamine B from aqueous solution using furfural industrial processing waste: furfural residue as an eco-friendly biosorbent. Colloids Surf., A 583, 123976.

Daneshvar, E., Kousha, M., Sohrabi, M.S., Khataee, A., Converti, A., 2012. Biosorption of three acid dyes by the brown macroalga Stoechospermum marginatum: isotherm, kinetic and thermodynamic studies. Chem. Eng. J. 195–196, 297–306.

Danish, M., Ahmad, T., Majeed, S., Ahmad, M., Ziyang, L., Pin, Z., Iqubal, S.M.S., 2018. Use of banana trunk waste as activated carbon in scavenging methylene blue dye: kinetic, thermodynamic, and isotherm studies. Biores. Technol. Rep. 3, 127–137.

Da Rosa, A.L.D., Carissimi, E., Dotto, G.L., Sander, H., Feris, L.A., 2018. Biosorption of rhodamine B dye from dyeing stones effluents using the green microalgae Chlorella pyrenoidosa. J. Clean. Prod. 198, 1302–1310.

Da Silva, D.C.C., Pietrobelli, J.M.T.A., 2019. Residual biomass of chia seeds (Salvia hispanica) oil extraction as low cost and eco-friendly biosorbent for effective reactive yellow B2R textile dye removal: characterization, kinetic, thermodynamic and isotherm studies. J. Environ. Chem. Eng. 7, 1038000.

Degermenci, G.D., Degermenci, N., Ayvaoglu, V., Durmaz, E., Cakir, D., Akan, E., 2019. Adsorption of reactive dyes on lignocellulosic waste; characterization, equilibrium, kinetic and thermodynamic studies. J. Clean. Prod. 225, 1220–1229.

Deniz, F., Karaman, S., 2011. Removal of Basic Red 46 dye from aqueous solution by pine tree leaves. Chem. Eng. J. 170, 67–74.

Deniz, F., Saygideger, S.D., 2010. Equilibrium, kinetic and thermodynamic studies of Acid Orange 52 dye biosorption by Paulownia tomentosa Steud. leaf powder as a low-cost natural biosorbent. Bioresour. Technol. 101, 5137–5143.

Deniz, F., Saygideger, S.D., 2011. Removal of a hazardous azo dye (Basic Red 46) from aqueous solution by princess tree leaf. Desalination 268, 6–11.

Deniz, F., Yildiz, H., 2019. Bioremediation potential of a widespread industrial biowaste as renewable and sustainable biosorbent for synthetic dye pollution. Int. J. Phytoremediation 21, 259–267.

Dogar, C., Gurses, A., Acıkyıldız, M., Ozkan, E., 2010. Thermodynamics and kinetic studies of biosorption of a basic dye from aqueous solution using green algae Ulothrix sp. Colloids Surf. B Biointerfaces 76, 279–285.

Dos Santos, K.J.L., dos Santos, G.E.D.S., de Sa, I.M.G.L., Ide, A.H., Duarte, J.L.D.S., Carvalho, S.H.V.D., Soletti, J.I., Meili, L., 2019. Wodyetia bifurcata biochar for methylene blue removal from the aqueous matrix. Bioresour. Technol. 293, 122093.

Du, C., Yang, H., Wu, Z., Ge, X., Cravotto, G., Ye, B.C., Kaleem, I., 2016. Microwave-assisted preparation of almond shell-based activated carbon for methylene blue adsorption. Green Process. Synth. 5, 395–406.

El Atouani, S., Belattmania, Z., Reani, A., Tahiri, S., Aarfane, A., Bentiss, F., Jama, C., Zrid, R., Sabour, B., 2019. Brown seaweed sargassum muticum as low-cost biosorbent of methylene blue. Int. J. Environ. Res. 13, 131–142.

Emrooz, H.B.M., Maleki, M., Shokouhimehr, M., 2019. Excellent adsorption of orange acid II on a water fern– derived micro- and mesoporous carbon. J. Taiwan Inst. Chem. Eng. 102, 99–109.

Erdoğdular, A.O., Apar, D.K., 2019. Bioremoval of reactive dye Remazol Navy by kefir grains. Appl. Biol. Chem. 62, 22.

Esmaeili, H., Foroutan, R., 2019. Adsorptive behavior of methylene blue onto sawdust of sour lemon, date palm, and Eucalyptus as agricultural wastes. J. Dispersion Sci. Technol. 40, 990–999.

Febrianto, J., Kosasih, A.N., Sunarso, J., Ju, Y.H., Indraswati, N., Ismadji, S., 2009. Equilibrium and kinetic studies in adsorption of heavy metals using biosorbent: a summary of recent studies. J. Hazard Mater. 162, 616–645.

Feng, Y., Yang, F., Wang, Y., Ma, L., Wu, Y., Kerr, P.G., Yang, L., 2011. Basic dye adsorption onto an agro-based waste material – sesame hull (Sesamum indicum L.). Bioresour. Technol. 102, 10280–10285.

Feng, Y., Dionysiou, D.D., Wu, Y., Zhou, H., Xue, L., He, S., Yang, L., 2013. Adsorption of dyestuff from aqueous solutions through oxalic acid-modified swede rape straw: adsorption process and disposal methodology of depleted bioadsorbents. Bioresour. Technol. 138, 191–197.

Feng, Y., Liu, Y., Xue, L., Sun, H., Guo, Z., Zhang, Y., Yang, L., 2017. Carboxylic acid functionalized sesame straw: a sustainable cost-effective bioadsorbent with superior dye adsorption capacity. Bioresour. Technol. 238, 675–683.

Fiorentin, L.D., Trigueros, D.E.G., Modenes, A.N., Espinoza-Quinones, F.R., Pereira, N.C., Barros, S.T.D., Santos, O.A.A., 2010. Biosorption of reactive blue 5G dye onto drying orange bagasse in batch system: kinetic and equilibrium modeling. Chem. Eng. J. 163, 68–77.

Ghaedi, M., Hajati, S., Barazesh, B., Karimi, F., Ghezelbash, G., 2013. *Saccharomyces cerevisiae* for the biosorption of basic dyes from binary component systems and the high order derivative spectrophotometric method for simultaneous analysis of Brilliant green and Methylene blue. J. Ind. Eng. Chem. 19, 227–233.

Ghariani, B., Hadrich, B., Louati, I., Mtibaa, R., Daassi, D., Rodriguez-Couto, S., Nasri, M., Mechichi, T., 2019. Porous heat-treated fungal biomass: preparation, characterization and application for removal of textile dyes from aqueous solutions. J. Porous Mater. 26, 1475–1488.

Gregory, P., 1990. Classification of Dyes by Chemical Structure. The Chemistry and Application of Dyes. Plenum Press, New York.

Guerrero-Coronilla, I., Morales-Barrera, L., Cristiani-Urbina, E., 2015. Kinetic, isotherm and thermodynamic studies of amaranth dye biosorption from aqueous solution onto water hyacinth leaves. J. Environ. Manag. 152, 99–108.

Gunduz, F., Bayrak, B., 2017. Biosorption of malachite green from an aqueous solution using pomegranate peel: equilibrium modelling, kinetic and thermodynamic studies. J. Mol. Liq. 243, 790–798.

Guo, F., Li, X., Jiang, X., Zhao, C., Rao, Z., 2018. Characteristics and toxic dye adsorption of magnetic activated carbon prepared from biomass waste by modified one-step synthesis. Colloid. Surface. Physicochem. Eng. Aspect. 555, 43–54.

Gupta, K., Khatri, O.P., 2019. Fast and efficient adsorptive removal of organic dyes and active pharmaceutical ingredient by microporous carbon: effect of molecular size and charge. Chem. Eng. J. 378, 122218.

Hajialigol, S., Masoum, S., 2019. Optimization of biosorption potential of nano biomass derived from walnut shell for the removal of Malachite Green from liquids solution: experimental design approaches. J. Mol. Liq. 286, 110904.

Hamitouche, A., Haffas, M., Boudjemaa, A., Benammar, S., Sehailia, M., Bachari, K., 2017. Efficient biosorption of methylene blue, malachite green and methyl violet organic pollutants on biomass derived from Anethum graveolens: an eco-benign approach for wastewater treatment. Desalin. Water Treat. 75, 225–236.

Huang, J., Liu, D., Lu, J., Wang, H., Wei, X., Liu, J., 2016. Biosorption of reactive black 5 by modified Aspergillus versicolor biomass: kinetics, capacity and mechanism studies. Colloid. Surface. Physicochem. Eng. Aspect. 492, 242–248.

Ismadji, S., Bhatia, S.K., 2001. Characterization of activated carbons using liquid-phase adsorption. Carbon 39, 1237–1250.

Jain, S.N., Gogate, P.R., 2019. Adsorptive removal of azo dye in a continuous column operation using biosorbent based on NaOH and surfactant activation of Prunus dulcis leaves. Desalin. Water Treat. 141, 331–341.

Javed, F., Feroze, N., Ikhlaq, A., Kazmi, M., Ahmad, S.W., Munir, H.M.S., 2019. Biosorption potential of Sapindus mukorossi dead leaves as a novel biosorbent for the treatment of Reactive Red 241 in aqueous solution. Desalin. Water Treat. 137, 345–357.

Karthik, V., Sivarajasekar, N., Padmanaban, V.C., Nakkeeran, E., Selvaraju, N., 2019. Biosorption of xenobiotic Reactive Black B onto metabolically inactive T. harzianum biomass: optimization and equilibrium studies. Int. J. Environ. Sci. Technol. 16, 3625–3636.

Kaushik, P., Malik, A., 2010. Alkali, thermo and halo tolerant fungal isolate for the removal of textile dyes. Colloids Surf. B Biointerfaces 81, 321–328.

Khan, A.A., Singh, R.P., 1987. Adsorption thermodynamics of carbofuran on Sn(IV) arsenosillicate in H^+, Na^+ and Ca^+ forms. Colloid. Surface. 24, 33–42.

Lebron, Y.A.R., Moreira, V.R., Santos, L.V.S., Jacob, R.S., 2018. Remediation of methylene blue from aqueous solution by Chlorella pyrenoidosa and Spirulina maxima biosorption: equilibrium, kinetics, thermodynamics and optimization studies. J. Environ. Chem. Eng. 6, 6680–6690.

Lebron, Y.A.R., Moreira, V.R., Santos, L.V.S., 2019. Studies on dye biosorption enhancement by chemically modified Fucus vesiculosus, Spirulina maxima and Chlorella pyrenoidosa algae. J. Clean. Prod. 240, 118197.

Lim, S.L., Chu, W.L., Phang, S.M., 2010. Use of Chlorella vulgaris for bioremediation of textile wastewater. Bioresour. Technol. 101, 7314–7322.

Lesmana, S.O., Febriana, N., Soetaredjo, F.E., Sunarso, J., Ismadji, S., 2009. Studies on potential applications of biomass for the separation of heavy metals from water and wastewater. Biochem. Eng. J. 44, 19–41.

Lin, Y., He, X., Han, G., Tian, Q., Hu, W., 2011. Removal of Crystal Violet from aqueous solution using powdered mycelial biomass of Ceriporia lacerata P2. J. Environ. Sci. 23, 2055–2062.

Liu, M., Li, X., Du, Y., Han, R., 2019. Adsorption of methyl blue from solution using walnut shell and reuse in secondary adsorption for Congo red. Bioreso. Technol. Rep. 5, 238–242.

Mahmoodi, N.M., Hayati, B., Bahrami, H., Arami, M., 2011. Dye adsorption and desorption properties of mentha pulegium in single and binary systems. J. Appl. Polym. Sci. 122, 1489–1499.

Mamani, A., Ramirez, N., Deiana, C., Gimenez, M., Sardella, F., 2019. Highly microporous sorbents from lignocellulosic biomass: different activation routes and their application to dyes adsorption. J. Environ. Chem. Eng. 7, 103148.

Mbarki, F., Selmi, T., Kesraoui, A., Seffen, M., Gaddoneix, O., Celzard, A., Fierro, V., 2019. Hydrothermal pre-treatment, an efficient tool to improve activated carbon performances. Ind. Crop. Prod. 140, 111717.

Meerbergen, K., Willems, K.A., Dewil, R., Impe, J.V., Appels, L., Lievens, B., 2018. Isolation and screening of bacterial isolates from wastewater treatment plants to decolorize azo dyes. J. Biosci. Bioeng. 125, 448–456.

Mishra, S., Mohanty, P., Maiti, A., 2019. Bacterial mediated bio-decolourization of wastewater containing mixed reactive dyes using jack-fruit seed as co-substrate: process optimization. J. Clean. Prod. 235, 21–33.

Módenes, A.N., Hinterholz, C.L., Neves, C.V., Sanderson, K., Trigueros, D.E.G., Espinoza-Quiñones, F.R., Borba, C.E., Steffen, V., Scheufele, F.B., Kroumov, A.D., 2019. A new alternative to use soybean hulls on the adsorptive removal of aqueous dyestuff. Biores. Technol. Rep. 6, 175–182.

Moghazy, R.M., Labena, A., Husien, S., 2019. Eco-friendly complementary biosorption process of methylene blue using micro-sized dried biosorbents of two macro-algal species (Ulva fasciata and Sargassum dentifolium): full factorial design, equilibrium, and kinetic studies. Int. J. Biol. Macromol. 134, 330–343.

Muntean, S.G., Todea, A., Bakardjieva, S., Bologa, C., 2017. Removal of non benzidine direct red dye from aqueous solution by using natural sorbents: beech and silver fir. Desalin. Water Treat. 66, 235–250.

Murthy, T.P.K., Gowrishankar, B.S., Prabha, M.N.C., Kruthi, M., Krishna, R.H., 2019. Studies on batch adsorptive removal of malachite green from synthetic wastewater using acid treated coffee husk: equilibrium, kinetics and thermodynamic studies. Microchem. J. 146, 192–201.

Nasrullah, A., Saad, B., Bhat, A.H., Khan, A.S., Danish, M., Isa, M.H., Naeem, A., 2019. Mangosteen peel waste as a sustainable precursor for high surface area mesoporous activated carbon: characterization and application for methylene blue removal. J. Clean. Prod. 211, 1190–1200.

Nausad, M., Alqadami, A.A., Alothman, Z.A., Alsohaimi, I.H., Algamdi, M.S., Aldawsari, A.M., 2019. Adsorption kinetics, isotherm and reusability studies for the removal of cationic dye from aqueous medium using arginine modified activated carbon. J. Mol. Liq. 293, 111442.

Ofomaja, A.E., Ukpebor, E.E., Uzoekwe, S.A., 2011. Biosorption of Methyl violet onto palm kernel fiber: diffusion studies and multistage process design to minimize biosorbent mass and contact time. Biomass Bioenergy 35, 4112–4123.

Ouhammou, M., Lahnine, L., Mghazli, S., Hidar, N., Bouchdoug, M., Jaouad, A., Mandi, L., Mahrouz, M., 2019. Valorisation of cellulosic waste basic cactus to prepare activated carbon. J. Saudi Soc. Agricul. Sci. 18, 133–140.

Oymak, T., Eruigur, M., 2019. Effective and rapid removal of cationic and anionic dyes from aqueous solutions using Elaeagnus angustifolia L. fruits as a biosorbent. Desalin. Water Treat. 138, 257–264.

Ozhan, A., Sahin, O., Kucuk, M.M., Saka, C., 2014. Preparation and characterization of activated carbon from pine cone by microwave-induced $ZnCl_2$ activation and its effects on the adsorption of methylene blue. Cellulose 21, 2457–2467.

Pandey, R.K., Tewari, S., Tewari, L., 2018. Lignolytic mushroom Lenzites elegans WDP2: laccase production, characterization, and bioremediation of synthetic dyes. Ecotoxicol. Environ. Saf. 158, 50–58.

Plazinski, W., Rudzinski, W., Plazinska, A., 2009. Theoretical models of sorption kinetics including a surface reaction mechanism: a review. Adv. Colloid Interface Sci. 152, 2–13.

Postai, D.L., Rodrigues, C.A., 2018. Adsorption of cationic dyes using waste from fruits of Eugenia umbelliflora Berg (myrtaceae). Arabian J. Sci. Eng. 43, 2425–2440.

Rehman, R., Farooq, S., Mahmud, T., 2019. Use of Agro-waste Musa acuminata and Solanum tuberosum peels for economical sorptive removal of Emerald green dye in ecofriendly way. J. Clean. Prod. 206, 819–826.

Russo, M.E., Di Natale, F., Prigione, V., Tigini, V., Marzocchella, A., Varese, G.C., 2010. Adsorption of acid dyes on fungal biomass: equilibrium and kinetics characterization. Chem. Eng. J. 162, 537–545.

Safa, Y., Bhatti, H.N., 2011. Biosorption of direct red-31 and direct orange-26 dyes by rice husk: application of factorial design analysis. Chem. Eng. Res. Des. 89, 2566–2574.

Sen, S.K., Raut, S., Bandyopadhyay, P., Raut, S., 2016. Fungal decolouration and degradation of azo dyes: a review. Fungal Biol. Rev. 30, 112–133.

Sewu, D.D., Lee, D.S., Nguyen, T.H., Woo, S.H., 2019. Effect of bentonite-mineral co-pyrolysis with macroalgae on physicochemical property and dye uptake capacity of bentonite/biochar composite. J. Taiwan Inst. Chem. Eng. 104, 106–113.

Shamsollahi, Z., Partovinia, A., 2019. Recent advances on pollutants removal by rice husk as a bio-based adsorbent: a critical review. J. Environ. Manag. 246, 314–323.

Shao, H., Li, Y., Zheng, L., Chen, T., Liu, J., 2017. Removal of methylene blue by chemically modified defatted brown algae Laminaria japonica. J. Taiwan Inst. Chem. Eng. 80, 525–532.

Shrivastava, V.S., 2010. The biosorption of Safranine onto Parthenium hysterophorus L: equilibrium and kinetics investigation. Desalin. Water Treat. 22, 146–155.

Sinha, S., Singh, R., Chaurasia, A.K., Nigam, S., 2016. Self-sustainable Chlorella pyrenoidosa strain NCIM2738 based photo bioreactor for removal of Direct Red-31 dye along with other industrial pollutants to improve the water-quality. J. Hazard Mater. 306, 386–394.

Spagnoli, A.A., Giannakoudakis, D.A., Bashkova, S., 2017. Adsorption of methylene blue on cashew nut shell based carbonsactivated with zinc chloride: the role of surface and structural parameters. J. Mol. Liq. 229, 465–471.

Sudaryanto, Y., Hartono, S.B., Irawaty, W., Hindarso, H., Ismadji, S., 2006. High surface area activated carbon prepared from cassava peel by chemical activation. Bioresour. Technol. 97, 734–739.

Sun, W., Sun, W., Wang, Y., 2019. Biosorption of Direct Fast Scarlet 4BS from aqueous solution using the green-tide-causing marine algae Enteromorpha prolifera. Spectrochim. Acta Mol. Biomol. Spectrosc. 223, 117347.

Sun, Z., Srinivasakannan, C., Liang, J., Duan, X., 2019. Preparation and characterization of shiitake mushroom-based activated carbon with high adsorption capacity. Arabian J. Sci. Eng. 44, 5443–5456.

Tan, C.Y., Li, G., Lu, X.Q., Chen, Z.L., 2010. Biosorption of basic orange using dried *A. filiculoides*. Ecol. Eng. 36, 1333–1340.

Wakkel, M., Khiari, B., Zagrouba, F., 2019. Textile wastewater treatment by agro-industrial waste: equilibrium modelling, thermodynamics and mass transfer mechanisms of cationic dyes adsorption onto low-cost lignocellulosic adsorbent. J. Taiwan Inst. Chem. Eng. 96, 439–452.

Yang, J.X., Hong, G.B., 2018. Adsorption behavior of modified Glossogyne tenuifolia leaves as a potential biosorbent for the removal of dyes. J. Mol. Liq. 252, 289–295.

Yu, X., Han, Z., Fang, S., Chang, C., Han, X., 2019. Optimized preparation of high value-added activated carbon and its adsorption properties for methylene blue. Int. J. Chem. React. Eng. 17, 20180267.

Zhou, Y., Lu, J., Zhou, Y., Liu, Y., 2019. Recent advances for dyes removal using novel adsorbents: a review. Environ. Pollut. 252, 352–365.

Chapter 5

Removal of dyes and pigments from industrial effluents

Ahmet Gürses, Kübra Güneş, Elif Şahin
Chemical Education Ataturk University, Erzurum, Turkey

1. Introduction

In addition to increasing the living standards of industrial and technological developments, the complex wastes they produce cause serious environmental problems. Industrial wastes emerge as undesirable products of industrial-scale processes such as textile, rubber, leather, paper, plastic, coal, food, petrochemical, pharmaceutical, paint industry, production of solvents and other chemicals, mills, and mining operations.

While industrial wastes are often produced as solid wastes from original materials, they can often contain liquid hazardous wastes. Liquid wastes are generally regarded as the most important industrial wastes associated with wastewater treatment systems. Moreover, many industrial processes can produce liquid wastes in huge quantities and varying qualities directly. Industrial effluents can also include in dissolved or dispersed forms value-added products such as feedstock materials, by-products, washing and cleaning agents, solvents and plasticizers. Effluents are characterized by flow volume and velocity, physical conditions, chemical and toxic components, and bacteriological contents, taking into account some basic parameters such as biological oxygen demand (BOD), chemical oxygen demand (COD), and dissolved oxygen (DO) (Chukwu, 2008).

Liquid wastes may contain nontoxic inorganic substances and toxic organic substances, and their increased amounts require higher DO. (Ali, 2004). However, while the inorganic materials which generally degrade more slowly cause blockages, the accumulated organic toxic compounds can cause serious chemical effects. Suspended solids, which are composed of various chemical and microbiological components in the effluents, primarily can lead to turbidity, and the sludge may also contain pathogenic microorganisms. Oils and solvents are also important fossil carbon sources in industrial liquid waste.

Green Chemistry and Water Remediation: Research and Applications.
https://doi.org/10.1016/B978-0-12-817742-6.00005-0

Colorants, dyes and pigments, are used to impart color to a wide range of materials, commonly referred to as substrates, for various purposes such as esthetics, protection, and art. They can be divided into two subgroups as dyes soluble in the application medium and pigments insoluble in the application medium. Dyes, which are organic compounds bound to give color to the fabric surface, are colored substances which, when applied to fibers, give a permanent color which is resistant to the negative effects of light, water, and detergent (Rai et al., 2005). They are water soluble, producing bright colors in water with acidic properties (Sharma and Kaur, 2018). Pigments are insoluble colorants used for fibers, plastics, and other polymeric materials (Wang and Yin, 2013). The most obvious distinction between dye and pigment is that the dyes are soluble in the substrate so that they are dispersed on a molecular scale and the pigments do not dissolve and are dispersed in particles (Gürses et al., 2016a).

2. Dyes and their classifications

Although dyes are classified in various ways, the most common classification is based on the source of the dyestuff and the nature of the chromophore groups.

According to the classification made according to the origin of dyes, dyes are divided into two basic groups as **natural** and **synthetic** dyes.

Natural dyes include all dyes obtained from natural sources such as plants, animals, and minerals (Benkhaya et al., 2017). Many plants and animals are known to have a very high potential in terms of their natural dyestuffs, and most of them have been used for natural dyeing for centuries. Various parts of plants, such as roots, stems, bark, leaves, fruit, and seeds, can contain a wide variety of coloring agents that can be extracted. Already natural dyes are commonly obtained from roots, stems, leaves, flowers, fruits of various plants, and dried stems of certain insects and minerals (Vankar, 2000).

Essentially, three basic colors are required to obtain any color or hue, which is much more critical for synthetic dyes, and since natural dyes cannot be easily blended to obtain the desired color, different dyeing procedures are used for different dyes to obtain various color or hues.

Today, due to the advantages of being cheaper, brighter, and faster and easily applicable, **synthetic dyes**, which form almost all of the dyes in the relevant market, have turned into one of the largest industrial production areas (Shindy, 2017).

According to the classification made according to chromophore group, the dyes are divided into subclasses such as **Nitro and Nitroso, Azo**, **Anthraquinonoid**, **Indigoid**, **Phthalocyanine,** and **Sulfur dyes**. Table 5.1 shows the subclasses made according to the chromophore group, their main chromophore groups except for the sulfur dyes, and their typical examples.

- **Nitro and nitroso dyes** constitute only a small part of the dye market and therefore have low commercial importance. Nitrogen dyes having simple chemical structures consist of small aromatic molecules containing one nitro group and one or more amino or hydroxyl groups (Raval et al., 2017).

TABLE 5.1 The subclasses made according to the chromophore group, their main chromophore groups except for the sulfur dyes, and their typical examples.

Dyes	Chromophore group	Typical example
Azo	-N=N-	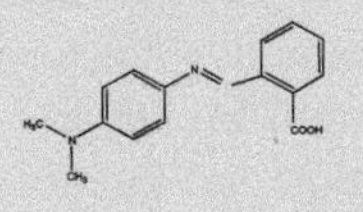 Acid Red 2
Nitro and Nitroso	 —N=O	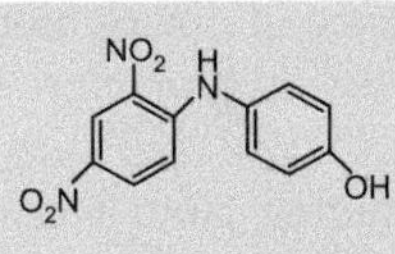 Disperse Yellow 1
İndigoid	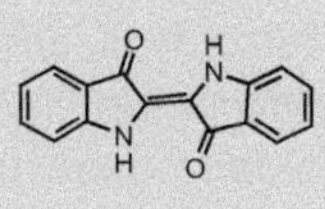	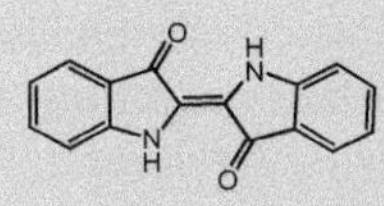 Indigo
Anthraquinonoid	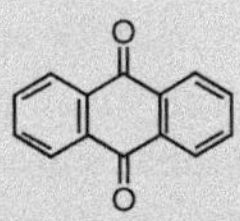	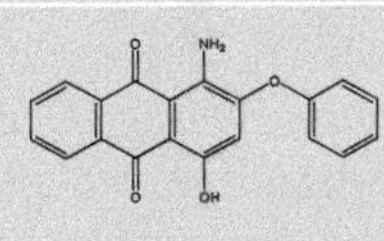 Disperse Red 60
Phthalocyanine	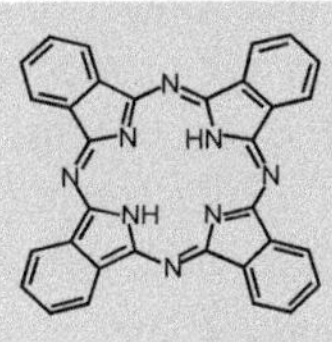	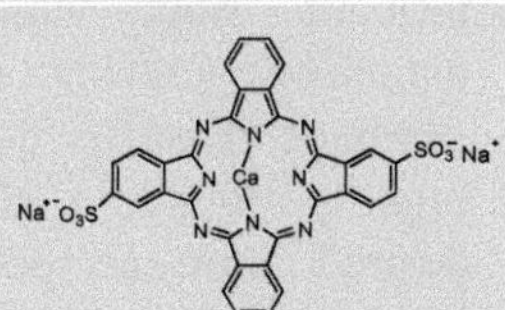 Direct Blue 86
Sulfur		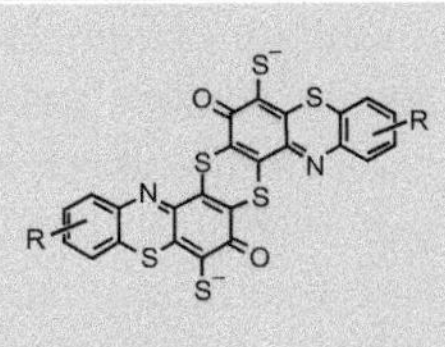 Sulfur Black 1

The first nitro dyes are acid dyes, which are mainly used for dyeing natural animal fibers such as wool and silk (Gregory, 1990). Nitroso compounds are not capable of dyeing alone, but they can form metal complexes with pigments (Gürses et al., 2016a).

Azo dyes, the most commonly used colorants, corresponding to more than 60% of dyes in use, are tinctorially strong. They cover the entire range of colors and often have some features such as cost-effective production, preparation from cheap and readily available starting materials, and good fastness (Gordon and Gregory, 1987; Gupta and Suhas, 2009; Shah, 2014a,b).

Azo dyes, which may contain at least one nitrogen–nitrogen (N = N) double bond, but may be of many different structures, are characterized by the presence of one or more azo chromophores (N = N) and the presence of bonds between two or more aromatic rings (Growther and Meenakshi, 2008; Costa et al., 2012). Azo dyes are stable in light and resistant to microbial degradation and fading from washing and are generally not easily removed from wastewater by conventional wastewater treatment methods (Puvaneswari et al., 2006; Mossvi et al., 2007).

- **Anthraquinonoid dyes**, which have been used for centuries and are the most common class of dyes after azo dyes, are characterized by their good fastness resulting from very stable anthraquinonoid structures (Saranraj, 2013; Jerič et al., 2013). In addition to their important advantages such as brightness and good fastness properties, anthraquinones can be also found in bacteria, fungi, lichens, and plants, unlike azo dyes that do not have natural counterparts (Han et al., 2001). Due to their high color fastness and stability, anthraquinonoid dyes are widely used in printing processes and textile dyeing (Gregory 1990; Han et al., 2001; Fleischmann et al., 2015)
- **Indigo dye**, which is one of the oldest known organic dyes and has been used for dyeing for approximately 5000 years in its natural form, is a unique blue-colored organic compound (Tölek and Kadem, 2016). Historically, indigo was a natural dye obtained from the leaves of certain plants, but almost all of the indigo dye produced in large quantities today is synthetic. Due to the different dye content, the natural indigo dyeing process requires a high level of expertize and experience to achieve the desired color and tone. It is widely used today in dyeing denim cloth and jackets (Ünlü, 2008; Saikhao et al., 2018).
- **Phthalocyanine dyes**, analogs of natural pigments, are known to be able to develop cheaper, brighter blue tones in cellulosics, and the shades produced are extremely resistant to all acid, alkali, and oxidizing and reducing agents (Chakraborty, 2014; Nemykin and Lukyanets, 2010). Phthalocyanines, which have an expanded p-electron conjugation system and which can exhibit absorption in the near IR region, are extremely stable bright and tinctorially strong dyes with various coordination properties and enhanced spectroscopic characteristics (Gregory, 1990; Dubinina et al., 2013; Durmuş and Nyokong, 2008; Sahin et al., 2015).

- **Sulfur dyes** are one of the most popular dye classes used to color cellulosic fibers in deep black shades and produce low-cost tertiary shades (Senior and Clarke, 1986).

These dyes are normally insoluble in water and are reduced to water--soluble leuco form for application to the substrate using sodium sulfite solution and then formed by atmospheric or chemical oxidation within the fiber pores (Hunger, 2003; Gürses et al., 2016a).

Sulfur dyes, which are complex organic compounds synthesized by heating simple amines or phenolic compounds in the presence of sulfur, cannot be characterized by an identifiable chromophore, unlike most other dye types. They may consist of phenothiazonethianthrone-type macromolecular structures in which sulfur exists as sulfur-bridging links and thiazine groups (Gregory, 1990; Atiq et al., 2019).

3. Pigments and their classifications

Pigments, which are inorganic or organic, colored, white or black substances that change the color of reflected or transmitted light as a result of wavelength-selective absorption, are dry colorants generally used as a fine powder in manufacturing and visual arts. For use in paints, powder pigments are added to a binder or carrier, which is a relatively neutral or colorless substance that suspends the pigment and adheres to the paint. In general, colorants that are insoluble (i.e., by forming suspension) in the medium are defined as pigments, whereas those which are soluble in the media (i.e., by forming solution) are defined as dyes. Thus, a colorant may function as a pigment or dye depending on the dispersion behavior in its medium. Moreover, in some cases, a soluble dye can be precipitated with a metallic salt to produce a pigment from the dye (Herbst and Hunger, 2004).

Although different ways of classifying pigments based on chemical structure and color characteristics are proposed, their classification as organic and inorganic pigments is considered a more appropriate way (Gürses et al., 2016a; Hunger and Schmidt, 2018). The classification of pigments together with their subgroups as organic and inorganic pigments is shown schematically in Fig. 5.1.

Organic pigments are organic compounds containing carbon, oxygen, nitrogen, and the like, which may form a chromophore, in part or in full, and therefore exhibit color. It is known that if the system has an optical absorption band in a visible wavelength region, a relatively large π-conjugation system containing these atoms can be a chromophore (Junji, 2015).

- **Azo pigments**, which are similar in the chemical structure to azo dyes but do not contain soluble groups, are not readily purified because they are virtually insoluble in any medium and therefore require highly purified precursors for their synthesis. Azo pigments containing a single $-N=N-$ group are defined as monoazo and those with two azo linkages are defined

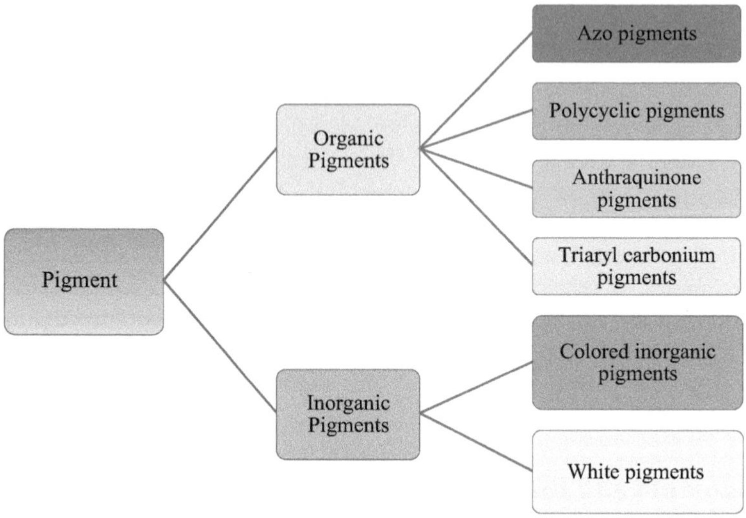

FIG. 5.1 The classification of pigments together with their subgroups as organic and inorganic pigments.

as disazo pigments. Most azo pigments, which are commonly used in various plastics, rubbers, and paints and can provide bright colors at low cost, are characterized by relatively poor light fastness or stability to heat and solvents (Berrie and Lomax, 1997).

- **Polycyclic pigments**, characterized by condensed aromatic or heterocyclic ring systems, are the second largest synthetic pigment group after the azo pigment group and are more expensive than azo pigments, except for phthalocyanine pigments (Ghelardi, 2014).
- **Anthraquinonoid pigments** are essentially a group of organic dyes with anthraquinonoid molecular structures and are subdivided according to the most suitable methods for their application to various fibers. Anthraquinonoid pigments are a large and important group that covers the tonal range from greenish yellow to bluish green, but are especially important for bluish red, violet, blue, and bluish green dyeing (Choudhury, 2011).
- **Triarylcarbonium pigments** have a triarylcarbonium structure and at least two of the amyl moieties carry amino groups that act as electron donating substituents. The compounds are therefore basic and can therefore combine with acids to form insoluble salts in the form in which they can be applied as pigments. Triarylcarbonium compounds can be identified in the quinoid form containing an ammonium ion or by a benzenoid structure containing a carbon ion and several mesomeric structures (Hunger, 2003).

Inorganic pigments derived from various natural minerals or ores such as iron oxide, cadmium sulfides, and titanium dioxide are chemical compounds that are not carbon-based and are generally metallic salts precipitated from solutions.

- **White pigments**, which are reflective in most wavelengths of solar radiation, provide optimal sun reflectance and have excellent hiding power and ultraviolet (UV) resistance (Comstock, 2016). The most common white pigments are titanium dioxide, zinc oxide, zinc sulfate, lithopone, and lead white.
- **Colored pigments** are complex inorganic pigments with mixed metal oxides such as bismuth vanadate, chromium (III) pigments, ultramarine pigments, cyanide iron blue, cadmium pigments, lead chromate, iron oxide and spinel, and retile and zircon pigments (Gürses et al., 2016a).

4. Industrial wastes

Wastes, which are an inevitable by-product of human activities, continue to be a life-threatening problem for centuries and a growing problem for humanity (Gürses et al., 2016b; Okwesili and Iroko, 2016). Although industrial growth plays an important role in the growth of this inevitable problem due to the economic development in the world, industrial wastes produced by factories and industrial plants have the biggest share in increasing pollution and negatively affecting the environment.

Moreover, this trend has led to drastic changes in pollution control and sustainability of environmental quality (Wei and Huang, 2001). The term industrial waste is commonly defined as undesirable residual materials arising from industrial processes or resulting from production processes, and food residues, ashes, rubbish, construction wastes, special wastes, and hazardous wastes are also considered industrial wastes (Gürses et al., 2016b). High amounts of industrial waste are potentially hazardous to the environment and extremely hazardous to living organisms, including humans (Makgae, 2011).

4.1 Classification of industrial wastes

A variety of chemical-, physical-, and biological-based industrial processes, nonrenewable raw materials such as petroleum, coal, ore, and minerals, and renewable raw materials such as plants and animals can convert into different products such as fuel, chemicals, plastics, pharmaceuticals, food, fertilizer, and feed. The structural properties of the products produced may vary due to chemical, electrochemical, biochemical, and biological reactions that occur during the process, and the processes may produce solid, liquid, and/or gaseous wastes as well as a certain amount of energy (Gürses et al., 2016b).

The current preclassification for waste types has been made by EPA as hazardous waste, restricted solid waste, and general solid waste (putrescible and nonputrescible) (EPA, 2014).

The impact of **hazardous wastes** on the environment, mainly due to industrial activities and their different modes of production, is a major concern with serious environmental risks, especially if it results in toxic contamination of soil, water, and air and is not well managed. Hazardous wastes with a wide variety of hazardous or potentially harmful properties to human health and the environment in general can be in the form of liquid, solid, or sludge and gas and also by-products of production processes or disposed commercial products (i.e., cleaning fluids or pesticides) (Evin, 2009).

Restricted solid waste includes wastes that the maximum possible levels of contaminants do not exceed the specific contaminant concentration and/or toxicity characteristics leaching procedure.

Wastes other than special wastes, liquid wastes, hazardous wastes, or limited solid wastes are classified as **general solid wastes (putrescible)** by EPA (e.g., household wastes containing putrescible organics, waste from litterbins, and food wastes). Rapid urbanization and changing lifestyles have changed the composition of waste from mainly organic species (putrescible), plastic, paper, and packaging materials with complex structures in nature. Storage and collection systems are becoming more complex and costly due to the increasing diversification of waste generated and the shrinkage of disposal sites (Idris et al., 2004). Wastes such as glass, plastic, rubber, gypsum board, ceramic, brick, concrete, metal, and paper or cardboard (other than special waste, liquid waste, hazardous waste, restricted solid waste, or putrescible general solid waste) constitute the **nonputrescible general solid wastes** (Gürses et al., 2016b).

Solid wastes are undesirable or useless solids resulting from combined housing, industrial and commercial activities in certain regions. They can be divided into three categories in terms of origin (domestic, industrial, commercial, construction, or institutional), content (organic material, glass, metal, plastic paper, etc.), and hazard potential (toxic, nontoxic, flammable, radioactive, infectious, etc.) (Sharma et al., 2014a).

Industrial solid wastes are also classified as organic, inorganic, solid, semisolid, hazardous, and nonhazardous wastes according to their components, species, and impacts (Gürses et al., 2016b).

The liquids originating from restaurants, cars, houses, washing machines, or tank cleaning operations, dye households and other industrial areas, as well as laboratory wastes that may be bulky, stagnant, or even completely liquid are considered as **liquid wastes**. Also, liquids from oil traps, septic tanks, surfactants, washing waters, typical medical or clinical wastes, solvents, dyes, resins, inks and paints, photographic wastes, insecticides, and other chemicals are included in the liquid waste category. Chemical wastes such as formaldehyde, alcohols, corrosive substances (acids and bases), contaminated soil, or quarantine wastes are considered as the most dangerous liquid wastes.

Gaseous wastes are waste products from various human activities such as manufacturing, processing, material consumption, or biological processes (Choppin et al., 2013). Gaseous emissions may contain organic and inorganic compounds that may have adverse effects on the population in an industrial environment or in nearby settlements (Gürses et al., 2016b).

5. Color removal technologies

Dyes and pigments, which are essential components in the dye industry, are highly visible substances, and even the minimum quantities released into the environment can lead to nonesthetic environmental pollution (Ejder-Korucu et al., 2015; Segneanu et al., 2013). Moreover, the dyeing industry is one of the largest water-consuming industries. Dye industry wastewater containing a wide range of chemicals and coloring compounds requires strictly fine treatment before discharge, but these wastewaters are highly difficult to treat because their composition is highly variable (Sivamani and Leena, 2009; Do and Chen, 1990).

Synthetic dyes, which are generally classified as acid, reactive, direct, basic, vat, disperse, metal complex, mordant, and sulfur dyes and produced around 1 million tonnes per year worldwide, are used in a variety of industries including textiles, paper printing, food, pharmaceuticals, leather, and cosmetics. More than 10,000 different dyes are used in textile and 70% of them are synthetic azo dyes. It is estimated that 2% of the dyes are discharged in effluent from manufacturing operations, while 10% was discharged from textile and associated industries (Hassaan and El Nemr, 2017). Among the industries, textile is one of the sectors that consumes the most water and chemicals. The effluents resulting from textile dyeing processes contain dyestuff (about 8%–20% of the total pollution load) and a large number of auxiliary chemicals with large amounts of water. For pigments, the loss rate in paper and leather dyeing is estimated to be 1%–10%. In parallel with the increasing demand for textile products, large increases in production and the use of synthetic dyes have caused dye wastewater to become an important source of serious pollution problems (Checker et al., 2013).

Processes that are used for the treatment of dye effluent, which is one of the biggest threats to the environment, causing esthetic pollution even at low concentrations and which also affect aquatic life and food chain as chemical pollution, are able to eliminate only half of the dissolved or suspended components in wastewater streams (Mukherjee et al., 1999; Allen et al., 2004; Nidheesh et al., 2018). Dyestuff-containing effluents cannot be effectively decolorized by conventional methods, since the dye molecules are resistant to aerobic digestion and stable to light, heat, and oxidizing agents (Hassaan et al., 2017). In addition, there are some disadvantages such as high cost and disposal problems in the large-scale treatment of colored effluents from the textile and paper industries (Ghoreishi and Haghighi, 2003).

In general, industrial color removal technologies can be divided into three categories as chemical, biological, and physical methods (Robinson et al., 2001).

5.1 Chemical treatment methods

The main chemical methods used in dye removal are coagulation or flocculation combined with flotation and filtration, precipitation-flocculation with di and trivalent cation salts, electroflotation, electrocoagulation(EC), ozonation, and irradiation. These methods, which generally have high operating costs and cause accumulation of colored sludge which are difficult to dispose, also have the potential to cause a secondary pollution problem due to excessive chemical use (Kyzas et al., 2013).

5.1.1 Coagulation—flocculation

The coagulation—flocculation method is carried out by the use of high amounts of expensive coagulants such as alum (aluminum sulfate, Al_2SO_4), iron chloride ($FeCl_3$), or iron sulfate (Fe_2SO_4), depending on the amount of effluent to be treated. In practice, coagulants are used to neutralize the charge of the particles to be removed so that they coagulate and precipitate, and therefore the process efficiency is strongly dependent on variables such as suspension pH, mixing speed, and time (Verma et al., 2012).

The most likely problem in the use of the chemical coagulation—flocculation method is that it is quite difficult to sufficiently reduce the solubility of the components capable of forming flocks to be removed from the effluent. For this reason, generally the coagulation—flocculation process is applied in combination with many other treatment methods in effluent treatment (Butler, 2013).

5.1.2 Electroflotation

Electroflotation has the ability to combine the advantages of methods such as coagulation, flotation, and EC used in the treatment of effluents. Compared with the conventional flotation method, electroflotation, which has many advantages, can produce uniformly dispersed fine bubbles, which can increase the contact surface between the flocks and the bubble and thus provide more efficient and faster clarification. Electroflotation, which is easily applicable with less sludge generation and voltage control, requires low maintenance and operating costs (El-Hosiny et al., 2017). It not only requires complex equipment but also has advantage of not using special chemicals, unlike chemical coagulation (Petsriprasit et al., 2010; Nandi and Patel, 2017).

The electroflotation process is based on the principle of electrochemically producing of cations from consumable iron or aluminum anodes and then coagulating contaminants in the aqueous medium (Chaturvedi, 2013; Naje et al., 2015). The process consists essentially of aggregation of the destabilized phase to form flocks for the formation of coagulates by destabilization of contaminants. Destabilization of the pollutants occurs by compressing the electrical double layer around the charged species with anodic counter ions, and counter ions further reduce the electrostatic repulsion by neutralizing the ionic species in the aqueous medium and thus coagulation occurs due to the attractive interactions between air bubbles and the neutralized species. (El-Hosiny et al., 2018).

5.1.3 Electrocoagulation process

EC, which is still in use today, is a popular electrochemical method used to treat contaminated water using electricity instead of expensive and more harmful chemical reagents. This method has been successfully applied for treatment of soluble or colloidal pollutants in various industrial effluents including effluents from food, textile, and dye industries, tanneries, mechanical operations, polymer manufacture, and industrial wastewater containing heavy metals, suspended solids, emulsified organics, and many other contaminants (Butler et al., 2011). EC, which has the advantage of removing even the smallest colloidal particles as compared with the conventional flocculation–coagulation method, has the potential of higher destabilization and, therefore, greater coagulation by the effect of the electric field that causes the movement of charged particles. EC also has advantages such as the need for shorter processing times as well as less secondary waste generation in wastewater treatment (Marol et al., 2017).

The process is based on the principle of producing coagulants using sacrificial electrodes, such as iron or aluminum, at appropriate current density. The metal ions dissolved in the electrode turn into polymetallic hydroxides at increasing pH based on electrolysis and initiate coagulation (Tien et al., 2017). The resulting poly metal hydroxides can provide the removal of organic components, settling by sweep coagulation and/or conversion of the colloidal particles present in the effluent to larger flocks by aggregation. While the formation of metal cations takes place at the anode, the hydrogen gas released at the cathode can provide an additional contribution to the removal of the flocculated particles in the medium through the electroflotation process. Anodic and cathodic reactions that occur when aluminum electrolytes are used in the process are given below:

$$\text{Anode: } Al \overrightarrow{x} Al^{3+}{}_{(aq)} + + 3\ e^{-}{}_{(aq)}$$

$$\text{Cathode: } 3H_2O + 3\ e^{3-} \overrightarrow{x} 3/2H_2\ (g) + 3\ OH^{-}{}_{(aq)}$$

Given these reactions, it can be argued that the EC process is in fact a combination of processes such as oxidation, flocculation, and flotation (Bazrafshan et al., 2015). Generally, in the EC process, the ions that produce coagulants are produced in situ, and the process consists of three basic steps: formation of cations and their hydrolysis products in the sacrificial electrode, destabilization of contaminants, particle suspension and breaking of emulsions and aggregation of destabilized phase to form flocks (Naje and Abbas, 2013).

5.1.4 Advanced oxidation process

The advanced oxidation process (AOP), which uses ozone for disinfection, water quality improvement, color and odor removal, and the degradation of refractory organic materials, is a serious alternative to conventional treatment methods for color removal (Chung and Kim, 2011). The process, which proceeds with a mechanism involving the production of hydroxyl radicals as oxidizing agents, leads to the production of organic peroxide radicals by interaction with chromogenic groups and initiates their conversion to CO_2, H_2O, and inorganic salts. (Sharma and Kaur, 2018). The radicals are produced using single oxidants or a combination of ozone, hydrogen peroxide, and UV radiation (Glaze and Kang, 1989). AOP are characterized by the production of OH radicals and their oxidative selectivity to target groups (Al-Kdasi et al., 2004).

5.1.5 Ozonation process

Although chlorine is gradually being used as an alternative to ozone in the treatment of most effluents, ozonation remains one of the most efficient methods to treat colored effluents, as ozone is one of the strongest commercially available oxidants (Mustapha, 2015). Moreover, as ozone can directly and indirectly oxidize unsaturated bonds and chromophore groups that produce color, it can degrade the structure of organic compounds and thus provide true color removal and lower COD (Konsowa, 2003; Saroj et al., 2005; Lee et al., 2007; Wijannarong et al., 2013; Sripiboon and Suwannahong, 2018).

The ozonation process is widely used for a variety of purposes including oxidation of inorganic pollutants such as iron and manganese, oxidation of organic micro pollutants such as fragrance compounds, phenolic pollutants and pesticides and organic macropollutants as well as disinfection, algae control, and coagulation. Especially the decolorization of colored effluents by ozonation process has been successfully applied today (Guendy, 2007).

5.1.6 Fenton process

Fenton process is based on the reactions of iron ions and peroxides (usually hydrogen peroxide (H_2O_2)) to form active oxygen species which can oxidize organic or inorganic compounds. The main reactions consist of the oxidation of ferrous iron to ferric iron simultaneously with the conversion of H_2O_2 to hydroxide ion and hydroxyl radicals. In the process, colored compounds are radically oxidized, while others are precipitated with ferric ions. The Fenton reaction has been used efficiently to remove dyes and many other hazardous organic substances from wastewater (Wang et al., 2011; Babuponnusami and Muthukumar 2014).

Essentially, the Fenton reaction is based on the addition of hydrogen peroxide to the acidic solution having a pH of about 2–3 containing Fe^{2+} ions to initiate the formation of hydroxyl radicals from H_2O_2. The Fenton reaction-based process, which is exothermic and generally carried out at a temperature higher than ambient temperature, is commonly applied as a pretreatment process for wastewater resistant to biological treatment and/or toxic to biomass. However, in large-scale applications, the reaction is generally carried out at ambient temperature using a greater amount of iron and hydrogen peroxide. In these conditions, since the ions do not act as catalysts, the possible reduction in the total COD may be largely due to the coprecipitation of contaminants with $Fe(OH)_3$. The main disadvantages of the method are that it requires high amounts of acid and alkali to reach the required pH and produces effluents containing high concentrations of iron ions, and in particular produces high amounts of sludge with dye and other contaminants (Lin and Chen, 1997).

5.1.7 Photocatalytic processes

Heterogeneous photocatalytic processes using semiconductor catalysts such as TiO_2, ZnO, Fe_20_3, CdS, GaP, and ZnS can easily decompose a wide variety of refractory organics into biodegradable compounds and eventually convert them into harmless carbon dioxide and water. Heterogeneous photocatalysis has become an attractive process in water treatment with a number of advantages, such as ambient operating temperature and pressure, complete mineralization of the main components and intermediate compounds without secondary pollution, and low operating costs (Chong et al., 2010). Also, UV irradiation is used in combination with solid catalysts (TiO_2) or H_2O_2 for color removal of dye solutions. In this process, hydroxyl radicals produced at high concentrations by UV irradiation in the presence of H_2O_2 can reduce dye molecules to CO_2 and H_2O. The ratio of dye removal may be influenced by the intensity of the UV radiation, medium pH, dye structure, and the bath composition of dye (Davis, 1994; Reeves et al., 1992).

The studies on the removal of various dyes using different chemical methods are given in Table 5.2.

5.2 Physical treatment methods

Many physical methods such as adsorption, ion exchange and irradiation, filtration and membrane filtration, nanofiltration(NF), reverse osmosis(RO), and electrodialysis are widely used in industrial scale especially in the treatment of textile wastewaters due to their high dye removal potential and low operating costs (Dabrowski, 2001 Iqbal and Ashiq, 2007). The main adsorbents used for the treatment of colored effluents are activated carbon (Pereira et al., 2003), coal, peat (McKay et al., 1986), fly ash (Khare et al., 1987; Singh and Rawat, 1994), silica, wood, clay, various agricultural wastes, and cotton waste (Theng and Wells, 1995; Juang et al., 1997).

5.2.1 Adsorption process

The adsorption process, defined as increasing the amount of another component at the solid–liquid or solid–gas interface and classified as physical and chemical adsorption, is one of the most effective and efficient methods used to remove dissolved colored organic pollutants such as dye and pigments from industrial effluents. In the adsorption process, the species to be adsorbed is

TABLE 5.2 The some studies on the removal of various dyes using different chemical methods.

Methods	Type	Name of Dye	References
Electrocoagulation	Acid	Marine Blue Erionly Mr Acid Black 194 Acid Red 97	Bellebia et al. (2009) Vidal et al. (2017) Patel et al. (2010)
	Direct	Direct Blue Dye Direct Blue 71 Direct Blue 86	Ghalib and Al-yaqoobi (2010) Esgair, (2017), ghali Attia, (2013)
	Azoic	Ostalan Black Sr Basic Red 18 Reactive Yellow 14	Malinović et al. (2017) Bazrafshan and Mahvi (2014) Yari et al. (2013)
	Disperse	Dispers Yellow 3 Disperse Red Disperse Yellow 218	Ghernaout et al. (2014) Merzouk et al. (2011) Yang and McGarrahan (2005)
	Vat	Indigo Dye Vat Yellow	Barrak et al. (2016) Chigozie and Joseph (2014)
	Basic	Basic Red 46 Basic Blue 3	Charoenlarp and Choyphan (2009) Daneshvar et al. (2006)
	Reactive	Reactive Red 120 Reactive Red 24	SenthilKumar et al. (2010) Ghalwa et al. (2016)
Advanced oxidation processes (AOPs)	Acid	Acid Red 17 Acid Orange 8	El Nemr et al. (2018) Abo-Farha (2010)
	Direct	Direct Blue 86 Direct Orange 39	Hassaan et al. (2017) Papić et al. (2006)
	Disperse	Disperse Yellow 126 Disperse Orange 288	Szpyrkowicz et al. (2001) Huang et al. (2011)
	Vat	Indanthrene Red FBB Vat Blue 1	Liu et al. (2007) Manu (2007)
	Basic	Methylene Blue Basic Orange 1 4	Masoumbeigi and Rezaee (2015) Abdullah Al-Hamedi (2007)
	Reactive	Reactive Yellow 16 Reactive Yellow 17	Sundararaman et al. (2009) Sennaoui et al. (2018)

TABLE 5.2 The some studies on the removal of various dyes using different chemical methods.—cont'd

Methods	Type	Name of Dye	References
Fenton process	Acid	Acid Red 18 Acid Red 14 Acid Blue 161	Barbusiński (2005) Sohrabi et al. (2017) Trovó et al. (2016)
	Direct	Direct Blue 71 Direct Red 80 Direct Red 81	Ertugay and Acar (2017) Jogani et al. (2017) Baghapour et al. (2016)
	Disperse	Disperse Yellow 23 Disperse Red 167	Wali (2015) Gökkuş and Ciner (2009)
	Vat	Vat Blue 20 Vat Green 01	Xu et al. (2004) Schrank et al. (2007)
	Basic	Malachite Green Basic Red 9 Methylene Blue	Thakur and Chauhan (2016) Güngör et al. (2014) Gowtham and Pauline (2015)
	Reactive	Reactive Red 198 Reactive Blue 19 Reactive Yellow 84 Reactive Red 2	Mansoorian et al. (2014) Moghadam and Kohbanan (2018) El Haddad et al. (2014) Mousavi et al. (2017)
Photo catalytic processes	Acid	Acid Blue 113 Acid Yellow 3 Acid Red 27 Acid Red 73	Mortazavian and James (2019) Bergamini et al. (2008) Shahmoradi et al. (2011) Vaez et al. (2012)
	Direct	Direct Yellow 86 Direct Green 99 Direct Blue 53	Mansouriieh et al. (2017) Wawrzyniak and Morawski (2006) Xu et al. (2010)
	Disperse	Rhodamine B Disperse Red 73	Rahman et al. (2013) Buscio et al. (2015)
	Vat	Vat Yellow 4 Vat Violet 3B	Kavitha and Palanisamy (2010) Putri et al. (2019)
	Basic	Methylene Blue Basic Yellow 15 Malachite Green	Salehi et al. (2012) Tang and An (1995) Kumar et al. (2014)
	Reactive	Reactive Blue 203 Reactive Yellow 84	Bagheri et al. (2019) Barka et al. (2010) Amiri et al. (2016)

Continued

TABLE 5.2 The some studies on the removal of various dyes using different chemical methods.—cont'd

Methods	Type	Name of Dye	References
		Reactive Yellow 81	
Electroflotation	Acid	Eriochrome Black Acid Red 14	Cestarolli et al. (2019) Hooshmandfar et al. (2015)
	Direct	Direct Red 81	Zodi et al. (2013)
	Disperse	Yellow Terasil 4G Disperse Red 1 Disperse Blue 14	Balla et al. (2010) Phalakornkule et al. (2010) El-Hosiny et al. (2018)
	Vat	Vat Yellow	Chen et al. (2000)
	Basic	Basic Violet 3 Basic Yellow 28	El-Hosiny et al. (2017) Ayhan Şengil and Özdemir (2012)
	Reactive	Yellow SPD Reactive Black 5 Levafix Orange E3 GA Reactive Black Dye RB-5	Chenik et al. (2013) Sadik (2019) Kobya et al. (2006) Ali et al. (2016)
Ozonation process	Acid	Acid Red 337 Congo Red Acid Black 1	Sevimli and Kinaci (2002) Gharbani et al. (2008) Kasiri et al. (2013)
	Direct	Direct Pink 3b Direct Blue 74 Direct Red 28	Guendy (2007) Tosik (2005) Pérez et al. (2016)
	Disperse	Disperse Blue 79 Disperse Red 74	Mohammed et al. (2019) Szpyrkowicz et al. (2001)
	Vat	Vat Blue 1	Strickland and Perkins (1995)
	Basic	Methylene Blue Basic Yellow 28	Turhan et al. (2012) Polat (2010)
	Reactive	Novacron Super Black G Reactive Orange 16 Reactive Yellow 17 Remazol Turquoize G	Wijannarong et al. (2013) Onder et al. (2018) Turhan (2018) Güneş et al. (2012)

defined as "adsorbate" and the adsorbing phase or substance is "adsorbent", and the nature of the adsorption is determined by the type and effectiveness of the interactions between adsorbate and adsorbent (Vital et al., 2016). Generally, physical adsorption occurs through weak physical interactions, and the adsorption enthalpy is about 10–20 kJ/mol. However, chemical adsorption occurs through interactions similar to chemical reactions between adsorbate and adsorbent, and the enthalpy of adsorption may be greater than 50–100 kJ/mol. Generally, another process that occurs in parallel with the adsorption process, but with a different interaction mechanism, is absorption. In most cases, it is difficult to differentiate between adsorption and absorption and therefore term "sorption" for many systems is commonly used to describe the transport of a certain substance between liquid and solid phases. (Ul-Islam, 2019).

Adsorption, which is economically feasible and capable of producing high-quality water, has become increasingly popular in wastewater treatment in recent years because it is more effective in removing contaminants that are highly resistant to biological methods (Allen and Koumanova, 2005). The adsorption properties and decolorization capacity of various inorganic and organic adsorbents were investigated. Inorganic adsorbents have advantages such as good mechanical and chemical stability, high specific area, and resistance to microbial degradation, while organic adsorbents can be renewable and noncommercial wastes or by-products of industrial processes (Forgacs et al., 2004; Ejder-Korucu et al., 2015).

Among the inorganic adsorbents, especially carbon-based ones are used extensively for the adsorption of various categories of dyes, the adsorption efficiency being highly dependent on the molecular size and solubility of the adsorbate, as well as the surface characteristics of the adsorbent as well as its pore size and distribution. (Joshi and Purwar, 2004).

In general, low adsorption on activated carbon is observed for acid and reactive dyes with low molar mass, whereas higher adsorption is achieved with basic and direct dyes having higher relative molecular mass and hydrophobic disperse dyes on the same adsorbent. Due to the low solubility and colloidal behavior of disperse dyes, vat dyes, and pigments, their adsorption on activated carbon at ambient temperature is relatively slow (Reife and Freeman, 1996).

As an organic adsorbent, dried biogas slurry and cross-linked chitosan fiber were used to remove some reactive dyes and direct dyes from wastewater (Yoshida and Takemori, 1997). However, activated carbons from many different precursors are still widely used in wastewater treatment systems, although they remain expensive adsorbents (Kaykhaii et al., 2018).

Active carbon, which is the most widely used adsorbent in the removal of dyes, is less effective in the adsorption of disperse dyes, direct, vat, pigment, and reactive dyes, but is highly effective in adsorption of cationic dyes, mordant, and acid dyes (Nasser and El-Geundi, 1991; Raghavacharya, 1997; Karaca et al., 2008; Rodriguez et al., 2009; Bao and Zhang, 2012). The efficiency and effectiveness of the adsorption process is strongly dependent on the type of activated carbon and the composition of the wastewater. The main reason for the high and nonselective adsorption effectiveness of powder and granular activated carbon is that it usually has a much larger surface area than other adsorbents (Ejder-Korucu et al., 2015; Guiza et al., 2004).

For the treatment of colored effluents, various adsorbent materials, which are sawdust (Parihar and Malaviya, 2013), banana and orange peels (Mane and Bhusari, 2012), wheat straw, corncobs, barley husks (Robinson et al., 2002), eucalyptus barks (Saliba et al., 2002), wood (M'hamdi et al., 2017), peat (Sepúlveda et al., 2004), rice husk (Ola et al., 2005), chitin (Shirsath and Shrivastava, 2012), algal biomass (Omar et al., 2018), metal hydroxide sludge (Golder et al., 2006), soil (Albanis et al., 2000), clay (Gürses et al., 2006), fly ash (Li et al., 2006), and coal (Gupta et al., 1990) are widely used.

Peat, having partially decomposed plant and animal components and containing various lignin and phenolic functional groups, is a widely used adsorbent for the adsorption of transition metals, polar organic compounds, and dyes from industrial effluents due to their superior adsorption properties resulting from cellular porous structure. Unlike activated carbon, peat, which does not require activation and is also much less costly, has the potential to be used for energy generation by burning after use or for protein enrichment as a substrate in solid-state fermentation (Poots and McKay, 1976).

Although wood chips have a high adsorption capacity against acid dyes, they are not as feasible as other absorbent materials due to their disadvantages such as hardness and long adsorption time (Ejder Korucu et al., 2015).

Fly ash is used as an adsorbent for dye adsorption from colored wastewater as a waste resulting from the combustion processes in thermal power plants, which is one of the biggest sources of environmental pollution (Gupta et al., 1990; Atul et al., 2013).

Silica gel is an amorphous inorganic adsorbent which is well known for its high surface area, high thermal and chemical stability, high mechanical and swelling resistance, reuse potential, and low cost and is also suitable for selective adsorption since it can interact with various organic materials, especially when grafted with silane compounds as an adsorbent (Farias et al., 2018). The commercial use of silica gel, which can be used as an effective adsorbent for the removal of basic dyes, is relatively limited due to its disadvantages such as interaction with particulate matter and air (Ejder-Korucu et al., 2015).

Similar to the behavior of substrates such as corn cob and rice husk, natural clays commonly used for dye removal have some basic advantages, such as prevalence, abundance, and cheapness (Nawar and Doma, 1989; Nasser and El-Geundi, 1991; Nigam et al., 2000). Natural clays have a negative surface in aqueous media and are highly effective in adsorption of cationic dyes (Fatiha and Belkacem, 2016).

Adsorption of dyestuffs consists essentially four consecutive steps: diffusion or transport of dyestuffs through solution bulk; diffusion of dyestuffs through a diffusion boundary layer (film diffusion); diffusion of dyestuffs from the surface to the pores of adsorbent materials; and the adsorption of the dyestuffs at the interface through various interactions (Al-Godah, 2000). Although parameters such as dye concentration and agitation speed are most effective in the second step, the third step is the rate determination step which will affect the adsorption of the dye molecules on the substrate. The fourth step depends mainly on the anionic and cationic nature of the dye molecules. In addition, the third step can be considered as two different steps, pore diffusion and surface diffusion (Seow and Lim, 2016).

5.2.2 Irradiation process

Most traditional techniques used in water and wastewater treatment can produce high-quality water, but there are some limitations in large-scale applications for these techniques for reclamation or disposal (high cost and low sustainability). In contrast, irradiation using gamma or electron beams is a simple and effective technology that removes a wide range of organic pollutants and disinfects harmful microorganisms. Irradiation, using catalysts such as titanium dioxide to improve the efficiency of the process, is more suitable for color removal over a wide range of dyes and at low volumes, but the degradation of dye in textile effluents requires higher amounts of DO (Robinson et al., 2001; Anjaneyulu et al., 2005).

5.2.3 Ion exchange

Ion exchange has great potential to remove dyes from textile effluents, but it has not been widely used for the treatment of wastewater containing dyes since essentially ion exchangers cannot accommodate a wide variety of dyes (Gosavi and Sharma, 2014). On the other hand, the process can produce high concentrations of waste that have to be disposed of carefully is not effective at low pHs. Furthermore, it is not effective for complex mixtures of metals, highly dissolved solids (TDS) levels, and high Fe, Mn, and Al concentrations. Therefore, this process requires pretreatment for most surface waters and removal of suspended solids prior to treatment and also includes additional limitations such as resin regeneration and high process costs (Sharma et al., 2014b). In the treatment of effluents containing dyes, the effluent is passed over the ion exchange resin until the existing exchange sites are saturated. This process is suitable for the successful removal of both cationic and anionic dyes from wastewater but is not very effective for disperse dyes. The advantages of this method are the absence of adsorbent loss in regeneration, recovery of solvent after use, and removal of soluble dyes (Anjaneyulu et al., 2005).

5.2.4 The membrane processes

Membrane processes are physical separation that can take place generally without the addition of chemicals to the feed stream, and also without phase change and therefore can be a good alternative to conventional processes such as distillation, ion exchange, precipitation, coagulation/flocculation, adsorption, and biological treatment. In most cases, low energy consumption, reduced number of processing steps, greater separation efficiency, and improved end product quality increase the attractiveness of these processes for applications in the chemical, pharmaceutical, biotechnology, and food industries (Mai, 2013). Membrane processes which offer an attractive alternative to the treatment of contaminated water, especially where the adsorbent is inexpensive and do not require an additional pretreatment step before application, have disadvantages such as a limited life due to membrane blockage and thus high periodic maintenance costs (Kyzas et al., 2013).

5.2.5 Membrane filtration

Membrane filtration, used to remove dissolved substances and fine particles from solutions, has the potential to continuously separate and concentrate dyes from effluents (Muro et al., 2012). Compared with other methods, this process has some specific properties, such as resistance to temperature rise, adverse chemical environments, and microbial attack (Giwa and Ogunribido, 2012). Membrane filtration is suitable for the treatment of effluents containing low concentrations of dyes and recycling of textile wastewater, but is not effective in reducing the dissolved solid content, which may make it difficult to reuse water (Atul et al., 2012). There are four common membrane separation processes: microfiltration (MF), ultrafiltration (UF), RO, and NF. The main advantages of membrane technology over other separation processes are low energy consumption, simplicity, and environmental friendliness (Igunnu and Chen, 2012; Wenten, 2002).

MF is similar to conventional raw filtration processes in textile wastewater treatment and has a limited application area. MF membranes generally have pore sizes in the range of $0.1-10\ \mu m$, and separation by MF is generally effective at low pressure differential within 2 bar. The process is mainly used for removal of particles suspension and colloidal dyes from exhausted dye bath and from discarded rinsing bath discharge. However, MF membranes may not prevent leakage of unconsumed auxiliary chemicals, dissolved organic pollutants, and other soluble contaminants (Juang et al., 2013).

UF is a membrane separation process that is mainly used to separate macromolecules and colloids from a solution and their application in the textile industry is limited. This is due to the fact that the molar masses of the dyes present in the highly colored textile discharge are much lower than the molar mass cut-off of UF membranes (Lafi et al., 2018).

NF, a newly developed membrane technology for various water treatment and purification purposes, is a membrane separation process where pressure is the main driving force responsible for the separation process and has performance characteristics between RO and UF (Lopes et al., 2005; Hassani et al., 2008). NF is a pressure-driven membrane-based separation process, using pressures from 4 to 20 MPa where particles with molar mass of 350−1000 Da and dissolved molecules can be retained (Mai, 2013).

RO is a process of reverse of the normal osmosis in which the solvent is pushed through a membrane from a zone with high solute concentration to a zone with low solute concentration, and its operating pressures are generally between 7 and 100 bar. RO processes can remove hardness, color, many types of bacteria and viruses, and organic contaminants such as agricultural chemicals and trihalomethane precursors simultaneously (Abid et al., 2012). The studies on the removal of various dyes using different physical methods are given in Table 5.3.

TABLE 5.3 The some studies on the removal of various dyes using different physical methods.

Methods	Type	Name of dye	References
Adsorption process	Acid	Acid Red 97 Acid Blue 74 Acid Blue 40	Gomez et al. (2007) Secula et al. (2011) Özacar and Şengil (2002)
	Direct	Direct Yellow 50 Direct Red-23 Direct Red 216	Yavuz and Aydin (2006) Abdelwahab et al. (2005) Ouslimani and Boureghda (2018)
	Disperse	Disperse Red 60 Disperse Blue 79 Disperse Blue 73	Sirianuntapiboon and Srisornsak (2007) Markandeya et al. (2017) Semeraro et al. (2017)
	Vat	Vat Blue 14 Vat Yellow 1 Vat Blue 4	Šmelcerović et al. (2010) Sirianuntapiboon et al. (2006) Chaari et al. (2009)
	Basic	Malachite Green Basic Violet 14 Basic Yellow 28 Basic Red 46	Seey and Kassim (2012) Suresh (2016) Albanis et al. (2000) Amin et al. (2017)
	Reactive	Reactive Red 222 Reactive Yellow 84 Reactive Red 31 Reactive Red 2	Chiou et al. (2003) Filipkowska (2006) Geethakarthi and Phanikumar (2011) Thitame and Shukla (2016)

Continued

TABLE 5.3 The some studies on the removal of various dyes using different physical methods.—cont'd

Methods	Type	Name of dye	References
Irradiation process	Acid	Acid Red 4092 Acid Orange 7	Dehghani and Mahdavi (2015) Talaiekhozani et al. (2019)
	Direct	Direct Blue Direct Blue 71 Direct Yellow 50	Mahmoud et al. (2017) Anvari et al. (2014) Lafta et al. (2015)
	Disperse	Disperse Violet 28 Disperse Yellow 119	Li et al. (2014) Çiner and Gökkuş (2013)
	Vat	Vat Green 03 Vat Red R1 Vat Blue 4	Rezig and Hadjel (2014) Moly and Shahul Hameed (2016) Adeel et al. (2018)
	Basic	Crystal Violet Malachite Green Basic Blue 9	Rehman et al. (2017) Mao et al. (2018) Dessouki and Abdel-Aab (1998)
	Reactive	Reactive Red-120 Reactive Orange 4 Apollofix Red Reactive Red 231	Paul et al. (2011) Sala et al. (2014) Solpan et al. (2007) Gutiérrez Bouzán and Buscio Olivera (2018)
Ion exchange	Acid	Acid Black 210 Acid Orange 7 Congo Red Acid Green 1	Deaconu et al. (2016) Akazdam et al., 2017 Khan et al. (2015) Elkhattabi et al. (2008)
	Direct	Direct Blue 71 Direct black 234	Wawrzkiewicz (2014) Deaconu et al. (2016)
	Disperse	Dispers Red S-R Disperse Blue 56	Yue et al. (2007) Rai et al. (2007)
	Vat	Vat Blue 14	Uddin et al. (2003)
	Basic	Methyl Violet 2B Basic Blue 3	Wu et al. (2008) Wawrzkiewicz (2013)
	Reactive	Procion Black 5b Remazol Black B Reactive Blue 4	Raghu and Basha (2007) Wawrzkiewicz and Hubicki (2011) Dhas et al. (2015)

TABLE 5.3 The some studies on the removal of various dyes using different physical methods.—cont'd

Methods	Type	Name of dye	References
Membrane processes	Acid	Acid Blue Victoria Blue Acid Orange 7	Kishore and Kamala (2015) Xuan Nguyen and Van der Bruggen (2015) Hammami et al. (2017)
	Direct	Direct Red 16 Direct Black	Zinadini et al. (2016) Majewska-Nowak et al. (2006)
	Disperse	Disperse Violet Disperse Red 73	Gholami et al. (2001) Buscio et al. (2015)
	Vat	Indigo Dye	Berradi et al. (2016)
	Basic	Basic Yellow 13	Gholami et al. (2001)
	Reactive	Reactive Blue Cibacron Yellow S-3R	Ahmad et al. (2002) Buscio et al. (2016a)
Membrane filtration	Acid	Acid Red Acid Black 210	Abid et al. (2012) Zahrim and Hilal (2013)
	Direct	Direct Red 80 Direct Blue	Akbari et al. (2007) Hassani et al. (2008)
	Disperse	Disperse Violet Disperse Orange 30	Gholami et al. (2003) Buscio et al. (2016b)
	Vat	Vat Yellow İndigo Dye	Chigozie and Joseph (2014) Ünlü (2008)
	Basic	Methyl Blue Basic Green 4	Johary et al. (2008) Kelewou et al. (2015)
	Reactive	Reactive yellow 145 Reactive Black 5 Reactive Red 24 Reactive Blue 5	Masmoudi et al. (2014) Kim et al. (2005) Petrinić et al. (2007) Vishnu and Joseph (2007)

5.3 Biological methods

In recent years, there has been a great interest in biological methods using some microorganisms such as fungi, bacteria, and algae which are highly effective in the adsorption and degradation of dyes from wastewater (Yuzhu and Viraraghavan, 2001). The efficacy of biological treatment processes in color removal from industrial effluents shows uncertainty and differences, and this is related to the fact that microorganisms do not use the color components as a food source and therefore the dyes themselves do not biodegrade. The most common methods known in biodegradation use aerobic microorganisms that consume molecular oxygen during the respiratory process (Birch et al., 1989; Anjaneyulu et al., 2005). Biological removal of dyes from textile industry effluents and dyestuff production industry wastewaters offers many advantages, such as the conversion of dyes to harmless compounds such as carbon dioxide and water and the formation of less harmless sludge compared with commonly used chemical and physicochemical methods (Rai et al., 2005; Forgacs et al., 2004).

The main goal of all biological wastewater treatment methods is to ensure that the microbial cells, which are fed with organic materials in the wastewater and thus reduce the BOD, come into contact with the wastes. BOD is a parameter that is used to determine the relative oxygen content of aerobic microorganisms required to decompose organic matter in wastewater or effluents and is widely used to measure waste load in treatment plants and to assess the effectiveness of BOD removal of such treatment systems. The test used to determine BOD measures the oxygen used during a particular incubation period for biochemical degradation of organic matter and for the oxidation of other inorganic materials such as sulfides, ferrous ions, and reduced nitrogen forms. In recent years, increasing awareness of environmental effects of organic loads determined by COD and BOD increases the demand for wastewater treatment before discharging into natural waters (Nayl et al., 2017). COD is used as a measure of the oxygen equivalent of the organic matter content of a sample susceptible to oxidation by a strong chemical oxidant (Kalel, 2006).

Total organic carbon (TOC) is one of the most important parameters that can be used in drinking water quality and wastewater treatment processes, and its measurement is therefore critical to water quality determination and optimization of treatment processes. TOC is indicative of the presence of many organic pollutants, including organic products such as petroleum products, humic and fulvic acids, pesticides, and various pathogens, and is a nonspecific but inclusive parameter for monitoring organics. In addition, TOC levels in plants can be used as a measure of treatment effectiveness and as an indicator of contamination. Unlike measurements such as BOD and COD, TOC contains all organic compounds and can be measured instrumentally in a very short time (Assmann et al., 2017).

The natural process of microbiological metabolism in the aquatic environment constitutes the basis of biological treatment of wastewater. Under special environmental conditions, soluble organic substances present in wastewater can be completely eliminated by converting them into harmless species by biological oxidation. Biological treatment systems usually include a biological reactor and a settling tank to remove the resulting biomass or sludge (Hedaoo et al., 2012).

Dyes are generally stable organic molecules and are therefore resistant to biological degradation. Also, the high solubility of dyes in water and their high molar mass inhibits their permeability through biological cell membranes and thus limit their biodegradability (Pearce et al., 2003). Biological removal of dyes in effluents from textile and dyestuff production industries is generally accomplished by aerobic treatment, anaerobic treatment, and a combination of both. (D'Antoni et al., 2017).

5.3.1 Aerobic treatment

The terms aerobic and anaerobic are used to characterize biological processes in the presence and absence of air or oxygen, respectively. Both terms are therefore directly related to the type of microorganisms involved in the degradation of organic impurities in a particular wastewater or effluent and the operating conditions of the bioreactor. Therefore, aerobic treatment, which can take place in the presence of air, utilizes microorganisms called erobes, which use molecular/free oxygen to convert organics into carbon dioxide, water, and biomass (Ejder-Korucu et al., 2015). Aerobic biological systems, which are not sufficiently effective in decolonization, have caused esthetic problems in the receiving waters and forced the dischargers to look for different alternatives. Since commercial dyes are structurally manufactured to be resistant to fading based on chemical and light-induced oxidation, the dyes themselves are also generally resistant to oxidative biodegradation. In addition to the toxicity problems of dye effluents, another problem with their aerobic biological treatment is the difficulty in acclimation into the substrates of microorganisms. Acclimation is also a serious problem in textile effluents due to continuous product changes and batch dyeing processes (Joshi et al., 2004). Bacteria and fungi are species of microorganisms commonly used to remove dyes under aerobic conditions (Balamurugan et al., 2011). Aerobic bacteria capable of degrading a number of dyes have been used for a long time (D'Antoni et al., 2017). Processes using bacteria and fungi for the complete removal and degradation of dyes from textile effluents have advantages such as being low cost and completing the mineralization of dyes by nontoxic by-products (Dawkar et al., 2010).

5.3.2 Anaerobic treatment

Anaerobic biological treatment is a process that enables the use of anaerobic bacteria to decompose organics under anaerobic conditions and was first used for sludge digestion. However, in recent years, it has been gradually used in wastewater treatment, which includes high and low concentrations of organic compounds (Arumai Dhas, 2008). Decolorization of dyes under anaerobic conditions is based on the transfer of electrons from an electron donor substrate oxidized by microorganisms to the final electron acceptor dyestuff and thus the reductive cleavage of the dye chromophore (Costa et al., 2012).

Anaerobic treatment is carried out in an oxygen-free environment and the organics present in the waste are converted into methane and CO_2, but ammonia and hydrogen sulfide are also produced when the waste contains nitrogen, and sulfide-containing contaminants. Anaerobic treatment also produces a significant amount of heat, and methane produced in particular is widely used for various energy-intensive purposes. When dyes are subjected to a biological reduction under anaerobic conditions, reductive cleavage occurs (Delée et al., 1998). Biological reduction of dyes under anaerobic conditions may include transformations leading to color removal as well as chemical decomposition. Anaerobic treatment is also applied for color removal from textile effluents (Dos Santos et al., 2007; Singh and Arora, 2011).

5.3.3 Combined anaerobic–aerobic treatment

Combined aerobic and anaerobic treatment is one of the best and most advantageous ways of removing dyestuffs from textile effluents, because complete removal can be achieved by the synergistic effect of different organisms. However, when toxic azo dyes decompose under anaerobic conditions, they can form carcinogenic and/or mutagenic aromatic amines which are extremely critical for the environment. Generally, under anaerobic conditions, azo dyes can be reduced by bacteria to colorless toxic aromatic amines, some of which are readily metabolized under erobic conditions (Naresh et al., 2013a). Therefore, in addition to the increasing attractiveness of azo dye removal by combined anaerobic–aerobic treatment, it is generally preferred to use an anaerobic step followed by erobic treatment in the treatment of dye wastewaters (Gupta and Suhas, 2009). In general, parameters such as initial dye concentration, initial pH, and waste water temperature are critical to color removal of dyes by this method, which is suitable for a wide range of dyes in terms of cost and applicability. However, biological treatment also exhibits some disadvantages such as low biodegradability, less flexibility in design and use, greater land area requirement, and longer time requirement for color removal (Uday et al., 2016).

5.3.4 Decolorization by bacteria

In biological treatment processes, various physicochemical operational parameters such as mixing speed, oxygen concentration, temperature, pH, dye structure, dye concentration, different carbon and nitrogen source supplementation, electron donor, and redox mediator can directly affect the bacterial decolorization performance of dye effluents. Therefore, in order to make the process more efficient, faster, and practically feasible, the effect of each factor on bacterial discoloration from dye effluents must be precisely determined (Junnarkar et al., 2006; Sriram and Reetha, 2015). Bacteria from different trophic groups can achieve a high degree of dye degradation as well as complete conversion of dyes to harmless products under optimal conditions (Roy et al., 2018).

Today, it has been determined that gram-negative bacteria species of different genera such as *Eromonas*, *Escherichia*, *Citrobacter*, *Pseudomonas*, and *Sphingomonas* can provide color removal in dye-containing solutions or simulated effluents. It was also found that gram-positive bacteria such as *Bacillus*, *Clostridium*, *Nocardia*, *Paenibacillus*, *Streptomyces*, and *Micrococcus* can degrade synthetic dyes (Forgacs et al., 2004).

The microbial degradation mechanism of azo dyes involves the reductive cleavage of azo bonds (−N = N−) with the aid of azo reductase under aerobic conditions, resulting in a colorless solution (Chang et al., 2001). For the reduction of azo dyes, it has been proposed that reduction to the anion radical occurs through a rapid electron transfer reaction followed by a second, slower electron transfer to produce stable dianion (Zimmermann et al., 1982). However, the functional group of azo dyes with higher electronic density may be inconvenient to the second electron transfer to form the dianion, in which case the color removal is either low or does not occur at all. Therefore, the sulfonated reactive group of azo dyes is normally considered to be more recalcitrant than carboxylated azo dyes in terms of degradation (Sultana, 2017).

5.3.5 Decolorization by fungi and yeast

Dyes can be removed by using fungi by biosorption, biodegradation, bioaccumulation, and enzymatic treatment with lignin peroxidase, manganese peroxidase, manganese-independent peroxidase, and laccase enzymes (Saranraj et al., 2010). Fungi, which can be excellent means of dye removal, often use an extracellular enzymatic system capable of converting aromatic substances such as lignin, polyaromatic hydrocarbons, or pesticides. Currently, fungal decolorization is widely studied, and the most popular fungi are white-rot fungi, which are the organisms capable of degrading lignin, the most complex structural polymer found in woody plants (Marimuthu et al., 2013). There are a wide variety of fungi available for decolorization, including string-shaped soil fungi such as *Fusarium*, and white-rot fungi such as *Hypoxylon, Ketomium, Pyulospora, Coriolis*, *Polaria*, and *Phenocyte* (Abbasi, 2017). Fungi are also used as sorbents and/or enzyme producers that may be involved in biological degradation/biotransformation (Przystas et al., 2015).

Because azo dyes are synthetically produced xenobiotic compounds, characterized by the presence of at least one chromophore group, the biological decolorization of effluents containing them is not very easy. However, effective decolorization including the cleavage of the azo bond can be achieved by using several microorganisms, notably white-rot fungi, other filamentous fungi, and bacterial species (Fu and Viraraghavan, 2001; Kirby et al., 1995). The fact that filamentous fungi, molds, growth rates are relatively slow compared with most yeast species provides advantages such as shorter reaction time and lower cost in terms of biotechnology. Also, like molds, yeasts are resilient microorganisms and are thereby resistant to unfavorable environments such as acidic pH, high salt concentration, and high-strength organic waste water such as textile effluents (Dias et al., 2010).

The effectiveness of the interactions between microorganisms such as yeast, bacteria and fungi, and dyes is strongly dependent on the chemical structure and properties of the interacting species. Yeasts, which have many advantages compared with bacteria and filamentous fungi, stand out as better bio sorbents for the removal of reactive dyes due to their unicellular nature and high growth rates. Yeast cells, which can be easily cultivated into relatively inexpensive growth media, are a readily available source of biomass having high bioremediation potential for wastes at lower pH values. (El-Rahim et al., 2009; Ertuğrul et al., 2009; El-Hennawi, 2013; Mahmoud, 2016). The some studies on the removal of various dyes using different biological methods are given in Table 5.4.

TABLE 5.4 The some studies on the removal of various dyes using different biological methods.

Methods	Type	Name of dye	References
Aerobic treatment	Acid	Acid Black 210 Acid Orange 7 Acid Orange 8 Acid Yellow 151	Mohan et al. (2009) Konsowa et al. (2011) Seshadri et al. (1994) Cerboneschi et al. (2015)
	Direct	Direct Red 81	Khalid et al. (2009)
	Disperse	Terasil Black Dye Dispersal Yellow C-4R	Pourbabaee et al. (2006) Li and Zhao (1997)
	Vat	Vat Black 25	Sirianuntapiboon and Saengow (2004)
	Basic	Basic Red 46	Cerboneschi et al. (2015)
	Reactive	Remazol Brilliant Blue Everzol Blue Bb Reactive Blue	Mirbolooki et al. (2017) Uysal and Bilgiç (2018) Agarry and Ajani (2011)

TABLE 5.4 The some studies on the removal of various dyes using different biological methods.—cont'd

Methods	Type	Name of dye	References
Anaerobic treatment	Acid	Acid Red 42 Acid Red 73	Gonçalves et al. (2000) Gonçalves et al. (2001)
	Direct	Direct Black 19 Direct Blue 71	Van der Zee et al. (2001) Jin et al. (2009)
	Disperse	Disperse Orange 3	Khalid et al. (2009)
	Vat	Vat Blue 1	Manu and Chaudhari (2003)
	Basic	Basic Yellow 28 Basic Blue 3 Basic Blue X-Grrl	Talarposhti et al. (2001) An et al. (1996) Deng et al. (2008)
	Reactive	Reactive Red 24 Reactive Red 22 Reactive Orange 107 Reactive Orange 96	Karataş et al. (2009) Chang and Lin (2000) Victral et al. (2016) Yoo et al. (2001)
Combined anaerobic −aerobic treatment	Acid	Acid Red 14 Acid Orange Acid Orange 12	Fitzgerald and Bishop (1995) Shah (2014a) Harrelkas et al. (2008)
	Direct	Direct Red 28 Direct Black 38 Direct Red 2	Işık and Sponza (2008) Sponza and Işık (2004) Rajaguru et al. (2000)
	Disperse	Disperse Blue 79	Melgoza and Buitron (2004)
	Vat	Indigo Dye	Dogdu and Yalcuk (2016)
	Basic	Basic Red 46	Assadi et al. (2017)
	Reactive	Remazol Black B Reactive Black 5 Reactive Red 180 Reactive Blue 13 Reactive Orange 16	Panswad et al. (2001) Sponza and Işık (2002) Pasukphun et al. (2010) Lin et al. (2010) Shah et al. (2014)

Continued

TABLE 5.4 The some studies on the removal of various dyes using different biological methods.—cont'd

Methods	Type	Name of dye	References
Decolorization by bacteria	Acid	Acid Blue 25 Acid Red 2 Acid Blue 40 Acid Yellow 36	Aruna et al. (2015) Joshi et al. (2015) Cerboneschi et al. (2015) Mahmood et al. (2015a)
	Direct	Direct Blue 71 Direct Red 28 Direct Red 31 Direct Brown 2 Direct Blue 201	Franciscon et al. (2009) Ajaz et al. (2019) Lalnunhlimi and Krishnaswamy (2016) Saratale et al. (2009) Manage (2017)
	Disperse	Disperse Brown 21 Disperse Red 78 Disperse Red 54 Disperse Yellow 211	Naresh et al. (2013b) Lade et al. (2015) Kurade et al. (2012) Sharma et al. (2009)
	Vat	Vat Blue 4 Vat Black Vat Violet 3	Olaganathan and Patterson (2012) Adebajo et al. (2017) Kulandaivel et al. (2014)
	Basic	Basic Red 46 Basic Blue 47 Crystal Violet Basic Green 4	Li et al. (2018) Cerboneschi et al. (2015) Kulandaivel et al. (2014) Sneha et al. (2014)
	Reactive	Reactive Black 5 Reactive Black B Reactive Orange 16 Reactive Blue 222	Franciscon et al. (2012) Rajeswari et al. (2011) Siti Zuraida et al. (2013) Mohamed (2016)

TABLE 5.4 The some studies on the removal of various dyes using different biological methods.—cont'd

Methods	Type	Name of dye	References
Decolorization by fungi and yeast	Acid	Acid Red B Acid Yellow 99 Acid Red 114 Indigo Carmine Acid Blue 74	Song et al. (2017) Park et al. (2004) Yang et al. (2017) Huang et al. (2014) Yildirim and Yesilada (2015)
	Direct	Direct Blue 1 Direct Blue 151 Direct Pink B Direct Blue 71 Direct Brown	Stolz (2001) Lalnunhlimi and Krishnaswamy (2016) Ngo et al. (2016) Jafari et al. (2013) Abd El-Rahim (2013)
	Disperse	Disperse Red 1 Disperse Yellow 5g Disperse Orange 30 Disperse Red 1 Dispers Blue 148	Yang et al. (2017) Ngo et al. (2016) Mahmood et al. (2015b) Yang et al. (2017) Schmitt et al. (2012)
	Vat	Vat Violet Cibanon Red 2b-Md Vat Red 10	Abioye et al. (2015) Asgher et al. (2008) Gurav et al. (2011)
	Basic	Basic Violet 3 Methylene Blue Malachite Green	Deivasigamani and Das (2011) Forootanfar et al. (2012) Rani et al. (2014)
	Reactive	Reactive Red 141 Reactive Black 5 Reactive Blue 38 Reactive Blue 4 Reactive Blue 19 Reactive Red 180	Telke et al. (2008) Prigione et al. (2012) Tehrani et al. (2015) Yemendzhiev et al. (2009) Omar (2016) Hartikainen et al. (2016)

References

Abbasi, B., 2017. Removal of dye by biological methods using fungi. Int. J. Med. Rev. 4 (4), 112–118.

Abd El-Rahim, W.M., Moawad, H., 2003. Enhancing bioremoval of textile dyes by eight fungal strains from media supplemented with gelatine wastes and sucrose. J. Basic Microbiol. 43 (5), 367–375.

Abdelwahab, O., El Nemr, A., El-Sikaily, A., Khaled, A., 2005. Use of rice husk for adsorption of direct dyes from aqueous solution: a case study of direct F. Scarlet. Egyp. J. Aquatic Res. 31 (1), 1–11.

Abdullah AI-Hamedi, F.H., 2007. Comparison of Efficiency of an Advanced Oxidation Process for Degrading Major Classes of Dyes. Master Thesis. United Arab Emirates University, United Arab Emirates.

Abid, M.F., Zablouk, M.A., Abid-Alameer, A.M., 2012. Experimental study of dye removal from industrial wastewater by membrane technologies of reverse osmosis and nanofiltration. Iran. J. Environ. Health Sci. Eng. 9 (1), 17–26.

Abioye, O.P., Aliyu, H.G., Aransiola, S.A., 2015. Bio-removal of vat dye from textile effluent by *Candida tropicalis* and Candida apis isolated from soil. Exp. Opin. Environ. Biol. 4 (3), 121–128.

Abo-Farha, S.A., 2010. Comparative study of oxidation of some azo dyes by different advanced oxidation processes: Fenton, Fenton-like, photo-Fenton and photo-Fenton-like. J. Am. Sci. 6 (10), 128–142.

Adebajo, S.O., Balogun, S.A., Akintokun, A.K., 2017. Decolourization of vat dyes by bacterial isolates recovered from local textile mills in Southwest, Nigeria. Microbiol. Res. J. Int. 18 (1), 1–8.

Adeel, S., Saeed, M., Abdullah, A., Rehman, F., Salman, M., Kamran, M., Zuber, M., Iqbal, M., 2018. Microwave assisted modulation of vat dyeing of cellulosic fiber: improvement in color characteristics. J. Nat. Fibers 15 (4), 517–526.

Agarry, S.E., Ajani, A.O., 2011. Evaluation of microbial systems for biotreatment of textile waste effluents in Nigeria: biodecolourization and biodegradation of textile dye. J. Appl. Sci. Environ. Manag. 15 (1), 79–86.

Ahmad, A.L., Harris, W.A., Ooi, B.S., 2002. Removal of dye from wastewater of textile industry using membrane technology. Jurnal Teknologi 36 (1), 31–44.

Ajaz, M., Rehman, A., Khan, Z., Nisar, M.A., Hussain, S., 2019. Degradation of azo dyes by Alcaligenes aquatilis 3c and its potential use in the wastewater treatment. Amb. Express 9 (1), 64.

Akazdam, S., Chafi, M., Yassine, W., Gourich, B., 2017. Removal of acid orange 7 dye from aqueous solution using the exchange resin amberlite FPA-98 as an efficient adsorbent: kinetics, isotherms, and thermodynamics study. J. Materials .Environ. Sci. 8 (8), 2993–3012.

Akbari, A., Desclaux, S., Rouch, J.C., Remigy, J.C., 2007. Application of nanofiltration hollow fibre membranes, developed by photografting, to treatment of anionic dye solutions. J. Membr. Sci. 297 (1–2), 243–252.

Al-Godah, Z., 2000. Adsorption of dyes using shale oil ash. Water Res. 34, 4295–4303.

Al-Kdasi, A., Idris, A., Saed, K., Guan, C.T., 2004. Treatment of textile wastewater by advanced oxidation processes—a review. Global NEST J. 6 (3), 222–230.

Albanis, T.A., Hela, D.G., Sakellarides, T.M., Danis, T.G., 2000. Removal of dyes from aqueous solutions by adsorption on mixtures of fly ash and soil in batch and column techniques. Global Nest: The Int. J 2 (3), 237–244.

Ali, J.K.A.K., 2004. The Impact of Industrial Waste on Human and Natural Resources: A Case Study of Khartoum North Industrial Area. Ph. D. Thesis. Omdurman Ahlia University, Sudan.

Ali, A.H., Razaq, Z.A., Hamaadi, N.J., 2016. Removal of reactive black dye Rb-5 from synthetic wastewater by electrocoagulation technique. J. Eng. Sust. Dev. 20 (6), 40–54.

Allen, S.J., Koumanova, B., 2005. Decolourisation of water/wastewater using adsorption. J. Uni. Chem. Technol. Metal. 40 (3), 175–192.

Allen, S.J., Mckay, G., Porter, J.F., 2004. Adsorption isotherm models for basic dye adsorption by peat in single and binary component systems. J. Colloid Interface Sci. 280 (2), 322–333.

Amin, G., Đorđević, D., Konstantinović, S., Jordanov, I., 2017. The removal of the textile basic dye from the water solution by using natural zeolite. Adv. Technol. 6 (2), 67–71.

Amiri, H., Ayati, B., Ganjidoust, H., 2016. Textile dye removal using a photocatalytic cascade disc reactor coated by ZnO nanoparticles. J. Mater. Sci. Chem. Eng. 4, 29–38.

An, H., Qian, Y., Gu, X., Tang, W.Z., 1996. Biological treatment of dye wastewaters using an anaerobic-oxic system. Chemosphere 33 (12), 2533–2542.

Anjaneyulu, Y., Chary, N.S., Raj, D.S.S., 2005. Decolourization of industrial effluents–available methods and emerging technologies–a review. Rev. Environ. Sci. Biotechnol. 4 (4), 245–273.

Anvari, F., Kheirkhah, M., Amraei, R., 2014. Treatment of synthetic textile wastewater by combination of coagulation/flocculation process and electron beam irradiation. J. Commun. Health Res. 3 (1), 31–38.

Arumai Dhas, J.P., 2008. Removal of Cod and Colour from Textile Wastewater Using Limestone and Activated Carbon. Ph. D. Thesis. University Sains Malaysia, Malaysia.

Aruna, B., Silviya, L.R., Kumar, E.S., Rani, P.R., Prasad, D.V., VijayaLakshmi, D., 2015. Decolorization of Acid Blue 25 dye by individual and mixed bacterial consortium isolated from textile effluents. Int. J. Curr. Microbiol. Appl. Sci. 4 (5), 1015–1024.

Asgher, M., Batool, S., Bhatti, H.N., Noreen, R., Rahman, S.U., Asad, M.J., 2008. Laccase mediated decolorization of vat dyes by Coriolus versicolor IBL-04. Int. Biodeterior. Biodegrad. 62 (4), 465–470.

Assadi, A., Naderi, M., Mehrasbi, M.R., 2017. Anaerobic–aerobic sequencing batch reactor treating azo dye containing wastewater: effect of high nitrate ions and salt. J. Water Reuse. Desalin. 8 (2), 251–261.

Assmann, C., Scott, A., Biller, D., 2017. Online total organic carbon (TOC) monitoring for water and wastewater treatment plants processes and operations optimization. Drink. Water Eng. Sci. 10 (2), 61–68.

Atiq, M.S., Rehman, A., Iqbal, K., Safdar, F., Basit, A., Ashraf, M., Maqsood, H.S., Khan, A., 2019. Salt free sulphur black dyeing of cotton fabric after cationization. Cellul. Chem. Technol. 53 (1–2), 155–161.

Atul, K., Pratibha, C., Poonam, V., 2012. A comparative study on the treatment methods of textile dye effluents. J. Chem. Pharmaceut. Res. 4 (1), 763–771.

Atul, K., Pratibha, C., Poonam, V., 2013. Adsorption of Reactive Red 194 dye from textile effluent by using class F Fly ash. Scholars J. Appl. Med. Sci. 1 (2), 111–116.

Ayhan Şengil, İ., Özdemir, A., 2012. Simultaneous decolorization of binary mixture of blue disperse and yellow basic dyes by electrocoagulation. Desalin. Water Treatment 46 (1–3), 215–226.

Babuponnusami, A., Muthukumar, K., 2014. A review on Fenton and improvements to the Fenton process for wastewater treatment. J.Environ. Chem. Eng. 2 (1), 557–572.

Baghapour, M., Dehghani, M., Elhamiyan, Z., 2016. Evaluation of Fenton process in removal of direct red 81. J. Health Sci. Surveill. Syst. 4 (1), 14–21.

Bagheri, M., Najafabadi, N.R., Borna, E., 2019. Removal of reactive blue 203 dye photocatalytic using ZnO nanoparticles stabilized on functionalized MWCNTs. J. King Saud Univ. Sci. (in press).

Balamurugan, B., Thirumarimurugan, M., Kannadasan, T., 2011. Anaerobic degradation of textile dye bath effluent using Halomonas sp. Bioresour. Technol. 102, 6365–6369.

Balla, W., Essadki, A.H., Gourich, B., Dassaa, A., Chenik, H., Azzi, M., 2010. Electrocoagulation/electroflotation of reactive, disperse and mixture dyes in an external-loop airlift reactor. J. Hazard Mater. 184 (1–3), 710–716.

Bao, Y., Zhang, G., 2012. Study of adsorption characteristics of methylene blue onto activated carbon made by *salix psammophila*. Energy Proc. 16, 1141–1146.

Barbusiński, K., 2005. The modified Fenton process for decolorization of dye wastewater. Pol. J. Environ. Stud. 14 (3), 281–285.

Barka, N., Qourzal, S., Assabbane, A., Nounah, A., Ait-Ichou, Y., 2010. Photocatalytic degradation of an azo reactive dye, Reactive Yellow 84, in water using an industrial titanium dioxide coated media. Arab. J. Chem. 3 (4), 279–283.

Barrak, N., Mannai, R., Zaidi, M., Kechida, M., Helal, A.N., 2016. Experimental design approach with response surface methodology for removal of indigo dye by electrocoagulation. J. Geosci. Environ. Protect. 4, 50–61.

Bazrafshan, E., Mahvi, A.H., 2014. Textile wastewater treatment by electrocoagulation process using aluminum electrodes. Iran. J. Health Sci. 2 (1), 16–29.

Bazrafshan, E., Mohammadi, L., Ansari-Moghaddam, A., Mahvi, A.H., 2015. Heavy metals removal from aqueous environments by electrocoagulation process—a systematic review. J. Environ. Health Sci. Eng. 13 (1), 74–90.

Bellebia, S., Kacha, S., Bouberka, Z., Bouyakoub, A.Z., Derriche, Z., 2009. Color removal from acid and reactive dye solutions by electrocoagulation and electrocoagulation/adsorption processes. Water Environ. Res. 81 (4), 382–393.

Benkhaya, S., El Harfi, S., El Harfi, A., 2017. Classifications, properties and applications of textile dyes: a review. Appl. J. Environ. Eng. Sci. 3 (3), 311–320.

Bergamini, R.B., Dezotti, M., de Araújo, L.R.R., Azevedo, E.B., 2008. Heterogeneous photocatalytic degradation of acid dyes in aqueous TiO_2 suspensions: kinetics and toxicity. J. Adv. Oxid. Technol. 11 (2), 308–315.

Berradi, M., Essamri, A., El Harfi, A., 2016. Discoloration of water loaded with vat dyes by membrane process of ultrafiltration. J. Mater. Environ. Sci. 7 (4), 1098–1106.

Berrie, B.H., Lomax, S.Q., 1997. Azo pigments: their history, synthesis, properties, and use in artists' materials. Stud. Hist. Art 57, 9–33.

Birch, R.R., Biver, C., Campagna, R., Gledhill, W.E., Pagga, U., Steber, J., Reust, H., Bontinck, W.J., 1989. Screening of chemicals for anaerobic biodegradability. Chemosphere 19 (10–11), 1527–1550.

Buscio, V., Brosillon, S., Mendret, J., Crespi, M., Gutiérrez-Bouzán, C., 2015. Photocatalytic membrane reactor for the removal of C.I. disperse Red 73. Materials 8, 3633–3647.

Buscio, V., García-Jiménez, M., Vilaseca, M., López-Grimau, V., Crespi, M., Gutiérrez-Bouzán, C., 2016a. Reuse of textile dyeing effluents treated with coupled nanofiltration and electrochemical processes. Materials 9 (6), 490–452.

Buscio, V., Crespi, M., Gutiérrez-Bouzán, C., 2016b. Application of PVDF ultrafiltration membranes to treat and reuse textile wastewater. Desalin. Water Treat. 57 (18), 8090–8096.

Butler, E.B., 2013. Electrochemical/electroflotation Process for Dye Wastewater Treatment. Master thesis. Cleveland State University, Ohio, USA.

Butler, E., Hung, Y.T., Yeh, R.Y.L., Suleiman Al Ahmad, M., 2011. Electrocoagulation in wastewater treatment. Water 3 (2), 495–525.

Cerboneschi, M., Corsi, M., Bianchini, R., Bonanni, M., Tegli, S., 2015. Decolorization of acid and basic dyes: understanding the metabolic degradation and cell-induced adsorption/precipitation by *Escherichia coli*. Appl. Microbiol. Biotechnol. 99 (19), 8235–8245.

Cestarolli, D.T., de Oliveira, A.D.G., Guerra, E.M., 2019. Removal of Eriochrome Black textile dye from aqueous solution by combined electrocoagulation–electroflotation methodology. Appl. Water Sci. 9 (4), 101–106.

Chaari, I., Feki, M., Medhioub, M., Fakhfakh, E., Jamoussi, F., 2009. Adsorption of a textile dye "Indanthrene Blue RS (CI Vat Blue 4)" from aqueous solutions onto smectite-rich clayey rock. J. Hazard Mater. 172 (2–3), 1623–1628.

Chakraborty, J.N., 2014. Dyeing with phthalocyanine dye. In: Chakraborty, J.N. (Ed.), Fundamentals and Practices in Colouration of Textiles. Woodhead Publishing, New Delhi, India, pp. 169–176.

Chang, J.S., Lin, Y.C., 2000. Fed-batch bioreactor strategies for microbial decolorization of azo dye using a pseudomonasluteola strain. Biotechnol. Prog. 16 (6), 979–985.

Chang, S.W., La, H.J., Lee, S.J., 2001. Microbial degradation of benzene, toluene, ethylbenzene and xylene isomers (BTEX) contaminated groundwater in Korea. Water Sci. Technol. 44 (7), 165–171.

Charoenlarp, K., Choyphan, W., 2009. Reuse of dye wastewater through colour removal with electrocoagulation process. Asian J. Energy Environ. 10, 250–260.

Chaturvedi, S.I., 2013. Electrocoagulation: a novel waste water treatment method. Int. J. Modern Eng. Res. 3 (1), 93–100.

Chen, G., Chen, X., Yue, P.L., 2000. Electrocoagulation and electroflotation of restaurant wastewater. J. Environ. Eng. 126 (9), 858–863.

Chenik, H., Elhafdi, M., Dassaa, A., Essadki, A.H., Azzi, M., 2013. Removal of real textile dyes by electrocoagulation/electroflotation in a pilot external-loop airlift reactor. J. Water Resour. Protect. 5 (10), 1000–1006.

Chequer, F.M.D., de Oliveira, G.A.R., Ferraz, E.R.A., Cardoso, J.C., Zanoni, M.V.B., de Oliveira, D.P., 2013. Textile dyes: dyeing process and environmental impact. In: Gunay, M. (Ed.), Eco-Friendly Textile Dyeing and Finishing. InTech, Crotia, pp. 151–176.

Chigozie, U.F., Joseph, N.T., 2014. Removal of orange-G, vat yellow, erythrosine dyes from synthetic wastewater by electrocoagulation and nanofiltration. J. Adv. Chem. Eng. 4 (3), 112–117.

Chiou, M.S., Kuo, W.S., Li, H.Y., 2003. Removal of reactive dye from wastewater by adsorption using ECH cross-linked chitosan beads as medium. J. Environ. Sci. Health. 38 (11), 2621–2631.

Chong, M.N., Jin, B., Chow, C.W., Saint, C., 2010. Recent developments in photocatalytic water treatment technology: a review. Water Research 44 (10), 2997–3027.

Choppin, G., Liljenzin, J.-O., Rydberg, J., Ekberg, C., 2013. Radiochemistry and Nuclear Chemistry, fourth ed. Academic Press, Oxford, UK.

Choudhury, A.R., 2011. Dyeing of synthetic fibres. In: Clark, M. (Ed.), Handbook of Textile and Industrial Dyeing. Woodhead Publishing, Cambridge, UK, pp. 40–128.

Chukwu, O., 2008. Impact of liquid wastes of food processing industries on quality of receiving streams. AU J. Technol. 12 (2), 121–125.

Chung, J., Kim, J.O., 2011. Application of advanced oxidation processes to remove refractory compounds from dye wastewater. Desalin. Water Treat. 25 (1–3), 233–240.

Çiner, F., Gökkuş, Ö., 2013. Treatability of dye solutions containing disperse dyes by Fenton and fenton-solar light oxidation processes. Clean 41 (1), 80–85.

Comstock, M., 2016. Complex Inorganic Colored Pigments: comparison of options and relative properties when faced with elemental restrictions. J. Surf. Coat. Australia 2016, 1–30.

Costa, M.C., Mota, F.S.B., Santos, A.B.D., Mendonça, G.L.F., Nascimento, R.F.D., 2012. Effect of dye structure and redox mediators on anaerobic azo and anthraquinone dye reduction. Quim. Nova 35 (3), 482–486.

D'Antoni, M., Iracà, F., Romero, M., 2017. Abstract Review on: Color Removal from Textile Effluents by Biological Processes, 10.13140/RG.2.2.35799.68009.

Dabrowski, A., 2001. Adsorption-from theory to practice. Adv. Colloid Interface Sci. 93, 135–224.

Daneshvar, N., Oladegaragoze, A., Djafarzadeh, N., 2006. Decolorization of basic dye solutions by electrocoagulation: an investigation of the effect of operational parameters. J. Hazard Mater. 129 (1–3), 116–122.

Davis, R.J., Gainer, J.L., O'Neal, G., Wu, I.W., 1994. Photocatalytic decolorization of wastewater dyes. Water Environ. Res. 66 (1), 50–53.

Dawkar, V.V., Jadhav, U.U., Tamboli, D.P., Govindwar, S.P., 2010. Efficient industrial dye decolorization by Bacillus sp. VUS with its enzyme system. Ecotoxicol. Environ. Saf. 73, 1696–1703.

Deaconu, M., Senin, R., Stoica, R., Athanasiu, A., Crudu, M., Oproiu, L., Ruse, M., Filipescu, C., 2016. Adsorption decolorization technique of textile/leather–dye containing effluents. Int. J. Waste Res. 6 (2), 1–7.

Dehghani, M.H., Mahdavi, P., 2015. Removal of acid 4092 dye from aqueous solution by zinc oxide nanoparticles and ultraviolet irradiation. Desalin. Water Treat. 54 (12), 3464–3469.

Deivasigamani, C., Das, N., 2011. Biodegradation of Basic Violet 3 by Candida krusei isolated from textile wastewater. Biodegradation 22 (6), 1169–1180.

Delée, W., Neill, C.O., Hawkes, F.R., Pinheiro, H.R.M., 1998. Anaerobic treatment of textile effluents. J. Chem. Technol. Biotechnol. 73, 323–335.

Deng, D., Guo, J., Zeng, G., Sun, G., 2008. Decolorization of anthraquinone, triphenylmethane and azo dyes by a new isolated Bacillus cereus strain DC11. Int. Biodeterior. Biodegrad. 62 (3), 263–269.

Dessouki, A.M., Abdel-Aab, S.E., 1998. Radiation degradation of some commercial dyes in wastewater. In: International Conference on Hazardous Waste: Source Effect and Management, pp. 1331–1345.

Dhas, P.G.T.N., Gulyas, H., Otterpohl, R., 2015. Impact of powdered activated carbon and anion exchange resin on photocatalytic treatment of textile wastewater. J. Environ. Protect. 6 (3), 191–203.

Dias, A.A., Lucas, M.S., Sampaio, A., Peres, J.A., Bezerra, R.M., 2010. Decolorization of azo dyes by yeasts. In: Atacag Erkut, H. (Ed.), Biodegradation of Azo Dyes. Springer, Berlin, pp. 183–193.

Do, J.S., Chen, M.L., 1990. Decolourization of dye containing solutions by electrocoagulation. J. Appl. Electrochem. 24, 785–790.

Dogdu, G., Yalcuk, A., 2016. Indigo dyeing wastewater treatment by eco-friendly constructed wetlands using different bedding media. Desalin. Water Treat. 57 (32), 15007–15019.

Dubinina, T.V., Tomilova, L.G., Zefirov, N.S., 2013. Synthesis of phthalocyanines with an extended system of π-electron conjugation. Rus. Chem. Rev. 82 (9), 865.

Durmuş, M., Nyokong, T., 2008. Photo physicochemical and fluorescence quenching studies of benzyloxyphenoxy-substituted zinc phthalocyanines. Spectrochim. Acta Mol. Biomol. Spectrosc. 69 (4), 1170–1177.

Ejder-Korucu, M., Gürses, A., Doğar, Ç., Sharma, S.K., Açıkyıldız, M., 2015. Removal of organic dyes from industrial effluents: an overview of physical and biotechnological applications. In: Sharma, S.K. (Ed.), Green Chemistry for Dyes Removal from Wastewater: Research Trends and Applications. Scrivener Publishing, Canada, pp. 1–34.

El Haddad, M., Regti, A., Laamari, M.R., Mamouni, R., Saffaj, N., 2014. Use of Fenton reagent as advanced oxidative process for removing textile dyes from aqueous solutions. J. Mater. Environ. Sci. 5 (3), 667–674.

El Nemr, A., Hassaan, M.A., Madkour, F.F., 2018. Advanced oxidation process (AOP) for detoxification of acid red 17 dye solution and degradation mechanism. Environ. Proc. 5 (1), 95–113.

El-Hennawi, H.M., Shahin, A.A., El-Thalouth, I.A., 2013. Evaluation of dried biomass of baker's yeast as reactive dye adsorbent using ultrasonic Technique. J. Appl. Sci. Res. 9 (3), 1401–1408.

El-Hosiny, F., AbdelKhalek, M.A., Selim, K., Osama, I., 2017. A designed electro-flotation cell for dye removal from wastewater. J. Appl. Res. Indus. Eng. 4 (2), 133–147.

El-Hosiny, F.I., Abdel-Khalek, M.A., Selim, K.A., Osama, I., 2018. Physicochemical study of dye removal using electro-coagulation-flotation process. Physicochem. Prob. Min. Proces. 54 (2), 321–333.

El-Rahim, W.M.A., El-Ardy, O.A.M., Mohammad, F.H., 2009. The effect of pH on bioremediation potential for the removal of direct violet textile dye by Aspergillus Niger. Desalination 249 (3), 1206–1211.

Elkhattabi, E.H., Lakraimi, M., Berraho, M., Legrouri, A., Hammal, R., El Gaini, L., 2018. Acid Green 1 removal from wastewater by layered double hydroxides. Appl. Water Sci. 8 (1), 45–56.

Environment Protection Authority (EPA), 2014. Waste Classification Guidelines – Part 1: Classification of Waste, NSW Environment Protection Authority, Sydney.

Ertugay, N., Acar, F.N., 2017. Removal of COD and color from Direct Blue 71 azo dye wastewater by Fenton's oxidation: kinetic study. Arab. J. Chem. 10, 1158–1163.

Ertuğrul, S., San, N.O., Dönmez, G., 2009. Treatment of dye (Remazol Blue) and heavy metals using yeast cells with the purpose of managing polluted textile wastewaters. Ecol. Eng. 35 (1), 128–134.

Esgair, K.K., 2017. A study on the removal of Direct Blue 71 dye from textile wastewater produced from state company of cotton industries by electrocoagulation using aluminum electrodes. J. Eng. 23 (2), 83–94.

Evin, H., 2009. Hazardous waste management in the world and Turkey: a comperative analysis. Fırat Univ. J. Soc. Sci. 19 (2), 197–208.

Farias, R.S.D., Buarque, H.L.D.B., Cruz, M.R.D., Cardoso, L.M.F., Gondim, T.D.A., Paulo, V.R.D., 2018. Adsorption of Congo red dye from aqueous solution onto amino-functionalized silica gel. Eng. Sanitária Ambient. 23 (6), 1053–1060.

Fatiha, M., Belkacem, B., 2016. Adsorption of methylene blue from aqueous solutions using natural clay. J. Mater. Environ. Sci. 7 (1), 285–292.

Filipkowska, U., 2006. Adsorption and desorption of reactive dyes onto chitin and chitosan flakes and beads. Adsorpt. Sci. Technol. 24 (9), 781–795.

Fitzgerald, S.W., Bishop, P.L., 1995. Two stage anaerobic/aerobic treatment of sulfonated azo dyes. J. Environ. Sci. Health. 30 (6), 1251–1276.

Fleischmann, C., Lievenbrück, M., Ritter, H., 2015. Polymers and dyes: developments and applications. Polymers 7 (4), 717–746.

Forgacs, E., Cserhati, T., Oros, G., 2004. Removal of synthetic dyes from wastewaters: a review. Environ. Int. 30, 953–971.

Forootanfar, H., Moezzi, A., Aghaie-Khozani, M., Mahmoudjanlou, Y., Ameri, A., Niknejad, F., Faramarzi, M.A., 2012. Synthetic dye decolorization by three sources of fungal laccase. Iran. J. Environ. Health Sci. Eng. 9 (1), 27.

Franciscon, E., Zille, A., Fantinatti-Garboggini, F., Silva, I.S., Cavaco-Paulo, A., Durrant, L.R., 2009. Microaerophilic–aerobic sequential decolourization/biodegradation of textile azo dyes by a facultative Klebsiella sp. strain VN-31. Process Biochem. 44 (4), 446–452.

Franciscon, E., Grossman, M.J., Paschoal, J.A.R., Reyes, F.G.R., Durrant, L.R., 2012. Decolorization and biodegradation of reactive sulfonated azo dyes by a newly isolated Brevibacterium sp. strain VN-15. SpringerPlus 1 (1), 37–47.

Fu, Y., Viraraghavan, T., 2001. Fungal decolorization of dye wastewaters: a review. Bioresour. Technol. 79, 251–262.

Geethakarthi, A., Phanikumar, B.R., 2011. Adsorption of reactive dyes from aqueous solutions by tannery sludge developed activated carbon: kinetic and equilibrium studies. Int. J. Environ. Sci. Technol. 8 (3), 561–570.

ghali Attia, H., 2013. Decolorization of direct blue dye by electrocoagulation process. J. Eng. Sust. Dev. 17 (1), 171–181.

Ghalib, A.M., Al-yaqoobi, A., 2010. Removal of direct blue dye in textile wastewater effluent by electrocoagulation. J. Eng. 16, 6198–6205.

Ghalwa, N.M.A., Saqer, A.M., Farhat, N.B., 2016. Removal of Reactive Red 24 dye by clean electrocoagulation process using iron and aluminum electrodes. J. Chem. Eng. Process Technol. 7 (1), 1–7.

Gharbani, P., Tabatabaii, S.M., Mehrizad, A., 2008. Removal of Congo red from textile wastewater by ozonation. Int. J. Environ. Sci. Technol. 5 (4), 495–500.

Ghelardi, E., 2014. A Multi-Analytical Approach for the Characterization of Contemporary Paint Materials. Ph. D. Thesis. Università degli Studi di Firenze, Italy.

Ghernaout, D., Al-Ghonamy, A.I., Naceur, M.W., Messaoudene, N.A., Aichouni, M., 2014. Influence of operating parameters on electrocoagulation of CI disperse yellow 3. J. Electrochem. Sci. Eng. 4 (4), 271–283.

Gholami, M., Nasseri, S., Fard, M.A., Mesdaghinia, A., Vaezi, F., Mahvi, A., Naddaffi, K., 2001. Dye removal from effluents of textile industries by ISO9888 method and membrane technology. Iran. J. Public Health 30 (1–2), 73–80.

Gholami, M., Nasseri, S., Alizadehfard, M.R., Mesdaghinia, A., 2003. Textile dye removal by membrane technology and biological oxidation. Water Quality Res. J. 38 (2), 379–391.

Ghoreishi, M., Haghighi, R., 2003. Chemical catalytic reaction and biological oxidation for treatment of non-biodegradable textile effluent. Chem. Eng. J. 95, 163–169.

Giwa, A., Ogunribido, A., 2012. The applications of membrane operations in the textile industry: a review. Br. J. Appl. Sci. Technol. 2 (3), 296–310.

Glaze, W.H., Kang, J.W., 1989. Advanced oxidation processes. Description of a kinetic model for the oxidation of hazardous materials in aqueous media with ozone and hydrogen peroxide in a semibatch reactor. Ind. Eng. Chem. Res. 28, 1573–1587.

Gökkuş, Ö., Çiner, F., 2009. Investigation of color and COD removal from wastewater containing disperse yellow 119 and disperse red 167 using Fenton oxidation process. J. Faculty Eng. Archit. Gazi Univ. 25, 49–55.

Golder, A., Samanta, A.N., Subhabrata, R., 2006. Anionic reactive dye removal from aqueous solution using a new adsorbent—sludge generated in removal of heavy metal by electrocoagulation. Chem. Eng. J. 122 (1–2), 107–115.

Gomez, V., Larrechi, M.S., Callao, M.P., 2007. Kinetic and adsorption study of acid dye removal using activated carbon. Chemosphere 69 (7), 1151–1158.

Gonçalves, I.M.C., Gomes, A., Bras, R., Ferra, M.I.A., Amorim, M.T.P., Porter, R.S., 2000. Biological treatment of effluent containing textile dyes. Color. Technol. 116 (12), 393–397.

Gonçalves, I.C., Ferra, M.I.A., Amorim, M.T., 2001. Anaerobic decolorization of an azo dye by a mixed culture. Toxicol. Environ. Chem. 79 (1–2), 81–93.

Gordon, P.F., Gregory, P., 1987. Azo dyes. In: Gordon, P.F., Gregory, P. (Eds.), Organic Chemistry in Colour. Springer, Berlin, Germany, pp. 95–162.

Gosavi, V.D., Sharma, S., 2014. A general review on various treatment methods for textile wastewater. J. Environ. Sci. Comput. Sci. Eng. Technol. 3, 29–39.

Gowtham, B., Pauline, S., 2015. Feasibility studies on application of photo-Fenton oxidation for methylene blue. Iran. J. Energy Environ. 6, 255–259.

Gregory, P., 1990. Classification of dyes by chemical structure. In: Waring, D.R., Hallas, G. (Eds.), The Chemistry and Application of Dyes. Plenum Press, New York, USA, pp. 17–47.

Growther, L., Meenakshi, M., 2008. Biotechnological approaches to combat textile effluents. Internet J. Microbiol. 7 (1), 1–7.

Guendy, H.R., 2007. Ozone treatment of textile wastewater relevant to toxic effect elimination in marine environment. Egyp. J. Aquatic Res. 33 (1), 98–115.

Guiza, S., Bagane, M., Al-Soudani, A.H., Amore, H.B., 2004. Adsorption of basic dyes onto natural clay. Adsorpt. Sci. Technol. 22 (3), 245–255.

Güneş, Y., Atav, R., Namırtı, O., 2012. Effectiveness of ozone in decolorization of reactive dye effluents depending on the dye chromophore. Textil. Res. J. 82 (10), 994–1000.

Gungor, A.A., Nadaroglu, H., Celebi, N., 2014. Fenton process for basic red 9 degradation: immobilized apolaccase on a nanomagnetite system. Environ. Experim. Biol. 12, 121–129.

Gupta, V.K., Suhas, 2009. Application of low-cost adsorbents for dye removal-A review. J. Environ. Manag. 90, 2313–2342.

Gupta, G.S., Prasad, G., Singh, V.H., 1990. Removal of chrome dye from aqueous solutions by mixed adsorbents: fly ash and coal. Water Res. 24, 45–50.

Gurav, A.A., Ghosh, J.S., Kulkarni, G.S., 2011. Decolorization of anthroquinone based dye Vat Red 10 by Pseudomonas desmolyticum NCIM 2112 and Galactomyces geotrichum MTCC 1360. Int. J. Biotechnol. Mol. Biol. Res. 2 (6), 93–97.

Gürses, A., Doğar, Ç., Yalçın, M., Açıkyıldız, M., Bayrak, R., Karaca, S., 2006. The adsorption kinetics of the cationic dye, methylene blue, onto clay. J. Hazard Mater. 131 (1–3), 217–228.

Gürses, A., Açıkyıldız, M., Güneş, K., Gürses, M.S., 2016a. Dyes and Pigments. Springer, Germany.

Gürses, A., Güneş, K., Ejder-Korucu, M., Açıkyıldız, M., 2016b. Industrial waste. In: Seidel, A. (Ed.), Kirk-Othmer Encyclopedia of Chemical Technology. John Wiley and Sons Inc, New York, USA, pp. 1–26.

Gutiérrez Bouzán, M.C., Buscio Olivera, V., 2018. Combining electrochemistry and UV for the simultaneous wastewater decolorization and reduction of salinity. AIMS Environ. Sci. 5 (2), 96–104.

Hammami, A., Charcosset, C., Amar, R.R.B., 2017. Performances of continuous adsorption-ultrafiltration hybrid process for AO7 dye removal from aqueous solution and real textile wastewater treatment. J. Membr. Sci. Technol. 7 (1), 1–8.

Han, Y.H., Robert Van der, H., Verpoorte, R., 2001. Biosynthesis of anthraquinones in cell cultures of the Rubiaceae. Plant Cell Tissue Organ Cult. 67, 201–220.

Harrelkas, F., Paulo, A., Alves, M.M., El Khadir, L., Zahraa, O., Pons, M.N., Van Der Zee, F.P., 2008. Photocatalytic and combined anaerobic–photocatalytic treatment of textile dyes. Chemosphere 72 (11), 1816–1822.

Hartikainen, E.S., Miettinen, O., Hatakka, A., Kähkönen, M.A., 2016. Decolorization of six synthetic dyes by fungi. Am. J. Environ. Sci. 12 (2), 77–85.

Hassaan, M.A., El Nemr, A., 2017. Health and environmental impacts of dyes: mini review. Am. J. Environ. Sci. Eng. 1 (3), 64–67.

Hassaan, M.A., El Nemr, A., Madkour, F.F., 2017. Testing the advanced oxidation processes on the degradation of Direct Blue 86 dye in wastewater. Egyp. J. Aquatic Res. 43 (1), 11–19.

Hassani, A.H., Mirzayee, R., Nasseri, S., Borghei, M., Gholami, M., Torabifar, B., 2008. Nanofiltration process on dye removal from simulated textile wastewater. Int. J. Environ. Sci. Technol. 5 (3), 401–408.

Hedaoo, M.N., Bhole, A.G., Ingole, N.W., Hung, Y.-T., 2012. Biological wastewater treatment. In: Hung, Y.-T., Vang, L.K., Shammas, N.K. (Eds.), Handbook of Environment and Waste Management: Air and Water Pollution Control. World Scientific Publishing, Singapore, pp. 431–473.

Herbst, W., Hunger, K., 2004. Industrial Organic Pigments. Wiley-VCH, Weinheim.

Hooshmandfar, A., Ayati, B., Khodadadi Darban, A., 2015. Optimization of material and energy consumption for removal of Acid Red 14 by simultaneous electrocoagulation and electroflotation. Water Sci. Technol. 73 (1), 192–202.

Huang, M., Zhao, K., Zhang, M., Chen, S., Xu, W., Ye, Y., Fu, W., Wei, Y., Qiu, Z., Sun, F., 2011. Removal of disperse dyes from wastewater by nano-iron modified goldmine waste-solid assisted AOPs. Proc. Eng. 18, 358–362.

Huang, J., Fu, Y., Liu, Y., 2014. Comparison of alkali-tolerant fungus *myrothecium* sp. IMER1 and white-rot fungi for decolorization of textile dyes and dye effluents. J. Biorem. Biodegrad. 5 (3), 221–226.

Hunger, K., 2003. Industrial Dyes: Chemistry, Properties, Applications. Wiley-VCH, Weinheim.

Hunger, K., Schmidt, M.U., 2018. Industrial Organic Pigments: Production, Crystal Structures, Properties, Applications. Wiley-VCH, , Weinheim.

Idris, A., Inanc, B., Hassan, M.N., 2004. Overview of waste disposal and landfills/dumps in Asian countries. J. Mater. Cycles Waste Manag. 6 (2), 104–110.

Igunnu, E.T., Chen, G.Z., 2012. Produced water treatment technologies. Int. J. Low Carbon Technol. 9 (3), 157–177.

Iqbal, M.J., Ashiq, M.N., 2007. Adsorption of dyes from aqueous solutions on activated charcoal. J. Hazard Mater. 139 (1–2), 57–66.

Işık, M., Sponza, D.T., 2008. Anaerobic/aerobic treatment of a simulated textile wastewater. Separ. Purif. Technol. 60 (1), 64–72.

Jafari, N., Kasra-Kermanshahi, R., Soudi, M.R., 2013. Screening, identification and optimization of a yeast strain, Candida palmioleophila JKS4, capable of azo dye decolorization. Iran. J. Microbiol. 5 (4), 434–440.

Jerič, T., Bisselink, R.J., Van Tongeren, W., Le Marechal, A.M., 2013. Decolorization and mineralization of reactive dyes, by the H_2O_2/UV process with electrochemically produced H_2O_2. Acta Chim. Slov. 60 (3), 666–672.

Jin, R., Yang, H., Zhang, A., Wang, J., Liu, G., 2009. Bioaugmentation on decolorization of CI Direct Blue 71 by using genetically engineered strain *Escherichia coli* JM109 (pGEX-AZR). J. Hazard Mater. 163 (2–3), 1123–1128.

Jogani, R., Bhervia, H., Kapoor, S., Singh, A., 2017. Optimization of different variables used in Fenton reagent process for removal of direct red 80 dye. Int. J. Advan. Agricul. Environ. Eng. 4 (1), 230–234.

Johary, F., Zaimin, F.F.M., Rohani, R., Hassan, A.R., Sharifuddin, S.S., Isa, M.H.M., 2018. Fabrication and characterisation of polyethersulfone/chitosan/non-ionic nanofiltration membrane for dyes removal. AIP Conf. Proc. 1972 (1), 1–8.

Joshi, M., Purwar, R., 2004. Developments in new processes for colour removal from effluent. Rev. Prog. Coloration Relat. Top. 34, 58–71.

Joshi, M., Bansal, R., Purwar, R., 2004. Colour removal from textile effluents. Indian J. Fiber Textil Res. 29, 239–259.

Joshi, P.A., Jaybhaye, S., Mhatre, K., 2015. Biodegradation of dyes using consortium of bacterial strains isolated from textile effluent. J. Exp. Biol. 5 (7), 36–40.

Juang, R.S., Wu, F.C., Tseng, R.L., 1997. The ability of activated clay for the adsorption of dyes from aqueous solutions. Environ. Technol. 18 (5), 525–531.

Juang, Y., Nurhayati, E., Huang, C., Pan, J.R., Huang, S., 2013. A hybrid electrochemical advanced oxidation/microfiltration system using BDD/Ti anode for acid yellow 36 dye wastewater treatment. Separ. Purif. Technol. 120, 289–295.

Junji, O., 2015. Study on Organic Pigments. Ph. D. Thesis. Ritsumeikan University, Japan.

Junnarkar, N., Murty, D.S., Bhatt, N.S., Madamwar, D., 2006. Decolorization of diazo dye Direct Red 81 by a novel bacterial consortium. World J. Microbiol. Biotechnol. 22 (2), 163–168.

Kalel, V., 2006. Biodegradation Kinetics of Azo Dye Amaranth by Free and Immobilized Bacteria. Master Thesis. University of Pune, Pune, India.

Karaca, S., Gürses, A., Açıkyıldız, M., Ejder (Korucu), M., 2008. Adsorption of cationic dye from aqueous solutions by activated carbon. Microporous Mesoporous Mater. 115 (3), 376–382.

Karatas, M., Dursun, S., Argun, M., 2009. Decolorization of reactive dyes under batch anaerobic condition by mixed microbial culture. Afr. J. Biotechnol. 8, 6856–6862.

Kasiri, M.B., Modirshahla, N., Mansouri, H., 2013. Decolorization of organic dye solution by ozonation; Optimization with response surface methodology. Int. J. Ind. Chem. 4 (1), 3–13.

Kavitha, S.K., Palanisamy, P.N., 2010. Photocatalytic degradation of vat yellow 4 using UV/TiO_2. Mod. Appl. Sci. 4 (5), 130.

Kaykhaii, M., Sasani, M., Marghzari, S., 2018. Removal of dyes from the environment by adsorption process. Chem. Mater. Eng. 6 (2), 31–35.

Kelewou, H., Lhassani, A., Merzouki, M., Drogui, P., 2015. Removal of textile-based dyes by nanofiltration: study of physicochemical parameters' effect on the retention by experimental designs methodology. Desalin. Water Treat. 54 (6), 1735–1746.

Khalid, A., Arshad, M., Crowley, D.E., 2009. Biodegradation potential of pure and mixed bacterial cultures for removal of 4-nitroaniline from textile dye wastewater. Water Res. 43 (4), 1110–1116.

Khan, M., Akhtar, S., Zafar, S., Shaheen, A., Luque, R., Rehman, A., 2015. Removal of Congo red from aqueous solution by anion exchange membrane (EBTAC): adsorption kinetics and themodynamics. Materials 8 (7), 4147–4161.

Khare, S.K., Panday, K.K., Srivastava, R.M., Singh, V.N., 1987. Removal of victoria blue from aqueous solution by fly ash. J. Chem. Technol. Biotechnol. 38 (2), 99–104.

Kim, T.H., Park, C., Kim, S., 2005. Water recycling from desalination and purification process of reactive dye manufacturing industry by combined membrane filtration. J. Clean. Prod. 13 (8), 779–786.

Kirby, N., Mc, M., Marchant, R., 1995. Decolorization of an artificial textile effluent by Phanerochaete chrysosporium. Biotechnol. Lett. 17, 761–764.

Kishore, K.A., Kamala, T., 2005. An experimental study on dye water treatment using nanoporous membrane process. Int. Conf. Chemical, Civil Environ. Eng. 124–127.

Kobya, M., Demirbas, E., Can, O.T., Bayramoglu, M., 2006. Treatment of levafix orange textile dye solution by electrocoagulation. J. Hazard Mater. 132 (2–3), 183–188.

Konsowa, A.H., 2003. Decolorization of wastewater containing direct dye by ozonation in a batch bubble column reactor. Desalination 158, 233–240.

Konsowa, A.H., El-Rahman, H.A., Moustafa, M.A., 2011. Removal of azo dye acid orange 7 using aerobic membrane bioreactor. Alexandria Eng. J. 50 (1), 117–125.

Kulandaivel, S., Kaleeswari, P., Mohanapriya, P., 2014. Decolorization and adsorption of dyes by consortium of bacteria with agriculture waste. Int. J. Curr. Microbiol. Appl. Sci. 3 (12), 865–882.

Kumar, A., Sharma, G., Naushad, M., Singh, P., Kalia, S., 2014. Polyacrylamide/Ni0. 02Zn0. 98O nanocomposite with high solar light photocatalytic activity and efficient adsorption capacity for toxic dye removal. Ind. Eng. Chem. Res. 53 (40), 15549–15560.

Kurade, M.B., Waghmode, T.R., Kagalkar, A.N., Govindwar, S.P., 2012. Decolorization of textile industry effluent containing disperse dye Scarlet RR by a newly developed bacterial-yeast consortium BL-GG. Chem. Eng. J. 184, 33–41.

Kyzas, G.Z., Kostoglou, M., Lazaridis, N.K., Bikiaris, D.N., 2013. Decolorization of dyeing wastewater using polymeric absorbents-an overview. In: Gunay, M. (Ed.), Eco-Friendly Textile Dyeing and Finishing. InTech, Crotia, pp. 177–206.

Lade, H., Kadam, A., Paul, D., Govindwar, S., 2015. Biodegradation and detoxification of textile azo dyes by bacterial consortium under sequential microaerophilic/aerobic processes. EXCLI J. 14, 158–174.

Lafi, R., Gzara, L., Lajimi, R.H., Hafiane, A., 2018. Treatment of textile wastewater by a hybrid ultrafiltration/electrodialysis process. Chem. Eng. Processing-Process Intensification 132, 105–113.

Lafta, A.J., Odeh, A.M., Abidalameer, H.A., Esmael, H.A., Mubark, I.J., Abid Alameeer, F.S., 2015. Photocatalytic removal of some textile dyes over suspension of titanium dioxide and irradiation with solar radiation. Int. J. Chem. Sci. 13 (4), 1755–1764.

Lalnunhlimi, S., Krishnaswamy, V., 2016. Decolorization of azo dyes (Direct Blue 151 and Direct Red 31) by moderately alkaliphilic bacterial consortium. Braz. J. Microbiol. 47 (1), 39–46.

Lee, B.H., Song, W.C., Manna, B., Yang, H.J., Kim, J.H., Kim, Y.S., 2007. Removal of color from wastewater using various ozonation techniques. Int. Forum .Strategic Technol. 2007, 120–124.

Li, X.Z., Zhao, Y.G., 1997. On-site treatment of dyeing wastewater by a bio-photoreactor system. Water Science and Technology 36 (2–3), 165–172.

Li, L., Wang, S., Zhu, Z., 2006. Geopolymeric adsorbents from fly ash for dye removal from aqueous solution. J. Colloid Interface Sci. 300 (1), 52–59.

Li, Y., Bi, H.Y., Jin, Y.S., Shi, X.Q., 2014. Removal of disperse violet 28 from water using self-assembled organo-layered double hydroxides through a one-step process. RSC Adv. 4 (102), 58307–58314.

Li, Q., Feng, X.L., Li, T.T., Lu, X.R., Liu, Q.Y., Han, X., Feng, Y.-J., Liu, Z.-Y., Zhang, X.-J., Xiao, X., 2018. Anaerobic decolorization and detoxification of cationic red X-GRL by Shewanella oneidensis MR-1. Environ. Technol. 39 (18), 2382–2389.

Lin, S.H., Chen, M.L., 1997. Treatment of textile wastewater by chemical methods for reuse. Water Res. 31 (4), 868–876.

Lin, J., Zhang, X., Li, Z., Lei, L., 2010. Biodegradation of reactive blue 13 in a two-stage anaerobic/aerobic fluidized beds system with a Pseudomonas sp. isolate. Bioresour. Technol. 101 (1), 34–40.

Liu, R., Chiu, H.M., Shiau, C.S., Yeh, R.Y.L., Hung, Y.T., 2007. Degradation and sludge production of textile dyes by Fenton and photo-Fenton processes. Dyes Pigments 73 (1), 1–6.
Lopes, C.N., Petrus, J.C.C., Riella, H.G., 2005. Color and COD retention by nanofiltration membranes. Desalination 172 (1), 77–83.
Mahmood, R., Sharif, F., Ali, S., Hayyat, M.U., 2015a. Enhancing the decolorizing and degradation ability of bacterial consortium isolated from textile effluent affected area and its application on seed germination. Scient World J 1–9.
Mahmood, R.T., Asad, M.J., Asgher, M., Gulfraz, M., Mukhtar, T., Akram, M., 2015b. Study of disperse dyes biodegradation and lignolytic enzymes production. Potential of indigenous coniophora puteana IBL-01, a Brown rot fungi. Adv. Environ. Biol. 9 (11), 139–150.
Mahmoud, M.S., 2016. Decolorization of certain reactive dye from aqueous solution using Baker's Yeast (*Saccharomyces cerevisiae*) strain. HBRC J. 12 (1), 88–98.
Mahmoud, G., Abdel-Aal, S.E., El-Kelesh, N.A., Alshafei, E., 2017. Effective removal of hazardous dyes from aqueous solutions using starch based hydrogel and gamma radiation. Environ. Ecol. Res. 5 (7), 480–488.
Mai, Z., 2013. Membrane Processes for Water and Wastewater Treatment: Study and Modeling of Interactions between Membrane and Organic Matter. Ph. D. Thesis. École Centrale Paris, Paris, France.
Majewska-Nowak, K., Kowalska, I., Kabsch-Korbutowicz, M., 2006. Ultrafiltration of aqueous solutions containing a mixture of dye and surfactant. Desalination 198 (1–3), 149–157.
Makgae, M., 2011. Key areas in waste management: a South African perspective. In: Kumar, S. (Ed.), Integrated Waste Management-Volume II. IntechOpen, Crotia, pp. 69–82.
Malinović, B.N., Pavlović, M.G., Djuričić, T., 2017. Electrocoagulation of textile wastewater containing a mixture of organic dyes by iron electrode. J. Electrochem. Sci. Eng. 7 (2), 103–110.
Manage, P.M., 2017. Decolorization of CI direct blue 201 textile dye by native bacteria. Int. J. Multidiscipl. Stud. 4 (1), 48–57.
Mane, R.S., Bhusari, V.N., 2012. Removal of colour (dyes) from textile effluent by adsorption using orange and banana peel. Int. J. Eng. Res. Appl. 2 (3), 1997–2004.
Mansoorian, H.J., Bazrafshan, E., Yari, A., Alizadeh, M., 2014. Removal of azo dyes from aqueous solution using Fenton and modified Fenton processes. Health Scope 3 (2), e15507.
Mansouriieh, N., Reza Sohrabi, M., Pouramir Vajargah, R., Roudbaraki, H., 2017. Photocatalytic degradation of Direct yellow 86 diazo dye using sulfanilic acid-modified TiO_2 in aqueous suspensions. Water Sci. Technol. 76 (8), 1992–2002.
Manu, B., 2007. Physico-chemical treatment of indigo dye wastewater. Color. Technol. 123 (3), 197–202.
Manu, B., Chaudhari, S., 2003. Decolorization of indigo and azo dyes in semicontinuous reactors with long hydraulic retention time. Process Biochem. 38 (8), 1213–1221.
Mao, Y., Yang, S., Xue, C., Zhang, M., Wang, W., Song, Z., Zhao, X., Sun, J., 2018. Rapid degradation of malachite green by $CoFe_2O_4$–SiC foam under microwave radiation. Royal Soci. Open Sci. 5 (6), 1–12.
Marimuthu, T., Rajendran, S., Manivannan, M., 2013. A review on fungal degradation of textile dye effluent. Acta Chimica Pharmaceutica Indica 3 (2), 192–200.
Markandeya, Shukla, S.P., Dhiman, N., Mohan, D., Kisku, G.C., Roy, S., 2017. An efficient removal of disperse dye from wastewater using zeolite synthesized from cenospheres. J. Hazard. Toxic Radioact. Waste 21 (4), 04017017.
Marol, C., Puneet, K., Sangeetha, K.V., 2017. Treatment of dairy industry waste water by electrocoagulation (EC) technique removal of BOD, COD, turbidity and color. Inter. J. Innov. Res. Sci. Technol. 4 (1), 246–251.

Masmoudi, G., Trabelsi, R., Ellouze, E., Amar, R.B., 2014. New treatment at source approach using combination of microfiltration and nanofiltration for dyeing effluents reuse. Int. J. Environ. Sci. Technol. 11 (4), 1007–1016.

Masoumbeigi, H., Rezaee, A., 2015. Removal of Methylene Blue (MB) dye from synthetic wastewater using UV/H_2O_2 advanced oxidation process. J. Health Policy and Sustai. Health 2 (1), 160–166.

McKay, G., Ramprasad, G., Mowli, P.P., 1986. Equilibrium studies for the adsorption of dyestuffs from aqueous solutions by low-cost materials. Water Air Soil Pollut. 29 (3), 273–283.

Melgoza, R., Buitron, C., 2004. Anaerobic/aerobic treatment of colorants present in textile effluents. Water Sci. Technol. 50, 149–155.

Merzouk, B., Gourich, B., Madani, K., Vial, C., Sekki, A., 2011. Removal of a disperse red dye from synthetic wastewater by chemical coagulation and continuous electrocoagulation. A comparative study. Desalination 272 (1–3), 246–253.

Mirbolooki, H., Amirnezhad, R., Pendashteh, A.R., 2017. Treatment of high saline textile wastewater by activated sludge microorganisms. J. Appl. Res. Technol. 15 (2), 167–172.

Moghadam, F., Kohbanan, N.N., 2018. Removal of reactive blue 19 dye using Fenton from aqueous solution. Avi. J. Environ. Health Eng. 5 (1), 50–55.

Mohamed, W.S.E.D., 2016. Isolation and screening of reactive dye decolorizing bacterial isolates from textile industry effluent. Int. J. Microbiol. Res. 7 (1), 1–8.

Mohammed, A.E., Hamed, H.H., Alabdraba, W.S., Ali, O.M., 2019. COD removal from disperse blue dye 79 in wastewater by using Ozone-Fenton process. In: IOP Conference Series: Materials Science and Engineering, vol. 518, p. 062015 (6).

Mohan, S.V., Rao, N.C., Sarma, P.N., 2009. Simulated acid azo dye wastewater treatment using suspended growth configured sequencing batch reactor (SBR) under anoxic-aerobic-anoxic microenvironment. Appl. Ecol. Environ. Res. 7 (1), 25–34.

Moly, A., Shahul Hameed, A., 2016. Adsorption of vat dye from aqueous solution using agricultural solid waste (ASW) after UV lamp irradiation with TiO_2: isotherm and kinetic studies. Int. J. Adv. Res. Sci. Eng. 5 (7), 174–184.

Mortazavian, S., Saber, A., James, D.E., 2019. Optimization of photocatalytic degradation of acid blue 113 and acid red 88 textile dyes in a UV-C/TiO_2 suspension system: application of response surface methodology (RSM). Catalysts 9 (4), 360–379.

Mossvi, S., Kher, X., Madamar, D., 2007. Isolation, characterization and decoloration of textile dyes by a mixed bacterial consortium. Dyes Pigments 7 (3), 723–729.

Mousavi, A., Mokhtari, M., Khashij, M., 2017. Evaluating the Fenton process efficiency in removal of reactive red 2 from aqueous solution. J. Environ. Health aSustain. Dev. 2 (2), 292–299.

Mukherjee, A.K., Gupta, B., Chowdhury, S.M.S., 1999. Separation of dyes from cotton dyeing effluent using cationic polyelectrolytes. Am. Dyest. Report. 88, 25–28.

Muro, C., Riera, F., del Carmen Díaz, M., 2012. Membrane separation process in wastewater treatment of food industry. In: Valdez, B. (Ed.), Food Industrial Processes-Methods and Equipment. InTech, Crotia, pp. 253–280.

Mustapha, A., 2015. Colour removal technology using ozone in textile industrial wastewater effluent: and overview. Int. J. Innov. Scient. Eng. Technol. Res. 3, 45–51.

M'hamdi, A.I., Kandri, N.I., Zerouale, A., 2017. Adsorption study of the methylene blue on sawdust beech and red wood. J. Mater. Environ. Sci. 8 (8), 2816–2831.

Naje, A.S., Abbas, S.A., 2013. Electrocoagulation technology in wastewater treatment: a review of methods and applications. Civ. Environ. Res. 3 (11), 29–42.

Naje, A.S., Chelliapan, S., Zakaria, Z., Abbas, S.A., 2015. Treatment performance of textile wastewater using electrocoagulation process under combined electrical connection of electrodes. Int. J. Electrochem. Sci. 10, 5924–5941.

Nandi, B.K., Patel, S., 2017. Effects of operational parameters on the removal of brilliant green dye from aqueous solutions by electrocoagulation. Arab. J. Chem. 10, 2961–2968.

Naresh, B., Jaydip, J., Prabhat, B., Rajkumar, P., 2013a. Recent biological technologies for textile effluent treatment. Int. Res. J. Biol. Sci. 2 (6), 77–82.

Naresh, B., Preethi, C., Sneha, S., Bhagyashree, R., Parizad, P., 2013b. Microbial decolorization of disperse textile dye brown 21 by Enterobacter gergoviae isolated from textile effluent. Int. Res. J. Environ. Sci. 2, 31–36.

Nasser, N.M., El-Geundi, M., 1991. Comparative cost of color removal from textile effluents using natural adsorbents. J. Chem. Technol. Biotechnol. 50, 257–264.

Nawar, S.S., Doma, H.S., 1989. Removal of dyes from effluents using low cost agricultural by-products. Sci. Total Environ. 79, 271–279.

Nayl, A.E.A., Elkhashab, R.A., El Malah, T., Yakout, S.M., El-Khateeb, M.A., Ali, M.M., Ali, H.M., 2017. Adsorption studies on the removal of COD and BOD from treated sewage using activated carbon prepared from date palm waste. Environ. Sci. Pollut. Control Ser. 24 (28), 22284–22293.

Nemykin, V.N., Lukyanets, E.A., 2010. Synthesis of substituted phthalocyanines. Rev. Accounts 2010 (1), 136–208.

Ngo, A.C.R., Devanadera, M.K.P., Dedeles, G.R., 2016. Decolorization of selected synthetic textile dyes by yeasts from leaves and fruit peels. J. Health and Pollut. 6 (10), 42–55.

Nidheesh, P.V., Zhou, M., Oturan, M.A., 2018. An overview on the removal of synthetic dyes from water by electrochemical advanced oxidation processes. Chemosphere 197, 210–227.

Nigam, P., Armour, G., Banat, I.M., Singh, D., Marchant, R., 2000. Physical removal of textile dyes and solid state fermentation of dye-adsorbed agricultural residues. Bioresour. Technol. 72, 219–226.

Okwesili, J., Iroko, C., 2016. Urban solid waste management and environmental sustainability in Abakaliki Urban, Nigeria. Eur. Sci. J. 12 (23), 155–183.

Ola, A., Ahmed, E.N., Amany, E.S., Azza, K., 2005. Use of rice husk for adsorption of direct dyes from aqueous solution: a case study of direct F. Scarlet. Egyp. J. Aquat. Res. 31 (1), 1–11.

Olaganathan, R., Patterson, J., 2012. Biodegradation of textile dye anthraquinone vat blue 4 by *Pseudomonas aeruginosa*. Int. J. Adv. Biotechnol. Res. 3, 743–751.

Omar, S.A., 2016. Decolorization of different textile dyes by isolated Aspergillus Niger. J. Environ. Sci. Technol. 9 (1), 149–156.

Omar, H., El-Gendy, A., Al-Ahmary, K., 2018. Bioremoval of toxic dye by using different marine macroalgae. Turk. J. Bot. 42 (1), 15–27.

Onder, I., Aydin, O., Guvenc, A., 2018. Decolorization of reactive orange 16 and reactive black 5 in aqueous solution by ozonation. Environ. Res. Technol. 1 (4), 6–13.

Ouslimani, N., Boureghda, M.Z.M., 2018. Removal of directs dyes from wastewater by cotton fiber waste. Int. J. Waste Res. 8 (2), 330–338.

Özacar, M., Şengil, İ.A., 2002. Adsorption of acid dyes from aqueous solutions by calcined alunite and granular activated carbon. Adsorption 8 (4), 301–308.

Panswad, T., Techovanich, A., Anotai, J., 2001. Comparison of dye wastewater treatment by normal and anoxic+ anaerobic/aerobic SBR activated sludge processes. Water Sci. Technol. 43 (2), 355–362.

Papić, S., Koprivanac, N., Božić, A.L., Vujević, D., Dragičević, S.K., Kušić, H., Peternel, I., 2006. Advanced oxidation processes in azo dye wastewater treatment. Water Environment Research 78 (6), 572–579.

Parihar, A., Malaviya, P., 2013. Textile wastewater treatment using sawdust as adsorbent. Int. J. Environ. Sci. 2, 110–113.

Park, C.H., Lee, Y.R., Kim, T.H., Lee, B.H., Lee, J.W., Kim, S.Y., 2004. Decolorization of three acid dyes by enzymes from fungal strains. J. Microbiol. Biotechnol. 14 (6), 1190–1195.

Pasukphun, N., Vinitnantharat, S., Gheewala, S., 2010. Investigation of decolorization of textile wastewater in an anaerobic/aerobic biological activated carbon system (A/A BAC). Pakistan J. Biol. Sci. 13 (7), 316–324.

Patel, N.B., Soni, B.D., Ruparelia, J.P., 2010. Studies on removal of dyes from wastewater using electro-coagulation process. Nirma Univ. J. Eng. Technol. 1 (2), 20–25.

Paul, J., Rawat, K.P., Sarma, K.S.S., Sabharwal, S., 2011. Decoloration and degradation of Reactive Red-120 dye by electron beam irradiation in aqueous solution. Appl. Radiat. Isot. 69 (7), 982–987.

Pearce, C.I., Lloyd, J.R., Guthrie, J.T., 2003. The removal of colour from textile wastewater using whole bacterial cells. Dyes Pigments 58, 179–196.

Pereira, M.F.R., Soares, S.F., Orfao, J.J.M., Figueiredo, J.L., 2003. A desorption of dyes on activated carbons: influence of surface chemical groups. Carbon 41 (4), 811–821.

Pérez, A., Poznyak, T., Chairez, I., 2016. Effect of inorganic additives in the textile dyes removal by ozonation. In: Akcakoca Kumbasar, E., Korlu, A. (Eds.), Textile Wastewater Treatment. InTech, Crotia, pp. 55–73.

Petrinić, I., Andersen, N.P.R., Šostar-Turk, S., Le Marechal, A.M., 2007. The removal of reactive dye printing compounds using nanofiltration. Dyes Pigments 74 (3), 512–518.

Petsriprasit, C., Namboonmee, J., Hunsom, M., 2010. Application of the electrocoagulation technique for treating heavy metals containing wastewater from the pickling process of a billet plant. Kor. J. Chem. Eng. 27 (3), 854–861.

Phalakornkule, C., Polgumhang, S., Tongdaung, W., Karakat, B., Nuyut, T., 2010. Electrocoagulation of blue reactive, red disperse and mixed dyes, and application in treating textile effluent. J. Environ. Manag. 91 (4), 918–926.

Polat, D., 2010. Catalytic Ozonation of Industrial Textile Wastewaters in A Three Phase Fluidized Bed Reactor. Ph. D. Thesis. Middle East Technical University, Ankara, Turkey.

Poots, V.J.P., McKay, J.J., 1976. The removal of acid dye from effluent using natural adsorbents – I Peat. Water Res. 10, 1061–1066.

Pourbabaee, A.A., Malekzadeh, F., Sarbolouki, M.N., Najafi, F., 2006. Aerobic decolorization and detoxification of a disperse dye in textile effluent by a new isolate of Bacillus sp. Biotechnol. Bioeng. 93 (4), 631–635.

Prigione, V., Grosso, I., Tigini, V., Anastasi, A., Varese, G., 2012. Fungal waste-biomasses as potential low-cost biosorbents for decolorization of textile wastewaters. Water 4 (4), 770–784.

Przystas, W., Zablocka-Godlewska, E., Grabinska-Sota, E., 2015. Efficacy of fungal decolorization of a mixture of dyes belonging to different classes. Braz. J. Microbiol. 46 (2), 415–424.

Putri, R.A., Safni, S., Jamarun, N., Septiani, U., Kim, M.-K., Zoh, K.-D., 2019. Degradation and Mineralization of violet-3B Dye Using C-N-Codoped TiO_2 Photocatalyst, Environmental Engineering Research, in Press.

Puvaneswari, N., Muthukrishnan, J., Gunasekkaren, P., 2006. Toxicity assessment and microbial degradation of azo dyes. Indian J. Exp. Biol. 44, 618–626.

Raghavacharya, C., 1997. Colour removal from industrial effluents - a comparative review of available technologies. Chem. Eng. World 32, 53–54.

Raghu, S., Basha, C.A., 2007. Chemical or electrochemical techniques, followed by ion exchange, for recycle of textile dye wastewater. J. Hazard Mater. 149 (2), 324–330.

Rahman, Q.I., Ahmad, M., Misra, S.K., Lohani, M., 2013. Effective photocatalytic degradation of rhodamine B dye by ZnO nanoparticles. Mater. Lett. 91, 170–174.

Rai, H.S., Bhattacharyya, M.S., Singh, J., Bansal, T.K., Vats, P., Banerjee, U.C., 2005. Removal of dyes from the effluent of textile and dyestuff manufacturing industry: a review of emerging techniques with reference to biological treatment. Crit. Rev. Environ. Sci. Technol. 35 (3), 219–238.

Rai, P.B., Banerjee, S.S., Jayaram, R.V., 2007. Removal of disperse dyes from aqueous solution using sawdust and BDTDA-sawdust. J. Dispersion Sci. Technol. 28 (7), 1066–1071.

Rajaguru, P., Kalaiselvi, K., Palanivel, M., Subburam, V., 2000. Biodegradation of azo dyes in a sequential anaerobic–aerobic system. Appl. Microbiol. Biotechnol. 54 (2), 268–273.

Rajeswari, K., Subashkumar, R., Vijayaraman, K., 2011. Biodegradation of mixed textile dyes by bacterial strains isolated from dyewaste effluent. Res. J. Environ. Toxicol. 5 (2), 97–107.

Rani, B., Kumar, V., Singh, J., Bisht, S., Teotia, P., Sharma, S., Kela, R., 2014. Bioremediation of dyes by fungi isolated from contaminated dye effluent sites for bio-usability. Braz. J. Microbiol. 45 (3), 1055–1063.

Raval, N.P., Shah, P.U., Shah, N.K., 2017. Malachite green "a cationic dye" and its removal from aqueous solution by adsorption. Appl. Water Sci. 7 (7), 3407–3445.

Reeves, P., Ohlhausen, R., Sloan, D., Pamplin, K., Scoggins, T., Clark, C., Hutchinson, B., Green, D., 1992. Photocatalytic destruction of organic dyes in aqueous TiO_2 suspensions using concentrated simulated and natural solar energy. Sol. Energy 48 (6), 413–420.

Rehman, F., Murtaza, S., Ali Khan, J., Khan, H.M., 2017. Removal of crystal violet dye from aqueous solution by gamma irradiation. J. Chil. Chem. Soc. 62 (1), 3359–3364.

Reife, A., Freeman, H.S., 1996. Environmental Chemistry of Dyes Pigments. Wiley-VCH, New York.

Rezig, W., Hadjel, M., 2014. Photocatalytic degradation of Vat Green 03 textile dye, using the Ferrihydrite-Modified Diatomite with TiO_2/UV process. Orient. J. Chem. 30 (3), 993–1007.

Robinson, T., McMullan, G., Marchant, R., Nigam, P., 2001. Remediation of dyes in textiles effluent: a critical review on current treatment technologies with a proposed alternative. Bioresour. Technol. 77, 247–255.

Robinson, T., Chandran, B., Nigam, P., 2002. Effect of pretreatments of three waste residues, wheat straw, corncobs and barley husks on dye adsorption. Bioresour. Technol. 85 (2), 119–124.

Rodriguez, A., Garcia, J., Ovejero, G., Mestanza, M., 2009. Adsorption of anionic and cationic dyes on activated carbon from aqueous solutions: equilibrium and kinetics. J. Hazard Mater. 172 (2–3), 1311–1320.

Roy, D.C., Biswas, S.K., Saha, A.K., Sikdar, B., Rahman, M., Roy, A.K., Prodhan, Z.H., Tang, S.S., 2018. Biodegradation of Crystal Violet dye by bacteria isolated from textile industry effluents. PeerJ 6, 1–15.

Sadik, M.A., 2019. Removal of reactive dye from textile mill wastewater by leading electrocoagulation process using aluminum as a sacrificial anode. Adv. Chem. Eng. Sci. 9, 182–193.

Sahin, S., Altun, S., Altındal, A., Odabas, Z., 2015. Synthesis of novel azo-bridged phthalocyanines and their toluene vapour sensing properties. Sensor. Actuator. B Chem. 206, 601–608.

Saikhao, L., Setthayanond, J., Karpkird, T., Bechtold, T., Suwanruji, P., 2018. Green reducing agents for indigo dyeing on cotton fabrics. J. Clean. Prod. 197, 106–113.

Sala, M., López-Grimau, V., Gutiérrez-Bouzán, C., 2014. Photo-electrochemical treatment of reactive dyes in wastewater and reuse of the effluent: method optimization. Materials 7 (11), 7349–7365.

Salehi, M., Hashemipour, H., Mirzaee, M., 2012. Experimental study of influencing factors and kinetics in catalytic removal of methylene blue with TiO_2 nanopowder. Am. J. Environ. Eng. 2 (1), 1–7.

Saliba, R., Gauthier, H., Gauthier, R., Petit-Ramel, M., 2002. The use of eucalyptus barks for the adsorption of heavy metal ions and dyes. Adsorpt. Sci. Technol. 20 (2), 119–129.

Dos Santos, A.B., Cervantes, F.J., Lier, J.B., 2007. Review paper on current technologies for decolourisation of textile wastewaters: perspectives for anaerobic biotechnology. Bioresour. Technol. 98, 2369–2385.

Saranraj, P., 2013. Bacterial biodegradation and decolourization of toxic textile azo dyes. Afr. J. Microbiol. Res. 7 (30), 3885–3890.

Saranraj, P., Sumathi, V., Reetha, D., Stella, D., 2010. Fungal decolourization of direct azo dyes and biodegradation of textile dye effluent. J. Ecobiotechnol. 2 (7), 12–16.

Saratale, R.G., Saratale, G.D., Kalyani, D.C., Chang, J.S., Govindwar, S.P., 2009. Enhanced decolorization and biodegradation of textile azo dye Scarlet R by using developed microbial consortium-GR. Bioresour. Technol. 100 (9), 2493–2500.

Saroj, D.P., Kumar, A., Bose, P., Tare, V., Dhopavkar, Y., 2005. Mineralization of some natural refractory organic compounds by biodegradation and ozonation. Water Res. 39 (9), 1921–1933.

Schmitt, S., de Souza, R., Bettin, F., Dillon, A.J.P., Valle, J.A.B., Andreaus, J., 2012. Decolorization of aqueous solutions of disperse textile dyes by oxidoreductases. Biocatal. Biotransform. 30 (1), 48–56.

Schrank, S.G., dos Santos, J.N.R., Souza, D.S., Souza, E.E.S., 2007. Decolourisation effects of Vat Green 01 textile dye and textile wastewater using H_2O_2/UV process. J. Photochem. Photobiol. Chem. 186 (2–3), 125–129.

Secula, M.S., Cagnon, B., Cretescu, I., Diaconu, M., Petrescu, S., 2011. Removal of an acid dye from aqueous solutions by adsorption on a commercial granular activated carbon: equilibrium, kinetic and thermodynamic study. Scient. Study Res. Chem. Chem. Eng. Biotechnol. Food Ind. 12 (4), 307–322.

Seey, T.L., Kassim, M.J.N.M., 2012. Acidic and basic dyes removal by adsorption on chemically treated mangrove barks. Int. J. Appl. 2 (3), 270–276.

Segneanu, A.E., Orbeci, C., Lazau, C., Sfirloaga, P., Vlazan, P., Bandas, C., Grozescu, I., 2013. Waste water treatment methods. In: Elshorbagy, W., Chowdhury, R.K. (Eds.), Water Treatment. InTech, Crotia, pp. 53–80.

Semeraro, P., Fini, P., D'Addabbo, M., Rizzi, V., Cosma, P., 2017. Removal from wastewater and recycling of azo textile dyes by alginate-chitosan beads. Int. J. Environ. Agricul. Biotechnol. 2 (4), 1835–1850.

Senior, C., Clarke, D.A., 1986. Dyeing with sulphur dyes. In: Preston, C. (Ed.), The Dyeing of Cellulosic Fibers. Dyers Company Publications Trust, Bradford, UK, pp. 256–289.

Sennaoui, A., Alahiane, S., Sakr, F., Assabbane, A., Addi, E.H.A., Hamdani, M., 2018. Advanced oxidation of reactive yellow 17 dye: a comparison between Fenton, photo-fenton, electro-fenton, anodic oxidation and heterogeneous photocatalysis processes. Port. Electrochim. Acta 36 (3), 163–178.

SenthilKumar, P., Umaiyambika, N., Gayathri, R., 2010. Dye removal from aqueous solution by electrocoagulation process using stainless steel electrodes. Environ. Eng. Manag. J. 9 (8), 1031–1037.

Seow, T.W., Lim, C.K., 2016. Removal of dye by adsorption: a review. Int. J. Appl. Eng. Res. 11 (4), 2675–2679.

Sepúlveda, L., Fernández, K., Contreras, E., Palma, C., 2004. Adsorption of dyes using peat: equilibrium and kinetic studies. Environ. Technol. 25 (9), 987–996.

Seshadri, S., Bishop, P.L., Agha, A.M., 1994. Anaerobic/aerobic treatment of selected azo dyes in wastewater. Waste Manag. 14 (2), 127–137.

Sevimli, M.F., Kinaci, C., 2002. Decolorization of textile wastewater by ozonation and Fenton's process. Water Sci. Technol. 45 (12), 279–286.

Shah, M.P., 2014a. Microbial degradation and decolorization of acid orange dye by anaerobic/aerobic sequential process. Int. J. Environ. Bioremed. Biodegrad. 2 (3), 112–116.

Shah, M., 2014b. Effective treatment systems for azo dye degradation: a joint venture between physico-chemical & microbiological process. Int. J. Environ. Bioremed. Biodegrad. 2 (5), 231–242.

Shah, M.P., Patel, K.,A., Nair, S.S., Darji, A.M., 2014. Microbial decolorization of remazol brilliant orange 3R, remazol black B & remazol brilliant violet dyes in a sequential anaerobic-aerobic system cite this article. Internat. J. Environ. Bioremed. Biodegrad. 1, 6–13.

Shahmoradi, B., Maleki, A., Byrappa, K., 2011. Photocatalytic degradation of Amaranth and Brilliant Blue FCF dyes using in situ modified tungsten doped TiO_2 hybrid nanoparticles. Catal. Sci. Technol. 1, 1216–1223.

Sharma, S., Kaur, A., 2018. Various methods for removal of dyes from industrial effluents - a review. Ind. J. Sci. Technol. 11 (12), 1–21.

Sharma, P., Singh, L., Dilbaghi, N., 2009. Optimization of process variables for decolorization of Disperse Yellow 211 by Bacillus subtilis using Box–Behnken design. J. Hazard Mater. 164 (2–3), 1024–1029.

Sharma, P., Dhanwantri, K., Mehta, S., 2014a. Solid waste management. Int. J. Environ. Res. Dev. 4 (1), 49–54.

Sharma, S., Saxena, R., Gaur, G., 2014b. Study of removal techniques for azo dyes by biosorption: a review. J. Appl. Chem. 7 (10), 06–21.

Shindy, A., 2017. Problems and solutions in colors, dyes and pigments chemistry: a review. Chem. Int. 3 (2), 97–105.

Shirsath, D.S., Shrivastava, V.S., 2012. Removal of hazardous dye Ponceau-S by using chitin: an organic bioadsorbent. Afr. J. Environ. Sci. Technol. 6 (2), 115–124.

Singh, K., Arora, S., 2011. Removal of synthetic textile dyes from wastewaters: a critical review on present treatment technologies. Crit. Rev. Environ. Sci. Technol. 41 (9), 807–878.

Singh, B.K., Rawat, N.S., 1994. Comparative sorption equilibrium studies of toxic phenols on fly ash and impregnated fly ash. J. Chem. Technol. Biotechnol. 61 (4), 307–317.

Sirianuntapiboon, S., Saengow, W., 2004. Removal of vat dyes from textile wastewater using biosludge. Water Qual. Res. J. 39 (3), 276–284.

Sirianuntapiboon, S., Srisornsak, P., 2007. Removal of disperse dyes from textile wastewater using bio-sludge. Bioresour. Technol. 98 (5), 1057–1066.

Sirianuntapiboon, S., Chairattanawan, K., Jungphungsukpanich, S., 2006. Some properties of a sequencing batch reactor system for removal of vat dyes. Bioresour. Technol. 97 (10), 1243–1252.

Siti Zuraida, M., Nurhaslina, C.R., Ku Halim, K.H., 2013. Removal of synthetic dyes from wastewater by using bacteria, lactobacillus delbruckii. Int. Ref. J. Eng. Sci. 2 (5), 1–7.

Sivamani, S., Leena, G.B., 2009. Removal of dyes from wastewater using adsorption-a review. Int. J. Biosci. Technol. 2 (4), 47–51.

Šmelcerović, M., Đorđević, D., Novaković, M., Mizdraković, M., 2010. Decolorization of a textile vat dye by adsorption on waste ash. J. Serb. Chem. Soc. 75 (6), 855–872.

Sneha, J., Gomashe, A.V., Sonal, A., 2014. Decolorization potential of Bacillus sp. for removal of synthetic textile dyes. Int. J. Curr. Microbiol. Appl. Sci. 3 (12), 83–88.

Sohrabi, M.R., Khavaran, A., Shariati, S., Shariati, S., 2017. Removal of Carmoisine edible dye by Fenton and photo Fenton processes using Taguchi orthogonal array design. Arab. J. Chem. 10, 3523–3531.

Solpan, D., Torun, M., Güven, O., 2007. Effect of gamma-irradiation on aqueous solutions of Apollofix dyes. Nukleonika 52 (3), 109–113.

Song, L., Shao, Y., Ning, S., Tan, L., 2017. Performance of a newly isolated salt-tolerant yeast strain Pichia occidentalis G1 for degrading and detoxifying azo dyes. Bioresour. Technol. 233, 21–29.

Sponza, D.T., Işik, M., 2002. Decolorization and azo dye degradation by anaerobic/aerobic sequential process. Enzym. Microb. Technol. 31 (1–2), 102–110.

Sponza, D., Isik, M., 2004. Decolorization and inhibition kinetic of Direct Black 38 azo dye with granulated anaerobic sludge. Enzym. Microb. Technol. 34, 147–158.

Sripiboon, S., Suwannahong, K., 2018. Color removal by ozonation process in biological wastewater treatment from the breweries. IOP Conf. Ser. Earth Environ. Sci. 167 (1), 1–6.

Sriram, N., Reetha, D., 2015. Isolation and characterization of dye degrading bacteria from textile dye effluents. Cent. Europ. J. Exp. Biol. 4 (2), 5–10.

Stolz, A., 2001. Basic and applied aspects in the microbial degradation of azo dyes. Appl. Microbiol. Biotechnol. 56 (1–2), 69–80.

Strickland, A.F., Perkins, W.S., 1995. Decolorization of continuous dyeing wastewater by ozonation. Text. Chem. Colorist 27 (5), 11–15.

Sultana, B., 2017. Degradation of Dyes Used in Textile Industries by Bacteria Isolated from Local Textile Effluents. Ph. D. Thesis. BRAC University, Dhaka Bangladesh.

Sundararaman, T.R., Ramamurthi, V., Partha, N., 2009. Decolorization and COD removal of reactive yellow 16 by Fenton oxidation and comparison of dye removal with photo Fenton and sono Fenton process. Mod. Appl. Sci. 3 (8), 15–22.

Suresh, S., 2016. Removal of Basic Violet 14 from aqueous solution using sulphuric acid activated materials. SpringerPlus 5 (1), 633–647.

Szpyrkowicz, L., Juzzolino, C., Kaul, S.N., 2001. A comparative study on oxidation of disperse dyes by electrochemical process, ozone, hypochlorite and Fenton reagent. Water Res. 35 (9), 2129–2136.

Talaiekhozani, A., Heydari Chaleshtori, A., Banisharif, F., Eskandari, Z., Nasiri, M., Aminsharei, F., Park, J., Rezania, S., Bazrafshan, M., 2019. Removal of Acid Orange 7 dye from wastewater using combination of ultraviolet radiation, ultrasonic method, and MgO nanoparticles. Environ. Health Eng. Manag. J. 6 (3). In (Press).

Talarposhti, A.M., Donnelly, T., Anderson, G.K., 2001. Colour removal from a simulated dye wastewater using a two-phase anaerobic packed bed reactor. Water Res. 35 (2), 425–432.

Tang, W.Z., An, H., 1995. UV/TiO_2 photocatalytic oxidation of commercial dyes in aqueous solutions. Chemosphere 31 (9), 4157–4170.

Tehrani, M.M., Assadi, M.M., Rashedi, H., 2015. Biodecolorization of textile effluents by autochthonous fungi. J. Appl. Biotechnol. Rep. 1 (4), 161–165.

Telke, A., Kalyani, D., Jadhav, J., Govindwar, S., 2008. Kinetics and mechanism of reactive red 141 degradation by a bacterial isolate rhizobium radiobacter MTCC 8161. Acta Chim. Slov. 55 (2), 320–329.

Thakur, S., Chauhan, M.S., 2016. Removal of malachite green dye from aqueous solution by Fenton oxidation. Int. Res. J. Eng. Technol. 3, 254–259.

Theng, B.K.G., Wells, N., 1995. Assessing the capacity of some New Zealand clays for decolorizing vegetable oil and butter. Appl. Clay Sci. 9, 321–326.

Thitame, P.V., Shukla, S.R., 2016. Adsorptive removal of reactive dyes from aqueous solution using activated carbon synthesized from waste biomass materials. Int. J. Environ. Sci. Technol. 13 (2), 561–570.

Tien, T.T., Linh, D.H., Vu, L.T., Hoa, P.T.T., Phuong, N.T.T., Le Luu, T., 2017. Electrochemical water treatment technology in Viet Nam: achievement & future development. Sci. J. Chem. 5 (6), 87–94.

Tölek, Ş., Kadem, F.D., 2016. An investigation on colour analysis and fastness properties of the denim fabric dyed with A different method. Textile Apparel 26 (2), 198–204.

Tosik, R., 2005. Dyes color removal by ozone and hydrogen peroxide: some aspects and problems. Ozone: Sci. Eng. 27 (4), 265–271.

Trovó, A.G., Hassan, A.K., Sillanpää, M., Tang, W.Z., 2016. Degradation of acid blue 161 by Fenton and photo-fenton processes. Int. J. Environ. Sci. Technol. 13 (1), 147–158.

Turhan, K., 2018. Decolorization of Reactive Yellow 17 in aqueous solution by ozonation in a semi-batch bubble column reactor. J. Turk. Chem. Soc. 5 (2), 835–844.

Turhan, K., Durukan, I., Ozturkcan, S.A., Turgut, Z., 2012. Decolorization of textile basic dye in aqueous solution by ozone. Dyes Pigments 92 (3), 897–901.

Uday, U.S.P., Bandyopadhyay, T.K., Bhunia, B., 2016. Bioremediation and detoxification technology for treatment of dye (s) from textile effluent. In: Kumbasar, E.A. (Ed.), Textile Wastewater Treatment. Crotia: InTech, pp. 75–92.

Uddin, A.B.M.H., Sujari, A.N.A., Asri, M.M.N., 2003. Effectiveness of peat coagulant for the removal of textile dyes from aqueous solution and textile wastewater. Malay. J. Chem. 5 (1), 34–43.

Ul-Islam, S., 2019. Intergrating Green Chemistry and Sustainable Engineering. John Wiley & Sons, USA.

Ünlü, M., 2008. Indıgo Dyeıng Wastewater Treatment by the Membrane Based Fıltratıon Process. Master Thesis. Middle East Technical University, Ankara, Turkey.

Uysal, Y., Bilgiç, M., 2018. Color removal from wastewater by using two-step (biological and chemical) aerobic filter reactors. Global NEST J. 20 (1), 7–13.

Vaez, M., Moghaddam, A.Z., Alijani, S., 2012. Optimization and modeling of photocatalytic degradation of azo dye using a response surface methodology (RSM) based on the central composite design with immobilized titania nanoparticles. Ind. Eng. Chem. Res. 51, 4199–4207.

Van der Zee, F.P., Lettinga, G., Field, J.A., 2001. Azo dye decolourisation by anaerobic granular sludge. Chemosphere 44 (5), 1169–1176.

Vankar, P.S., 2000. Chemistry of natural dyes. Resonance 5 (10), 73–80.

Verma, A.K., Dash, R.R., Bhunia, P., 2012. A review on chemical coagulation/flocculation technologies for removal of colour from textile wastewaters. J. Environ. Manag. 93, 154–168.

Victral, D.M., Aquino, S.F.D., Silva, S.D.Q., Baeta, B.E.L., 2016. Application of residual yeast as a source of redox mediators for the anaerobic decolorization of a model azo dye. Braz. J. Chem. Eng. 33 (4), 705–711.

Vidal, J., Espinoza, C., Contreras, N., Salazar, R., 2017. Elimination of industrial textile dye by electrocoagulation using iron electrodes. J. Chil. Chem. Soc. 62 (2), 3519–3524.

Vishnu, G., Joseph, K., 2007. Nanofiltration and ozonation for decolorisation and salt recovery from reactive dyebath. Color. Technol. 123 (4), 260–266.

Vital, R.K., Saibaba, K.V.N., Shaik, K.B., Gopinath, R., 2016. Dye removal by adsorption: a review. J. Biorem. Biodegrad. 7 (6), 371–374.

Wali, F.K., 2015. Color removal and COD reduction of dyeing bath wastewater by Fenton reaction. Int. J. Waste Res. 5 (1), 1–6.

Wang, C., Yin, Y., 2013. Preparation, characterization and application of ultra-fine modified pigment in textile dyeing. In: Gunay, M. (Ed.), Eco-Friendly Textile Dyeing and Finishing. InTech, Crotia, pp. 77–101.

Wang, Z., Xue, M., Huang, K., Liu, Z., 2011. Textile dyeing wastewater treatment. Adv. Treat. Textile Effl. 5, 91–116.

Wawrzkiewicz, M., 2013. Removal of C.I. Basic Blue 3 dye by sorption onto cation exchange resin, functionalized and non-functionalized polymeric sorbents from aqueous solutions and wastewaters. Chem. Eng. J. 217, 414–425.

Wawrzkiewicz, M., 2014. Anion-exchange resins for CI Direct Blue 71 removal from aqueous solutions and wastewaters: effects of basicity and matrix composition and structure. Ind. Eng. Chem. Res. 53 (29), 11838–11849.

Wawrzkiewicz, M., Hubicki, Z., 2011. Remazol Black B removal from aqueous solutions and wastewater using weakly basic anion exchange resins. Open Chem. 9 (5), 867–876.

Wawrzyniak, B., Morawski, A.W., 2006. Solar-light-induced photocatalytic decomposition of two azo dyes on new TiO_2 photocatalyst containing nitrogen. Appl. Catal. B Environ. 62, 150–158.

Wei, M.-S., Huang, K.-H., 2001. Recycling and reuse of industrial wastes in Taiwan. Waste Manag. 21, 93–97.

Wenten, I.G., 2002. Recent development in membrane science and its industrial applications. Songklanakarin J. Sci. Technol. 24 (Suppl. 1), 1010–1024.

Wijannarong, S., Aroonsrimorakot, S., Thavipoke, P., Sangjan, S., 2013. Removal of reactive dyes from textile dyeing industrial effluent by ozonation process. APCBEE Proc. 5, 279–282.

Wu, J.S., Liu, C.H., Chu, K.H., Suen, S.Y., 2008. Removal of cationic dye methyl violet 2B from water by cation exchange membranes. J. Membr. Sci. 309 (1–2), 239–245.

Xu, X.R., Li, H.B., Wang, W.H., Gu, J.D., 2004. Degradation of dyes in aqueous solutions by the Fenton process. Chemosphere 57 (7), 595–600.

Xu, W.G., Liu, S.F., Lu, S.X., Kang, S.Y., Zhou, Y., Zhang, H.F., 2010. Photocatalytic degradation in aqueous solution using quantum-sized ZnO particles supported on sepiolite. J. Colloid Interface Sci. 351, 210–216.

Xuan Nguyen, H., Van der Bruggen, B., 2015. Nanofiltration of synthetic and industrial dye baths: influence of temperature on rejection and membrane fouling. J. Memb. Sci. Res. 1 (1), 34–40.

Yang, C.L., McGarrahan, J., 2005. Electrochemical coagulation for textile effluent decolorization. J. Hazard Mater. 127 (1–3), 40–47.

Yang, S.O., Sodaneath, H., Lee, J.I., Jung, H., Choi, J.H., Ryu, H.W., Cho, K.S., 2017. Decolorization of acid, disperse and reactive dyes by *Trametes versicolor* CBR43. J. Environ. Sci. Health. 52 (9), 862–872.

Yari, A.R., Alizadeh, M., Hashemi, S., Biglari, H., 2013. Efficiency of electrocoagulation for removal of reactive yellow 14 from aqueous environments. Arch. Hyg. Sci. 2 (1), 7–15.

Yavuz, Ö., Aydin, A.H., 2006. Removal of direct dyes from aqueous solution using various adsorbents. Pol. J. Environ. Stud. 15 (1), 155–161.

Yemendzhiev, H., Alexieva, Z., Krastanov, A., 2009. Decolorization of synthetic dye reactive blue 4 by mycelial culture of white-rot fungi *Trametes versicolor* 1. Biotechnol. Biotechnol. Equip. 23 (3), 1337–1339.

Yildirim, S.C., Yesilada, O., 2015. A comparative study on decolorization of reactive azo and indigoid dyes by free/immobilized pellets of *Trametes versicolor* and Funalia trogii. J. Environ. Biol. 36 (6), 1393–1400.

Yoo, E.S., Libra, J., Adrian, L., 2001. Mechanism of decolorization of azo dyes in anaerobic mixed culture. J. Environ. Eng. 127 (9), 844–849.

Yoshida, H., Takemori, T., 1997. Adsorption of direct dye on cross-linked chitosan. Water Sci. Technol. 35, 29–37.

Yue, Q.Y., Li, Q., Gao, B.Y., Wang, Y., 2007. Kinetics of adsorption of disperse dyes by poly-epicholorohydrin-dimethylamine cationic polymer/bentonite. Separ. Purif. Technol. 54 (3), 279–290.

Yuzhu, F., Viraraghavan, T., 2001. Fungal decolorization of dye wastewaters: a review. Bioresour. Technol. 79, 251–262.

Zahrim, A.Y., Hilal, N., 2013. Treatment of highly concentrated dye solution by coagulation/flocculation–sand filtration and nanofiltration. Water Res. Ind. 3, 23–34.

Zimmermann, T., Kulla, H.G., Leisinger, T., 1982. Properties of purified Orange II azoreductase, the enzyme initiating azo dye degradation by Pseudomonas KF46. Eur. J. Biochem. 129 (1), 197–203.

Zinadini, S., Moradi, M., Zinatizadeh, A.A.L., 2016. Influence of operating variables on performance of nanofiltration membrane for dye removal from synthetic wastewater using response surface methodology. Int. J. Eng. 29 (12), 1650–1658.

Zodi, S., Merzouk, B., Potier, O., Lapicque, F., Leclerc, J.P., 2013. Direct red 81 dye removal by a continuous flow electrocoagulation/flotation reactor. Separ. Purif. Technol. 108, 215–222.

Chapter 6

Membrane processes for removal of polyaromatic hydrocarbons from wastewater

S.R. Barman[a,b], U. Roy[c], P. Das[c], A. Mukhopadhayay[b]
[a]*Department of Environmental Science, Asutosh College, Kolkata, West Bengal, India;*
[b]*Department of Environmental Science, University of Calcutta, Kolkata, West Bengal, India;*
[c]*Department of Chemical Engineering, Jadavpur University, Kolkata, West Bengal, India*

1. Introduction

Pollution of waterbodies with various organic and inorganic substances has become an increasing concern in the present time. Effluents from various sources are mostly discharged into waterbodies resulting in building up of pollutants (Sun et al., 2017). Moreover, leachates from landfill containing toxic organic pollutants further add to pollution load of waterbodies (Paxéus, 2000). Increasing levels of emergent persistent pollutants such as polyaromatic hydrocarbons (PAHs) in the aquatic environment are adversely affecting its flora and fauna. PAHs are a group of ubiquitous lipophilic pollutants with fused aromatic rings in their molecular structure (Barman et al., 2018; Behera et al., 2018). With increasing number of aromatic rings, the toxicity of PAHs increases (Kuppusamy et al., 2017).

PAHs enter the environment mainly due to incomplete combustion of carboncontaining fuels, indiscriminate use of pesticides, and oil spills (Barman et al., 2017; Behera et al., 2018). Discharges from industries like paper making, wood preserving, oil refining, and dye production are responsible for high PAH concentrations in wastewater (Gong et al., 2017; Barman et al., 2017). Factors such as the weather conditions, type of industry, and effectiveness of wastewater treatment plant determine the level of PAH in water in an area. The average PAH load in industrial wastewater is estimated to be approximately 100 μg/L (Smol and Włodarczyk-Makuła, 2012) which may, however, vary due to the type of industry.

Green Chemistry and Water Remediation: Research and Applications.
https://doi.org/10.1016/B978-0-12-817742-6.00006-2

PAHs are sparingly soluble in water, and being a persistent pollutant, they can remain in water for a long period of time (Gutiérrez et al., 2018). PAHs from aquatic environment can enter the food chain and bioaccumulate causing a great deal of damage due to their carcinogenic and mutagenic properties (Li et al., 2019).

PAHs often interact with other compounds available in the environment and produce even more toxic derivatives (Qiao et al., 2019; Wlodarczyk-Makula, 2005).

In the case of their presence in the environment, PAHs may react with other components of the matrix, which can lead to the formation of their derivatives, often more toxic.

It has been observed that even at very low concentrations, continuous exposure to PAH can cause considerable damage to tissues leading to harmful health effects (Nwaichi and Ntorgbo, 2016). Due to their lipophilic property, PAHs can get absorbed from the intestinal tract (Barman et al., 2017) and can cross membrane barriers (Celebioglu et al., 2019). They can form DNA adducts which ultimately lead to carcinogenesis (Pratt et al., 2011). Considering the health impact associated with PAHs, the USEPA has listed 16 of them as priority pollutant (Barman et al., 2018). The molecular structure and physical attributes of these 16 PAHs are listed in Table 6.1. The maximum permissible concentration of PAH in industrial wastewater that can be discharged into the sewage system is approximately around 0.2 mg C/L (Smol and Włodarczyk-Makuła, 2012; Smol et al., 2014).

Thus, it can be concluded that the mitigation of improperly disposed PAH-containing effluent can be a costly affair for many industries (Zeledon-Toruno et al., 2007). It is crucial to develop cost-effective and efficient techniques for the elimination of PAHs from various types of wastewater before they are discharged.

The removal of PAHs from wastewater can be achieved by biological decomposition as well as physical processes. Very often these processes are combined for generating better results (Pérez-Gregorio et al., 2010). PAH degradation using microorganisms such as bacteria fungi and algae has been extensively studied in the past (Smol and Włodarczyk-Makuła, 2017; Barman et al., 2018). These biological processes, however, have their drawbacks especially when industrial wastewater is considered. The high salinity or other attribute of industrial wastewater may become toxic for the microorganism which ultimately results in the process being rendered ineffective.

Adsorption techniques using various materials have been conducted widely in the past. Activated carbon and biochars have been used extensively as adsorbents. However, in recent times, the use of various membranes for filtration of water has received popularity. Membrane techniques have been broadly applied in the wastewater treatment because of its ability in terms of its structural integrity and stability, hydrophobicity, hydrophilicity, permeability (higher as well as lower), less energy requirements, and mechanical and

TABLE 6.1 Properties and molecular structure of the 16 PAHs enlisted as priority pollutant by the US EPA.

Order	Compound name	Abbreviation	CAS number	Formula	Mol wt(g/mol $^{-1}$)	Structure
1	Naphthalene	Nap	91-20-3	$C_{10}H_8$	128.17	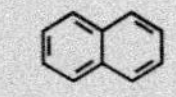
2	Acenaphthylene	AcPy	208-96-8	$C_{12}H_8$	152.19	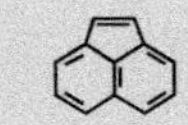
3	Acenaphthene	Acp	83-32-9	$C_{12}H_{10}$	154.21	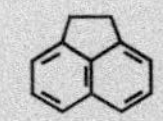
4	Fluorene	Flu	86-73-7	$C_{13}H_{10}$	166.22	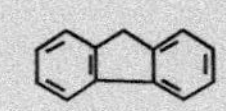
5	Phenanthrene	PA	85-01-8	$C_{14}H_{10}$	178.23	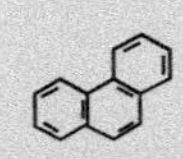
6	Anthracene	Ant	120-12-7	$C_{14}H_{10}$	178.23	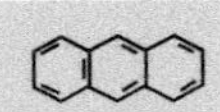

Continued

TABLE 6.1 Properties and molecular structure of the 16 PAHs enlisted as priority pollutant by the US EPA.—cont'd

Order	Compound name	Abbreviation	CAS number	Formula	Mol wt(g/mol $^{-1}$)	Structure
7	Fluoranthene	FL	206-44-0	$C_{16}H_{10}$	202.25	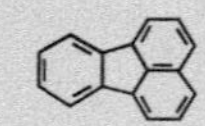
8	Pyrene	Pyr	129-00-0	$C_{16}H_{10}$	202.25	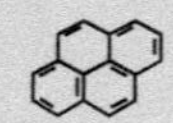
9	Benzo[a]anthracene	BaA	56-55-3	$C_{18}H_{12}$	228.29	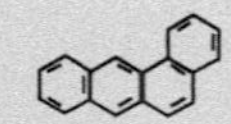
10	Chrysene	CHR	218-01-9	$C_{18}H_{12}$	228.29	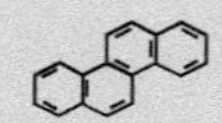
11	Benzo[b]fluoranthene	BbF	205-99-2	$C_{20}H_{12}$	252.31	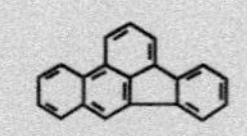
12	Benzo[k]fluoranthene	BkF	207-08 9	$C_{20}H_{12}$	252.31	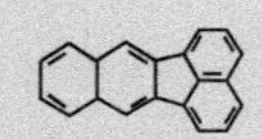

13	Benzo[a]pyrene	BaP	50-32-8	$C_{20}H_{12}$	252.31	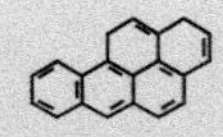
14	Indeno[l,2,3-cd]pyrene	IND	193-39-5	$C_{22}H_{12}$	276.33	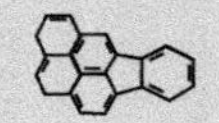
15	Dibenzo[a,h]anthracene	DBA	3-0-3	$C_{22}H_{14}$	278.35	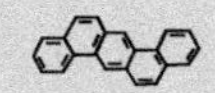
16	Benzo[g,h,i]perylene	BghiP	191-24-2	$C_{22}H_{12}$	276.33	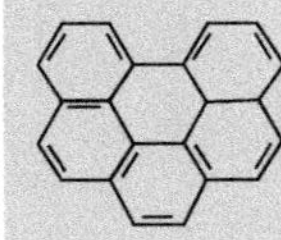

Adapted from Ke, S., Liu, Q., Yao, Y., Zhang, X., Sui, G., 2018. An in vitro cytotoxicities comparison of 16 priority polycyclic aromatic hydrocarbons in human pulmonary alveolar epithelial cells HPAEpiC. Toxicol. Lett. 290, 10–18.

thermal stability (Fan et al., 2008; Rajesh et al., 2013). Membrane filtration may be ultafiltration, microfiltration, or nanofiltration depending on the characteristic of the water being treated and the particular pollutant needing removal (Zularisam et al., 2006). In a membrane filtration system, the basic idea is that the membrane acts as a specific sieve, allowing only molecules of a particular size, corresponding to the pore diameter, to pass through. The selectivity of the membrane is chosen with respect to the pollutant under concern and its average molecular size.

2. Polyaromatic hydrocarbon remediation using membrane techniques

The elimination of PAHs from aqueous solutions applying various membrane techniques to reduce treatment time as well as increase cost-effectiveness is being evaluated extensively in recent times. Several studies reported that a variety of commercially accessible ultrafiltration (UF) membranes, nanofiltration membrane, low-pressure reverse osmosis (RO) membranes, RO membrane, etc., can be applied to eliminate PAHs from wastewaters (Gong et al., 2017; Bodzek and Konieczny, 2011; Kowal and Świderska-Bróż, 2007; Bodzek and Konieczny, 2006).

UF membrane processes are widely used for the removal of organic contaminants like PAHs, trihalomethanes, phthalates, etc., from industrial effluents. Industrial UF membranes are increasingly utilized in order to choose the best PAHs removal approach because of their thermal and structural stability, lower processing time, and energy efficiency.

The basic applications of UF membrane techniques include treatment of natural water, brackish water, refining of macromolecular solutions, and industrial effluents contaminated with oil derivatives (Dudziak et al., 2003). The PAH concentration of raw wastewater nearing about 240 μg/L has been successfully removed through filtration and UF techniques to up to almost 78% in a work by Włodarczyk-Makuła et al. (2011).

The application of UF membranes for PAH removal in presence of humic materials has indicated that this combination has higher rejection efficiency as compared with UF alone. It was found that anthracene removal via UF was increased up to 97% in presence of humic material. (Smol and Wlodarczyk-Makula, 2012). Dudziak et al. (2003) also established in their work that UF technique was able to remove more than 80% of PAHs from wastewaters. UF membranes are usually classified by the molecular weight cut off (MWCO) value which can be described as the lowest molecular weight that is 90% retained by the membrane pores. Most UF membranes generally have a MWCO values ranging between 1000 and 100,000 Da (Mielczarek et al., 2011; Dudziak et al., 2003). The properties of these membranes include high hydraulic efficacy, proficient assorting properties, and high resistance to different factors rendering UF membranes more suitable for application in the removal of PAHs (Smol and Wlodarczyk-Makula, 2012).

It has been found that even when the molecular weight of the PAH under study is lesser than the UF membrane cut off, rejection percentage is high. This result may be explained by adsorption of PAH compounds onto the surface of the membrane (Smol and Wlodarczyk-Makula, 2017).

Modification of UF membrane to enhance adsorption ability is another widely studied avenue of contemporary research. Mntambo et al. (2019) formulated a C18-incorporated UF membrane which was found to be able to remove higher percentage of PAHs as compared with unmodified UF membranes. Gong et al. (2017) in another study coupled electrocoagulation techniques with membrane filtration to achieve almost 90% removal of PAHs from wastewater.

Pressure-dependent filtration processes such as RO had previously attained huge popularity in sustainable wastewater treatment. Conventional RO has been efficient in removing various organic as well as inorganic contaminants, including PAHs, from wastewater, but at a comparatively higher filtration pressures. Energy-efficient low-pressure RO membranes were later designed and formulated to reduce operational cost without compromising its permeability and filtration efficiency (Gong et al., 2017).

Submerged membrane bioreactor (MBR) is an alternative membrane application for PAH removal from wastewater. This technology incorporates the biologically activated sludge method with physical membrane separation of the solid or liquid thereby eliminating the need for a previous sedimentation phase. Thus, an innovative system has evolved that merges first, second, and third treatment steps, permitting the construction of more concise plants capable of forming reusable effluent (Wisniewski, 2007). In such system, various configurations may be utilized regarding final water quality. Therefore, mechanisms for removing PAHs in MBR can defer from traditional activated sludge (González et al., 2012).

Sorption of PAHs mainly depends on usage of hydrophobic substances that can interact with the PAHs. On this context, Eeshwarasinghe et al., in a study, utilized a granular activated carbon (GAC) for elimination of aqueous PAHs within few hours (Eeshwarasinghe et al., 2018). Recently, the development of PAH removal systems occurs through the mechanism of intercalation, inclusion complexation, $\pi-\pi$ interaction, etc., In this regard, Topuz et al. described a bioinspired model of DNA nanogel scavenger that interacts with PAHs and subsequently removes them from wastewater (Topuz et al., 2016). Short dispersion pathway and interconnected DNA molecules provided a rapid and sufficient removal of PAHs in this method. Furthermore, in this system, the removal approach does not depend on surface interaction, but on absolute swollen network (Eeshwarasinghe et al., 2018).

Nanofiltration treatment is also a widely accepted process of organic micropollutants removal from potable water. Due to its higher efficiency of PAH removal and comparatively lower removal of monovalent salts, it can be efficiently used in water treatment. Size exclusion is considered the major mechanism of the PAH removal using nanofiltration (Zhang et al., 2012). Several research works have applied polyamide nanofiltration or RO membrane for removal of PAHs (Ren et al., 2018; Pal and Kumar 2013). In a study by Wu et al. (2016b), the potency of membrane nanofiltration and ozonation (O_3) was tested for purification of drinking water contaminated with micropollutants. It was reported that membrane filters had higher efficiency in removing trace amount of PAHs from water as compared with the O_3 technique. Moreover, it was also found that the organic compounds with higher molecular weight had higher removal rate by nanofiltration as compared with lesser molecular weight organic compound.

Carbon nanotube (CNT)-based membranes are another widely suggested potential adsorbent for several organic compounds—detergents, fats, different pesticides, grease, volatile organic compounds, perchlorate, different medicines, coloring agents, and other organohalides as well as inorganic substances including heavy metals, ammonia, salts, etc, (Jame and Zhou, 2016; Upadhyayula et al., 2009). These membranes have also been reported as efficient adsorbing agents for biological components including bacteria, viruses, fungi, and algae. (Ali et al., 2014; Rao et al., 2007). Doped CNTs for the desalination of sea and brackish water have also been examined (Das et al., 2014a). Aligned CNTs can be used in engineered membranes as cone or pores (Das et al., 2014b). CNTs comprise of porous structure which are hydrophobic in nature, thus facilitating easy capture of PAHs while allowing smooth flow of water (Akinpelu et al., 2019). CNT membranes are also considered to be more energy efficient and they can also be specialized with incorporation of various functional groups to facilitate target-specific pollution removal.

2.1 Composition and formulation

Recent studies indicate that there is an incessant requirement of efficient and cheap membrane formulation for abatement of pollutants in wastewater treatment.

Basically membranes are composed of different biopolymers along with inorganic materials.

Organic compounds utilized for synthesis of different types of membranes include polysulfone (Chen et al., 2016), polyethersulfone (Idris et al., 2007), polyacrylonitrile, polyamide, polyimide (Wang et al., 2009), polyvinylidene fluoride (PVDF) (Mahlambi et al., 2014), and polytetrafluorethylene (Grevstad and Leknes, 1993). PVDF is the most frequently and effectively utilized organic polymers in recent decade. Moreover, Ma et al. (2017) and Shen et al. (2017) have also described the efficiency of PVDF membrane and its possible large-scale implementation.

Langmuir monolayer formed from bacterial phospholipids was studied by Kallimanis et al. (2007) in their work. This was then applied as model membranes for eliminating PAHs from wastewaters. The phospholipid found in membrane of *Arthrobacter* sp. (pyrene, and benzo(a)pyrene-degrading bacterial species) was utilized to design the filtration membranes. The major components of these bacteria-derived membrane were cardiolipin (65%), phosphatidylglycerol (30%), and phosphatidylethanolamine (5%) (Broniatowski and Urbaś, 2017).

In another study, Shevate et al. (2018) described an easy and eco-friendly way of synthesizing a nanoporous polysterene membrane using block copolymer as the parent material.

Various nanomaterials have also been incorporated previously into the membrane matrix to enhance their performance. Among the various nanomaterials used, CNTs have significantly attracted scientific interest as they have higher flux-induced specificity and decreased fouling. Other membrane processes reported so far have usually offered superior flux or better specificity, thus CNT-incorporated membranes have emerged as highly suitable membrane technology for separation purposes (Kim et al., 2012).

Thin-film composite (TFC) CNT membranes are another vastly discussed nanomaterial incorporated membrane which are mainly designed by poly condensation with amine and acyl chloride monomers. Thus, polyamide skin layers are prepared at the oil or water interfaces (Wu et al., 2016a). TFC membranes consist of a macroporous matrix coated with a thin layer of polyamide bed. The breadth of the polyamide bed varies between 100 nm and 500 nm (Lee et al., 2011). These membranes have high specificity and reusability while also being energy efficient and are thus considered to be highly efficient in desalination processes (Xu et al., 2011). CNT membranes are also highly favored in RO desalination techniques for their greater water fluxes and increased tolerance to extreme pH alterations (Dume'e et al., 2013).

Another area of growing interest in recent times is the modifications and pretreatment of membrane filters to mitigate fouling effects. Liu et al. in their work pretreated their UF membrane using polyacrylamide flocculation along with alum coagulation which resulted in higher efficiency of the filtration system (Liu et al 2017).

The schematic diagram of a membrane filtration unit is shown in Fig. 6.1.

2.2 Classification

The four major membrane filtration techniques viz. RO, nanofiltration, UF, and microfiltration can be classified with respect to the particle size retained by the respective pores (Cui et al., 2010). The membrane filtration methods along with their respective particle retention size and operational pressure required is shown in Table 6.2.

Membrane filtration processes requiring transmembrane pressure conventionally are conducted in either one of the two flow models viz. dead end and crossflow model. The two models are shown in Fig. 6.2.

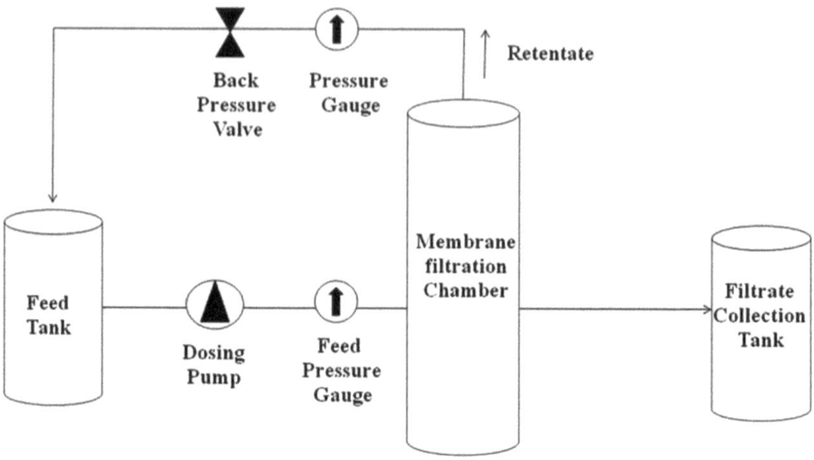

FIG. 6.1 Schematic diagram of a membrane filtration System.

Basically if the feed flow is vertical to the membrane filter and there is no continuous outflow of retentate, then the filtration model is called "dead end", whereas when the feed flow is tangential to the membrane filter and there is a constant flow of retentate, then the model is known as "crossflow" model. Dead end model has higher recovery as compared with crossflow model as only a portion of feedwater is converted to permeate in the latter model. However, the pore blocking due to formation of filtrate cake on the membrane filter reduces recovery for the dead end model over time which demands frequent cleaning and increased transmembrane pressure. Crossflow model is advantageous in this aspect as the sheering flow of the feedwater carries out the retentate which reduces fouling and provides a stable system over a longer period of time (Kirschner et al., 2019).

TABLE 6.2 Classification of membrane filtration techniques with respect to their particle retention size and cross membrane pressure requirement.

Particle size (nm)	Filtration techniques	Required pressure across the membrane (bar)
<1	Reverse osmosis	35–100
1–10	Nanofiltration	10–30
1–100	Ultrafiltration	1–10
100–10000	Microfiltration	1–10

Adapted from Cui, Z. F., Jiang, Y., Field, R. W., 2010. Fundamentals of pressure-driven membrane separation processes. Memb. Technol., 1–18. (Editor Butterworth-Heinemann).

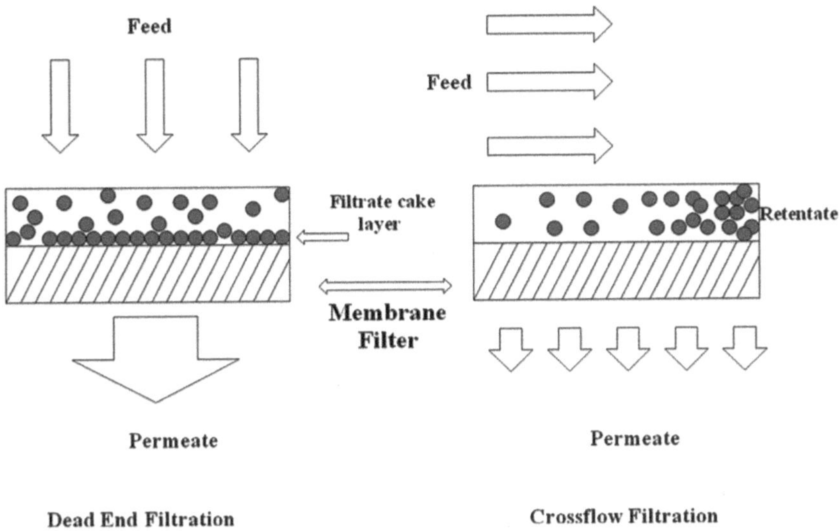

FIG. 6.2 Graphical representation of dead end and crossflow model of membrane filtration.

Membrane filtration can also be classified on basis of pressure requirement for the filtration processes. From Table 6.2, it is evident that both microfiltration and UF processes require lower pressure as compared with higher pressure required for nanofiltration and RO processes.

2.3 Membrane fouling

Several membrane techniques have been designed and applied to scavenge pollutants in different waterbodies, especially focusing on potable water sources (Celebioglu et al., 2019). A variety of commercially accessible membranes like UF membranes, nanofiltration membrane, low-pressure RO membranes, PVDF membrane, etc., are being widely used to eliminate the PAHs from aqueous solutions.

A major problem faced during membrane filtration is the fouling of membrane due to particles present in the feed water. Previous studies have established that the pollutant's physicochemical properties such as polarity, size, hydrophobicity, and charge play an important role in its interaction with the membrane filter and its subsequent removal (Li et al., 2019). Understanding the fouling mechanisms of particular membrane used for removal of pollutants is important for increasing the efficiency of the filtration process. It can also provide important insight for future membrane fabrication, modification, and fouling mitigation.

There are four major models of membrane fouling described by Hermia for dead end filtration which include standard, intermediate, complete pore blocking, and cake filtration (Kirschner et al., 2019). The various pore blocking models are shown in Fig. 6.3. These four proposed models assume that the pores in the membrane under consideration are uniform in size and distribution and the filtrate also consists of particles with similar shape and are not deformed due to the filtration process (Kirschner et al., 2019).

The pore blocking mechanism usually may not be constant for a filtration process. Two pore blocking mechanisms might occur simultaneously as well as there may be constant interchanging among the four fouling mechanism (Koonani and Amirinejad, 2019; Salazar-Peláez et al., 2017).

The resultant decline in flux due to the above-mentioned fouling mechanism is dependent on various factors including process temperature, salinity, molecular weight of the contaminant, and membrane characteristics (Mondal and De, 2009).

In complete pore blocking membrane fouling, it is assumed that the pores on the surface of the membrane filter are blocked completely by individual contaminant particles. There is no further particle deposition on top of already deposited particle (Hermia, 1982). In intermediate blocking of membrane pore, the contaminant particles can either deposit on an unblocked pore or can deposit on a previously blocked pore on top of the previous contaminant particle. Both complete and intermediate pore blocking considers only the surface of the semipermeable membrane. In this aspect, standard pore blocking differs mainly as in this model, the subsurface area of the pores is considered. Contaminant particles are considered to deposit in the inner parts of the pores thus reducing the total pore area. Finally, in cake filtration module of fouling, contaminant particles cover the total surface of the semipermeable membrane in layers. The flux reduces proportionately to the thickness of the total cake layer formed during filtration process.

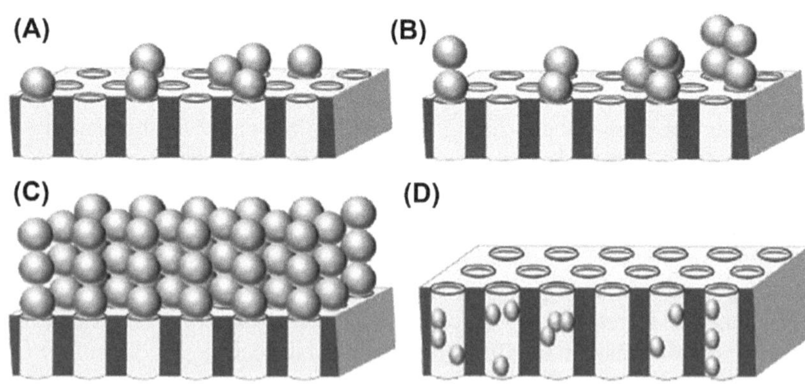

FIG. 6.3 Four dead end membrane fouling mechanism. (A) Complete pore blocking, (B) Intermediate pore blocking, (C) Cake filtration, and (D) Standard pore blocking. *Adapted from Kirschner, A. Y., Cheng, Y. H., Paul, D. R., Field, R. W., Freeman, B. D., 2019. Fouling mechanisms in constant flux crossflow ultrafiltration. J. Memb. Sci. 2019, 574, 65–75.*

Fouling of membrane filters has various negative impacts on membrane performance. It reduces efficiency of the membrane due to reduced flux which leads to change in separation properties. Further, it necessitates recurrent cleaning of membrane to recover flux, reducing the cost-effectiveness of the system and also shortening of membrane lifetime. In most previous work, alkaline agents were used for cleaning of organic matter and acid agents were used for cleaning of inorganic fouling agent (Rudolph et al., 2017).

Hwang et al. (2008) studied the fouling of membrane in UF, microfiltration, and nanofiltration of natural water. It was seen that the mechanisms of rejection of the different membrane processes varied. Pore plugging due to particles in feed water and subsequent cake formation reducing filtration surface was the primary cause of microfilter fouling. In UF process, reduction of internal pore size due to pore adsorption of calcium-organic flocs was majorly responsible for amplified rejection. In nanofiltration process, the deposited particle charge was the factor affecting the fouling mechanism. Thus, it was concluded from the study that the characteristics of rejection in membranes depend majorly on the nature of the foulants present in the feed water as well as the fouling state of the membrane under study rather than on the original characteristics of the membrane. Ahmadiannamini et al. (2015) used polyelectrolyte multilayer (PEM)-coated polysulfone UF membranes for treatment of methylene blue (MB) in water. It was found that permeability of water increased after dismantling of PEMs on UF membranes, while also decreasing MB rejection. The modified membrane was able to recover 80% of water permeability as compared with unmodified UF membrane.

Li et al. (2019) assessed the removal of PAHs using nanofiltration and found that fouling of membrane was chiefly associated with the decrease in porosity due to PAHs adsorption.

Further, this chapter also revealed that nanofiltration membrane with smaller pore size and finer pore distribution led to random pore blocking after PAH adsorption; therefore, there was very little change of average pore distribution and size after fouling. However, in nanofiltration membrane with comparatively larger pore size and spacious pore distribution, pore contraction was observed after PAH adsorption. Therefore, it was concluded that PAH removal using nanofiltration membrane having narrow pore size distribution was more efficient compared with membrane with broader pore size distribution.

Bengani-Lutzet al. 2017 formulated a novel zwitterion copolymer membrane with high resistance to fouling for treatment of wastewater consisting of oil and textile waste. The formulated membrane demonstrated zero percent flux decline after a 7-day dead end oil emulsion filtration. Use of ultrasonic technologies for mitigation of membrane fouling is another field of extensive research (Hou et al., 2017; Qasim et al., 2018; Arefi-Oskoui et al., 2019) which has been found successful in many cases.

Physicochemical characteristics of PAHs and nanofiltration membranes such as charge, hydrophilicity, and polarity play an important role in determining the degree of rejection and membrane fouling mechanism (Albergamo et al., 2018). Therefore, to design an energy-efficient high-performance filtration system, it is crucial to study the rejection and fouling mechanisms of nanofiltration process.

Reduction of membrane fouling is a primary area of research because fouling not only reduces productivity of the membrane but also decreases lifespan of the membrane. Due to reduced flux, the pressure requirement also increases leading to an increase in energy requirement. The cost-effectiveness is majorly affected due to fouling. Therefore, research needs to focus on formulation of various foul mitigation technologies. Cheaper and more efficient fouling mitigation strategies needs to be developed for any kind of membrane filtration.

2.4 Calculations

There are a number of formulas used to access the productivity of a membrane process as well as the efficiency of newly formulated filtration membranes.

The flux is calculated using the equation

$$J = \frac{Q_p}{S_m} \tag{6.1}$$

where J is the flux expressed in L/m^{-2}/h, Q_p is the flow rate of permeate expressed in L/h, and S_m is the surface area of membrane.

The pressure across the membrane filter which is the difference between the feed and permeate pressure is an important factor in estimating the pressure required for the filtration process. It can be calculated by

$$\Delta P = P_f - P_p \tag{6.2}$$

The percentage recovery, which is the amount of feed that is transformed into permeate, can be calculated by

$$R\% = \frac{Q_p}{Q_f} * 100 \tag{6.3}$$

where Q_p and Q_f represent the permeate and feed flow, respectively.

The percentage rejection of the pollutant under study is calculated by

$$R = 1 - \frac{C_p}{C_f} * 100 \tag{6.4}$$

where C_p and C_f represent the concentration of pollutant under study in permeate and feed water, respectively. Percentage rejection is important as it is a direct indicator of the efficiency of a particular membrane with respect to the contaminant requiring removal.

As discussed earlier, the understanding of fouling mechanism of a membrane is essential to design an efficient filtration system.

Fouling of membrane due to pore blocking can be calculated by Hermia's equation

$$\frac{d^2t}{dV^2} = K\left(\frac{dt}{dV}\right)^n \tag{6.5}$$

where V represents the permeate flux, K is filtration constant, t is time, and n has values of 0, 1, 1.5, and 2 for cake filtration, intermediate, standard, and complete pore blocking, respectively.

3. Conclusion

Membrane filtration for PAH laden wastewater is a promising new avenue. With increasing pollution of waterbodies as well as increasing complexities related to removal of PAHs from water, membrane filtration can be a cost-effective, compact, and time-efficient solution.

Membrane filtration can also be implemented in large-scale industrial use. Further research is required for specialization of membranes to reduce fouling and increase efficiency. Studying the membrane fouling mechanism and its relation to process parameters is required to optimize the process under study. Promising new fouling-resistant membrane filter is the requirement for the future and thus extensive study is required for their formulation, structure, and function.

References

Ahmadiannamini, P., Bruening, M.L., Tarabara, V.V., 2015. Sacrificial polyelectrolyte multilayer coatings as an approach to membrane fouling control: disassembly and regeneration mechanisms. J. Membr. Sci. 491, 149–158.

Akinpelu, A.A., Ali, M.E., Johan, M.R., Saidur, R., Qurban, M.A., Saleh, T.A., 2019. Polycyclic aromatic hydrocarbons extraction and removal from wastewater by carbon nanotubes: a review of the current technologies, challenges and prospects. Process Saf. Environ. Protect. 122, 68–82.

Albergamo, V., Blankert, B., Cornelissen, E., Hofs, B., Knibbe, W.J., Meer, van derW., Voogt, DeP., 2018. Removals of polar organic micropollutants by pilot-scale reverse osmosis drinking water treatment. Water Res. 148, 535–545.

Ali, M., Das, R., Maamor, A., Hamid, S.B.A., 2014. Multifunctional carbon nanotubes (CNTs): a new dimension in environmental remediation. Adv. Mater. Res. 832, 328–332.

Arefi-Oskoui, S., Khataee, A., Safarpour, M., Orooji, Y., Vatanpour, V., 2019. A review on the applications of ultrasonic technology in membrane bioreactors. Ultrason. Sonochem. 104633.

Barman, S.R., Banerjee, P., Mukhopadhayay, A., Das, P., 2017. Biodegradation of acenapthene and naphthalene by Pseudomonas mendocina: process optimization, and toxicity evaluation. J. Environ. Chem. Eng. 5 (5), 4803–4812.

Barman, S.R., Banerjee, P., Das, P., Mukhopadhayay, A., 2018. Urban wood waste as precursor of activated carbon and its subsequent application for adsorption of polyaromatic hydrocarbons. Int. J. Energy and Water Res. 2 (1–4), 1–13.

Behera, B.K., Das, A., Sarkar, D.J., Weerathunge, P., Parida, P.K., Das, B.K., Bansal, V., 2018. Polycyclic Aromatic Hydrocarbons (PAHs) in inland aquatic ecosystems: perils and remedies through biosensors and bioremediation. Environ. Pollut. 241, 212–233.

Bengani-Lutz, P., Zaf, R.D., Culfaz-Emecen, P.Z., Asatekin, A., 2017. Extremely fouling resistant zwitterionic copolymer membranes with ~ 1 nm pore size for treating municipal, oily and textile wastewater streams. J. Membr. Sci. 543, 184–194.

Bodzek, M., Konieczny, K., 2011. Membrane techniques in the removal of inorganic anionic micropollutants from water environment- State of the art. Arch. Environ. Protect. 37 (2), 15–22.

Bodzek, M., Konieczny, K., 2006. Membrane processes in water treatment – state of art. Engineering and Protection of Environment 9, 129–159.

Broniatowski, M., Urbaś, M., 2017. Interactions of two structurally related anionic phospholipids cardiolipin and phosphatidylglycerol with phospholipase A2, Langmuir monolayer studies. Biochim. Biophys. Acta 1859 155–166.

Celebioglu, A., Topuz, F., Yildiz, Z.I., Uyar, T., 2019. Efficient removal of polycyclic aromatic hydrocarbons and heavy metals from water by electrospun nanofibrous polycyclodextrin membranes. ACS Omega 4 (4), 7850–7860.

Chen, X., Qiu, M., Ding, H., Fu, K., Fan, Y., 2016. A reduced graphene oxide nanofiltration membrane intercalated by well-dispersed carbon nanotubes for drinking water purification. Nanoscale 8 (10), 5696–5705.

Cui, Z.F., Jiang, Y., Field, R.W., 2010. Fundamentals of pressure-driven membrane separation processes. Memb. Technol. 1–18. Editor Butterworth-Heinemann.

Das, R., Ali, M.E., Hamid, S.B.A., Ramakrishna, S., Chowdhury, Z.Z., 2014. Carbon nanotube membranes for water purification: a bright future in water desalination. Desalination 336, 97–109.

Das, R., Hamid, S.B.A., Ali, M.E., Ismail, A.F., Annuar, M.S.M., Ramakrishna, S., 2014. Multifunctional carbon nanotubes in water treatment: the present, past and future. Desalination 354, 160–179.

Dumée, L., Lee, J., Sears, K., Tardy, B., Duke, M., Gray, S., 2013. Fabrication of thin film composite poly(amide)-carbon-nanotube supported membranes for enhanced performance in osmotically driven desalination systems. J. Membr. Sci. 427, 422–430.

Dudziak, M., Luks-Betlej, K., Bodzek, M., 2003. Removal of PAHs from water using membrane processes. Eng. Protect. Environ. 3 (4), 299–311.

Eeshwarasinghe, D., Loganathan, P., Kalaruban, M., Sounthararajah, D.P., Kandasamy, J., Vigneswaran, S., 2018. Removing polycyclic aromatic hydrocarbons from water using granular activated carbon: kinetic and equilibrium adsorption studies. Environ. Sci. Pollut. Res. 25, 13511–13524.

Fan, Z., Wang, Z., Sun, N., Wang, J., Wang, S., 2008. Performance improvement of polysulfone ultrafiltration membrane by blending with polyaniline nanofibers. J. Membr. Sci. 320 (1), 363–371.

Gong, C., Huang, H., Qian, Y., Zhang, Z., Wu, H., 2017. Integrated electrocoagulation and membrane filtration for PAH removal from realistic industrial wastewater: effectiveness and mechanisms. RSC Adv. 7 (83), 52366–52374.

González, D., Ruiz, L.M., Garralón, G., Plaza, F., Arévalo, J., Parada, J., Pérez, J., Moreno, B., Gómez, M.Á., 2012. Wastewater polycyclic aromatic hydrocarbons removal by membrane bioreactor. Desalin. Water Treat. 42 (1–3), 94–99.

Grevstad, H.J., Lekens, K.N., 1993. Ultrastructure of plaque associated with polytetrafluoroethylene (PTFE) membranes used for guided tissue regeneration. J. Clin. Periodontol. 20 (3), 193–198.

Gutiérrez, M.S., Duel, P., Hierro, F., Morey, J., Piña, M.N., 2018. A very highly efficient magnetic nanomaterial for the removal of PAHs from aqueous media. Small 14 (8), 1702573.

Hermia, J., 1982. Constant pressure blocking filtration laws – application to power-law non-Newtonian fluids. Trans. Am. Inst. Chem. Eng. 60, 183–187.

Hou, D., Lin, D., Zhao, C., Wang, J., Fu, C., 2017. Control of protein (BSA) fouling by ultrasonic irradiation during membrane distillation process. Separ. Purif. Technol. 175, 287–297.

Hwang, K.J., Liao, C.Y., Tung, K.L., 2008. Effect of membrane pore size on the particle fouling in membrane filtration. Desalination 234 (1–3), 16–23.

Idris, A., Zain, N.M., Noordin, M., 2007. Synthesis, characterization and performance of asymmetric polyethersulfone (PES) ultrafiltration membranes with polyethylene glycol of different molecular weights as additives. Desalination 1–3, 324–339.

Jame, S.A., Zhou, Z., 2016. Electrochemical carbon nanotube filters for water and wastewater treatment. Nanotechnol. Rev. 5 (1), 41–50.

Kallimanis, A., Frillingos, S., Drainas, C., Koukkou, A.I., 2007. Taxonomic identification, phenanthrene uptake activity, and membrane lipid alterations of the PAH degrading *Arthrobacter* sp. strain Sphe3. Appl. Microbiol. Biotechnol. 76, 709–717.

Ke, S., Liu, Q., Yao, Y., Zhang, X., Sui, G., 2018. An in vitro cytotoxicities comparison of 16 priority polycyclic aromatic hydrocarbons in human pulmonary alveolar epithelial cells HPAEpiC. Toxicol. Lett. 290, 10–18.

Kim, E.S., Hwang, G., El-Din, M.G., Liu, Y., 2012. Development of nanosilver and multi-walled carbon nanotubes thin-film nanocomposite membrane for enhanced water treatment. J. Membr. Sci. 394, 37–48.

Kirschner, A.Y., Cheng, Y.H., Paul, D.R., Field, R.W., Freeman, B.D., 2019. Fouling mechanisms in constant flux crossflow ultrafiltration. J. Memb. Sci. 574, 65–75.

Koonani, H., Amirinejad, M., 2019. Combined three mechanisms models for membrane fouling during microfiltration. J. Memb. Sci. Res. 5 (4), 274–282.

Kowal, A., Świderska-Bróż, L., 2007. Water Treatment. Theoretical Background and Technological Processes and Devices. Polish Scientific Publishers, Warszawa.

Kuppusamy, S., Thavamani, P., Venkateswarlu, K., Lee, Y.B., Naidu, R., Megharaj, M., 2017. Remediation approaches for polycyclic aromatic hydrocarbons (PAHs) contaminated soils: technological constraints, emerging trends and future directions. Chemosphere 168.

Lee, K.P., Arnot, T.C., Mattia, D., 2011. A review of reverse osmosis membrane materials for desalination-development to date and future potential. J. Membr. Sci. 370, 1–22.

Li, S., Luo, J., Hang, X., Zhao, S., Wan, Y., 2019. Removal of polycyclic aromatic hydrocarbons by nanofiltration membranes: rejection and fouling mechanisms. J. Membr. Sci. 582, 264–273.

Liu, T., Lian, Y., Graham, N., Yu, W., Rooney, D., Sun, K., 2017. Application of polyacrylamide flocculation with and without alum coagulation for mitigating ultrafiltration membrane fouling: role of floc structure and bacterial activity. Chem. Eng. J. 307, 41–48.

Ma, Z., Lu, X., Wu, C., Gao, Q., Zhao, L., Zhang, H., Liu, Z., 2017. Functional surface modification of PVDF membrane for chemical pulse cleaning. J. Membr. Sci. 524, 389–399.

Mahlambi, M.M., Vilakati, G.D., Mamba, B.B., 2014. Synthesis, characterization, and visible light degradation of rhodamine B dye by carbon-covered alumina supported Pd-TiO_2/polysulfone membranes. Separ. Sci. Technol. 49 (14), 2124–2134.

Mielczarek, K., Bohdziewicz, J., Kwarciak-Kozłowska, A., 2011. Comparing the effectiveness of treatment coking industry wastewater with applying for the coagulation processes and advanced oxidation processes. Ecol. Eng. 4, 184–194.

Mntambo, S.A., Mdluli, P.S., Mahlambi, M.M., Onwubu, S.C., Nxumalo, N.L., 2019. Synthesis and characterisation of ultrafiltration membranes functionalised with C18 as a modifier for adsorption capabilities of polyaromatic hydrocarbons. WaterSA 45 (1), 131–140.

Mondal, S., De, S., 2009. Generalized criteria for identification of fouling mechanism under steady state membrane filtration. J. Membr. Sci. 344 (1–2), 6–13.

Nwaichi, E.O., Ntorgbo, S.A., 2016. Assessment of PAHs levels in some fish and seafood from different coastal waters in the Niger Delta. Toxicol. Rep. 3, 167–172.

Pal, P., Kumar, R., 2013. Treatment of coke wastewater: a critical review for developing sustainable management strategies. Separ. Purif. Rev. 43, 89–123.

Paxéus, N., 2000. Organic compounds in municipal landfill leachates. Water Sci. Technol. 42 (7–8), 323–333.

Pérez-Gregorio, M.R., García-Falcón, M.S., Martínez-Carballo, E., Simal-Gándara, J., 2010. Removal of polycyclic aromatic hydrocarbons from organic solvents by ashes wastes. J. Hazard Mater. 178 (1–3), 273–281.

Pratt, M.M., John, K., MacLean, A.B., Afework, S., Phillips, D.H., Poirier, M.C., 2011. Polycyclic aromatic hydrocarbon (PAH) exposure and DNA adduct semi-quantitation in archived human tissues. Int. J. Environ. Res. Publ. Health 8 (7), 2675–2691.

Qiao, M., Fu, L., Cao, W., Bai, Y., Huang, Q., Zhao, X., 2019. Occurrence and removal of polycyclic aromatic hydrocarbons and their derivatives in an ecological wastewater treatment plant in South China and effluent impact to the receiving river. Environ. Sci. Pollut. Control Ser. 26 (6), 5638–5644.

Qasim, M., Darwish, N.N., Mhiyo, S., Darwish, N.A., Hilal, N., 2018. The use of ultrasound to mitigate membrane fouling in desalination and water treatment. Desalination 443, 143–164.

Rajesh, S., Senthilkumar, S., Jayalakshmi, A., Nirmala, M., Ismail, A.F., Mohan, D., 2013. Preparation and performance evaluation of poly (amide-imide) and TiO_2 nanoparticles impregnated polysulfone nanofiltration membranes in the removal of humic substances. Colloids Surf., A: Physicochemistry Engineering Aspects 418, 92–104.

Rao, G.,P., Lu, C., Su, F., 2007. Sorption of divalent metal ions from aqueous solution by carbon nanotubes: a review. Separ. Purif. Technol. 58, 224–231.

Ren, J., Li, J., Chen, Z., Cheng, F., 2018. Fate and wetting potential of bio-refractory organics in membrane distillation for coke wastewater treatment. Chemosphere 208, 450–459.

Rudolph, G., Thuvander, J., Jönsson, A.S., Lipnizki, F., 2017. Fouling and Cleaning of Membranes in Biorefineries.

Salazar-Peláez, M.L., Morgan-Sagastume, J.M., Noyola, A., 2017. Fouling layer characterization and pore-blocking mechanisms in an UF membrane externally coupled to a UASB reactor. WaterSA 43 (4), 573–580.

Shen, L., Feng, S., Li, J., Chen, J., Li, F., Lin, H., Yu, G., 2017. Surface modification of polyvinylidene fluoride (PVDF) membrane via radiation grafting: novel mechanisms underlying the interesting enhanced membrane performance. Scitific Rep. 7 (1), 2721.

Shevate, R., Kumar, M., Karunakaran, M., Canlas, C., Peinemann, K.V., 2018. Surprising transformation of a block copolymer into a high performance polystyrene ultrafiltration membrane with a hierarchically organized pore structure. J. Mater. Chem. 6 (10), 4337–4345.

Smol, M., Maria, W.M., Bohdziewicz, J., 2014. The use of integrated membrane systems in the removal of selected pollutants from pre-treated wastewater in coke plant. Mono. Environ. Eng. Commit. 119, 143–152.

Smol, M., Włodarczyk-Makuła, M., 2017. The effectiveness in the removal of PAHs from aqueous solutions in physical and chemical processes: a review. Polycycl. Aromat. Comp. 37 (4), 292–313.

Smol, M., Włodarczyk-Makuła, M., 2012. Effectiveness in the removal of Polycyclic Aromatic Hydrocarbons from industrial wastewater by ultrafiltration technique. Arch. Environ. Protect. 38 (4), 49–58.

Sun, H., Jiang, J., Xiao, Y., Du, J., 2017. Efficient removal of polycyclic aromatic hydrocarbons, dyes, and heavy metal ions by a homopolymer vesicle. ACS Appl. Mater. Interfaces 10 (1), 713–722.

Topuz, F., Singh, S., Albrecht, K., Möller, M., Groll, J., 2016. DNA nanogels to snare carcinogens: a bioinspired generic approach with high efficiency. Angew. Chem. 55, 12210–12213.

Upadhyayula, V.K.K., Deng, S.G., Mitchell, M.C., Smith, G.B., 2009. Application of carbon nanotube technology for removal of contaminants in drinking water: a review. Sci. Total Environ. 408, 1–13.

Wang, Y., Goh, S.H., Chung, T.S., Na, P., 2009. Polyamide-imide/polyetherimide dual-layer hollow fibre membranes for pervaporation dehydration of C1–C4 alcohols. J. Membr. Sci. 326 (1), 222–233.

Wisniewski, C., 2007. Membrane bioreactor for water reuse. Desalination 203, 15–19.

Włodarczyk-Makuła, M., Kalaga, K., Kipigroch, M., Smol, M., 2011. Determination of PAHs in coke wastewater. Eng. Protect. Environ. 3, 266–274.

Wlodarczyk-Makula, M., 2005. The loads of PAHs in wastewater and sewage sludge of municipal treatment plant. Polycycl. Aromat. Comp. 25 (2), 183–194.

Wu, M.B., Lv, Y., Yang, H.C., Liu, L.F., Zhang, X., Xu, Z.K., 2016. Thin film composite membranes combining carbon nanotube intermediate layer and microfiltration support for high nanofiltration performances. J. Membr. Sci. 515, 238–244.

Wu, Y.C., Chen, L.J., Lan, Y.Q., Liu, R., 2016. Performance of nanofiltration for improving the drinking water quality in a water supply plant with micropolluted water resource. Huan jing ke xue= Huanjing kexue. 37 (9), 3466–3472.

Xu, R., Wang, J., Kanezashi, M., Yoshioka, T., Tsuru, T., 2011. Development of robust organosilica membranes for reverse osmosis. Langmuir 27, 13996–13999.

Zhang, W., Wei, C., Chai, X., He, J., Cai, Y., Ren, M., Yan, B., Peng, P., Fu, J., 2012. The behaviours and fate of polycyclic aromatic hydrocarbons (PAHs) in a coking wastewater treatment plant. Chemosphere 88, 174–182.

Zularisam, A.W., Ismail, A.F., Salim, R., 2006. Behaviours of natural organic matter in membrane filtration for surface water treatment—a review. Desalination 194 (1–3), 211–231.

Zeledon-Toruno, Z.C., Lao-Luque, C., de las Heras, F.X.C., Sole-Sardans, M., 2007. Removal of PAHs from water using an immature coal (leonardite). Chemosphere 67 (3), 505–512.

Chapter 7

Hybrid application of advanced oxidation processes to dyes′ removal

Süheyda Atalay, Gülin Ersöz
Department of Chemical Engineering, Ege University, İzmir, Turkey

1. Introduction

The dyes are organic compounds that have unsaturated groups such as chromophores in their structures. These unsaturated groups are responsible to give the color and functional groups are responsible for their fixation on the fibers (Habib et al., 2012). Depending on the chromophore structures of the dyes, they can be categorized as acidic, basic, reactive, azo, disperse, diazo, and anthraquinone dyes. In addition subject to their differences in structure, they can be named as cationic, anionic, or nonionic dyes (Su et al., 2016; Robinson et al., 2001).

Commercial dyes are used as raw materials at various industries such as textiles, paper, plastics, food, and pharmaceuticals. Discharges from these industries contain various toxic-colored contaminants which are not easily degraded by the traditional methods. Some of dye groups, mainly azo dyes, may be transformed into mutagenic and carcinogenic aromatic amines which are very dangerous to nature and human health if they are not treated properly (He et al., 2013). Hence, a complete removal of these effluents before discharging to the environment is absolutely required (Nguyen and Juang, 2019). However, it is difficult to treat dye containing wastewaters by traditional methods due to their recalcitrant nature and resistance to biodegradation.

In order to decrease toxicity and to achieve complete mineralization of organic pollutants advanced oxidation processes (AOPs) can be used. Thus, AOPs for the degradation of nonbiodegradable organic compounds are becoming very promising alternatives in comparison with conventional wastewater treatment processes (Orak et al., 2017). Common AOPs are the Fenton process, ozonation, photocatalytic oxidation, catalytic wet air oxidation (WAO), or the proper combination of them (Ribeiro et al., 2015). However,

Green Chemistry and Water Remediation: Research and Applications.
https://doi.org/10.1016/B978-0-12-817742-6.00007-4

especially when the wastewater is highly toxic or contains high concentration of suspended solids, most of these oxidation processes individually have their drawbacks. To solve this problem, the combined systems can be effective to treat dyeing wastewater (Kim et al., 2004).

The physicochemical processes such as adsorption, flocculation, coagulation, and reverse osmosis have the ability of just transferring the pollutants from one phase to another without essentially solving the color problem in the effluents (Torrades et al., 2004; Solmaz et al., 2006).

Generally, it can be concluded that none of the processes applied individually are always advantageous for the degradation of the complex dye containing effluents. These methods can be used in combination to handle the drawbacks of the individual method.

In recent days, considerable attention has been focused on the removal of pollutants from wastewater by combined processes. Using wastewater treatment techniques in combination can considerably enhance the treatment performance especially when dealing with high strength industrial wastewaters (Hai et al., 2007; Orak et al., 2017). For example, a combination of different AOPs may increase the efficiency of treatment when compared with individual applications by generating higher quantum of desired radicals and high degree of energy efficiency. Combined application of traditional physicochemical methods with the AOPs has also seemed to be an interesting alternative (Hai et al., 2007; Su et al., 2016).

In this chapter, a detailed summary of the potential hybrid techniques mentioned in literature for the treatment of dye containing wastewaters together with their advantages and disadvantages has been given. Before emphasizing on the combined methods, the basis and limitations of relevant sole method are mentioned briefly. This chapter is divided into two main categories, namely Decolorization and Degradation of Dyes by Combination among Advanced Oxidation Processes and Decolorization and Degradation of Dyes by Combination of Advanced Oxidation Processes and Other Physicochemical Processes. These two main categories are then further divided into subcategories.

2. Decolorization and degradation of dyes by combination among advanced oxidation processes

AOPs are well known to be competitive methods for the treatment of dye containing wastewaters which are hardly treated by traditional techniques due to their high chemical stability, toxicity, and low biodegradability. Hence, in recent years, AOPs have been considered to be powerful and promising alternative processes over the traditional technologies.

AOPs such as Fenton and Fenton-like processes, ozonation, photocatalysis, catalytic WAO, and sonolysis have emerged as useful methods for the treatment of contaminants in aqueous media. AOPs involve many effective oxidative agents to treat water such as ultraviolet (UV) radiation, ozone, chlorine, hydrogen peroxide, and oxygen. Although the methods have different mechanisms, they have similar chemical characteristic which is using a very high oxidizing agent (hydroxyl radical) (Krishnakumar and Swaminathan, 2011; Asghar et al., 2015).

2.1 Individual advanced oxidation processes

In order to remove recalcitrant contaminants successfully, various AOPs have been tested and studied in the previous years.

2.1.1 Fenton and Fenton-like processes

Fenton Reagent Oxidation is the most commonly used method in advanced oxidation methods. The Fenton Oxidation process includes the activation of H_2O_2 by ferrous (Fe^{+2}) ions to produce hydroxyl radicals (•OH) over a complex reaction mechanism under acidic condition (Bokare and Choi, 2014). Besides the classical Fenton, there are three more Fenton-based processes. These are photo-Fenton, photo-Fenton like, and electro-Fenton systems (Ribeiro et al., 2015).

Photo-Fenton process is the same as classical Fenton oxidation; it occurs under visible or UV light. (Sreeja and Sosamony, 2016). For Fenton-like oxidation, regarding Fe other transition metals can be used to produce hydroxyl radicals from hydrogen peroxide (Krishnan et al., 2017). Electro-Fenton process mechanism is related on the generation of electrons continuously by hydrogen peroxide on an appropriate cathode by the reduction of dissolved O_2 or air in the presence of an iron catalyst in the treated solution to produce hydroxyl radicals (Pushpalatha and Krishna, 2017).

2.1.2 Ozonation

Ozonation is one of the advanced treatment processes that is in common use for the degradation of different kinds of dyes in wastewater. Ozone may selectively destroy the conjugated double bonds that are directly related with color (Solmaz et al., 2006). Ozone oxidizes the organic molecules by the help of the combined radical and molecular reactions. It can be called as a green process since no sludge is formed during the reaction. The process has the potential to realize the decolorization and the degradation of dye containing effluents in a single step. When compared with the other processes, less space is required for equipment installation and it is said to be less hazardous since no stock of H_2O_2 is desired for the oxidation process. Therefore, it can be strongly considered to be an efficient method for dye degradation (Asghar et al., 2015; Oğuz et al., 2005).

2.1.3 Photocatalysis

Among the AOP, photocatalytic oxidation has received great attention as it is one of the most efficient and environmentally friendly technologies and inexpensive in comparison to many conventional processes and AOP. The photocatalytic method has many benefits such as application procedure is not complicated, easily adaptable to industrial scales, investment cost is less, and solar light can be used as a sustainable energy source. The main steps of heterogeneous photocatalysis comprise the reactants transfer in the bulk liquid to the surface of the catalyst, adsorption of the reactants on the catalyst surface, reaction between the adsorbed reactants, and desorption of the reaction products. Finally, the products are removed in the bulk liquid. In detailed, when photocatalyst absorbs, UV light electrons are promoted from the valence band to the conduction band and electron−hole pairs $\left(h_{vb}{}^{+}\text{ and }e_{cb}{}^{-}\right)$ are generated. Mostly, $h_{vb}{}^{+}$ reacts with surface bound H_2O to produce hydroxyl radicals, whereas $e_{cb}{}^{-}$ reacts with O_2 to produce superoxide radical anion $^{\cdot}O_2{}^{-}$. These radicals react with the organic molecules to mineralize them (Park et al., 2013). In photocatalysis, semiconducting materials such as TiO_2, Fe_2O_3, ZnO, etc., are used to generate holes and free radicals for the removal of organic pollutants.

2.1.4 Catalytic wet air oxidation

Other eco-friendly advanced oxidation method known as WAO is one of the most appropriate techniques to eliminate highly concentrated, persistent, toxic, and hazardous organic compounds. It converts them into CO_2 and H_2O at high temperature and pressure in the presence of oxygen. No emissions of NO_x, SO_2, HCl, dioxins, furans, and fly ash occur (Jing et al., 2016). However, the need of high temperature and pressure limits the applications of WAO in practical. The disadvantage of the WAO can be prevented by using appropriate catalysts. The catalytic WAO processes succeed better performances at milder conditions when compared with WAO (Hua et al., 2013). The common catalysts used in catalytic WAO processes are transition metals oxides, noble metals, and the homogeneous catalysts. Though these catalysts are effective in organic removal, they have significant disadvantages such as deactivation of the metal oxides due to the leaching of the active sites, expensive costs of noble metal catalysts, and the difficulty in separation of homogeneous catalysts (Keav et al., 2010). In this frame, development of innovative and sustainable catalysts to be used in advanced oxidation of the resistant organic compounds has become a significant area of studies recently.

2.1.5 Sonolysis

Sonolysis (or sonication) is another promising AOP. The method does not use any chemical and it can mineralize the contaminants by thermolysis in which the effluent is subjected to ultrasound irradiation. By cavitation, the ultrasound shows its physical and chemical effects (Chakma et al., 2015; Chakma and Moholkar, 2011, 2013).

In sonolysis, various radicals such as •H,•OH,•O, and $\bullet HO_2$ are generated by the help of the water vapor caught in the bubble. This bubble of cavitation undergoes thermal dissociation. •H,•OH,•O, and $\bullet HO_2$ are released with distruption of the bubble and react with the contaminant molecules. But, if the pollutant concentration is low, the chance of interaction of these radicals and the contaminant molecule is very low and instead the radicals may recombine. Combination of •OH radicals induces the generation of H_2O_2. Since H_2O_2 has lower oxidation potential than •OH radicals, a loss of oxidation potential may be faced (Chakma and Moholkar, 2014).

2.2 Limitation of advanced oxidation processes

Degradation and decolorization of dye containing effluents can be accomplished successfully by using AOPs depending mainly on hydroxyl radicals production at moderate conditions. But most of the AOPs if operated individually results in problem such as limited applicability and high costs.

One of the major drawbacks faced in application of AOP is that usually AOPs may bring along by-products that are toxic and undesired. In addition, complete degradation may not be possible. The efficiency of the processes may be reduced by the presence of possible radical scavengers (Hai et al., 2007).

Below some limitations of the AOP in the treatment of dye containing wastewaters are summarized.

2.2.1 Fenton and Fenton-like processes

Even though Fenton process is an efficient method for treating dye containing wastewaters, there are various factors that prevent the process to have higher efficiencies. This process may be evaluated as economically nonfeasible because of using high amounts of catalyst and hydrogen peroxide. In addition to these, hazards related with the transport and storage of large amounts of H_2O_2 have made the technique hazardous and expensive (Asghar et al., 2015; Hassan et al., 2012; Sheriff et al., 2007).

Some of the limitations of Fenton oxidation method can be listed in detail as below:

- Using the oxidant amount beyond a particular limit is not possible and hence the oxidation degree is limited (Bagal and Gogate, 2014; Gogate and Pandit, 2004).
- High iron salt concentrations (10–500 mgL^{-1}) which are above acceptable discharge limit (2 mgL^{-1}) are normally required to gain remarkable degradation (Adityosulindro et al., 2017).
- Additional water pollution is caused by the usage of homogeneous catalyst (iron salt).
- The processes are effective within narrow pH range of <3.5 especially when homogeneous catalyst is used. Hence, the acidification of the inlet stream and neutralization of the treated solution before disposal is required (Brillas et al., 2009; Hai et al., 2007).

- A longer reaction time is required when compared with the other methods (Hai et al., 2007).
- Huge amounts of small flocs are consistently observed during the process involving sludge generation. Hence, an extra treatment step to remove the iron sludge is required (Brillas et al., 2009; Hai et al., 2007).
- Relatively high cost of hydrogen peroxide makes the process expensive.
- There is also a risk in transportation and to store H_2O_2 (Brillas et al., 2009).

2.2.2 Ozonation

Ozonation is a thoroughly tested process for the degradation of dye in the wastewaters; however, it is not efficient for complete mineralization of complex effluents to ensure current discharge standards (Ince and Tezcanlı, 2001). The applications of ozonation have been limited by the following situations:

- Decolorization by ozonation mainly depends on dye concentration. If the initial concentration is high, more ozone is required (the production cost of ozone is high) (Sheng and Chi,1993; Sevimli and Sarıkaya, 2002; Konsowa et al., 2010).
- The ozone cannot be utilized efficiently due to poor mass transfer rate, low solubility, and the instability of the ozone (Konosowa et al., 2010).

2.2.3 Photocatalytic oxidation

Photocatalytic oxidation when applied indivually has the following disadvantages (Buscio et al., 2015; Hai et al., 2007):

- Light penetration limitation
- Fouling of catalysts
- Catalyst separation from the treated stream

2.2.4 Wet air oxidation

WAO process when applied indivually has the following disadvantages:

- Complete mineralization cannot be usually achieved.
- WAO of effluents with concentration higher than 100 gL^{-1} of chemical oxygen demand (COD) produces by-products which are toxic such as dioxins and furans (Sina and Mohsen, 2017; Tyagi and Lo, 2011).
- Compounds with lower molecular weights are not compliant to WAO.
- The high operating pressures and temperatures applied cause higher costs (both capital and operating).

2.2.5 Sonolysis

The low degradation performance and a high energy requirement are the main limitations of sonolysis process when it is used as single treatment. If the pollutant concentration is low, the chance of interaction of the radicals and the contaminant molecule is very low and instead the radicals may recombine with each other and generate H_2O_2. Since H_2O_2 has lower oxidation potential than •OH radicals, a loss of oxidation potential may be faced (Chakma and Moholkar, 2014).

Generally, it can be concluded that each of the method operating individually cannot always be advantageous for the effective removal of the complex dye containing effluents. The combination of these techniques can be tested to overcome the drawbacks of the individual method.

2.3 Combination among advanced oxidation processes

Recently, many studies have been presented on the use of hybrid methods among AOPs for the degradation of different kinds of pollutants. Results show that important enhancement in the removal performance (when compared to single application of AOPs) has been obtained showing the good synergism of the hybrid process of advanced oxidation methods.

In this section, synergistic mechanisms of various hybrid AOPs with combination of Fenton reagent oxidation, UV light irradiation, ozonation, and sonolysis for the treatment of dye containing wastewaters will be investigated and the studies performed in literature will be given.

The combinations that are mainly focused on are as follows:

- Combination of Fenton process with
  - UV light
  - Sonolysis
  - UV light and sonolysis
  - UV light and ozonation
- Combination of ozonation with
  - UV light
  - Sonolysis
- Combination of sonolysis and UV light

The Fenton process combined with other AOP such as UV light, ultrasound, and ozone has been applied for dye containing wastewaters.

2.3.1 Combination of Fenton process and ultraviolet light

The combination of Fenton process with UV light is named as photo-Fenton oxidation. In the combined process, the efficiency of the process is enhanced by the presence of UV irradiation by increasing the rate of the oxidation process. The UV light promotes the oxidation of Fe^{+2} to Fe^{+3}, while generating hydroxyl radicals.

Hydrolysis of Fe^{+3} in solution ensures the generation of some complexes of Fe^{+3}–hydroxy. By the help of UV light irradiation these formed complexes may be broken down and Fe^{+2} is regenerated by the help of generation of extra hydroxyl radicals. In classical Fenton reagent oxidation, there is a high sludge volume and hence the degradation is less compared with photo-Fenton reaction. The operating cost of the photo-Fenton process is significantly lower, since the consumption of chemicals is lower. The photo-Fenton process leads to negligible formation of sludge. In the photo Fenton oxidation processes, light irradiation (visible light or UV) effects hydroxyl radical generation. In the process hydrogen peroxide is degraded into two OH^- ions, $Fe(OH)^{+2}$ is decomposed into ferrous ion and OH molecules (Oancea and Meltzer, 2013).

One of the most preferred homogeneous methods proposed for treatment of organic contaminants is homogeneous Fenton oxidation in which hydroxyl radicals are generated with Fe^{+2} and hydrogen peroxide. But the classical homogeneous system needs an additional expensive and complicated removal system to separate the reacted iron ions. To overcome this problem, alternatively heterogeneous Fenton process can be applied. In heterogeneous catalysis concept, the solid–liquid separation problem can be prevented and also operating pH can be shifted from acidic to near neutral (Palas et al., 2016; Shih et al., 2013; Chen and Zhu, 2011; Ji et al., 2011; Muthukumari et al., 2009).

Tekin et al. focused on the treatment of tartrazine (food dye), by photo-Fenton oxidation with novel catalysts using visible light. They prepared bismuth oxyhalide catalysts and metal-doped bismuth oxyhalide catalysts. Cu–BiOCl had a flower-like nanostructure with 2.53 eV band gap. This property enhanced the catalyst's photocatalytic activity. The optimum conditions were determined as 0.25 gL^{-1} for photocatalyst amount, 100 W for visible light power, ph = 6, $[H_2O_2] = 6$ mM, T = 70 °C. At the determined optimum conditions, approximately 91% degradation, 95% color removal, and 59% TOC removal were achieved (Tekin et al., 2018).

In 2017, Palas et al. tested the effect of visible and UV light irradiation in the catalytic activity of $LaCuO_3$ in photo-Fenton–like oxidation of Tartrazine. The effects of the solution pH, concentration of H_2O_2, the catalyst loading, and the light source for irradiation on the food dye degradation and color removal were evaluated. The optimum reaction conditions (Tartrazine concentration of 100 mgL^{-1}) were found as 50 °C, pH = 3, $[H_2O_2]_o = 8$ mM, and 0.25 gL^{-1} of catalyst loading. At these conditions under visible light irradiation 46.6% degradation was achieved. Also the color removal was 83.9%. Under UV light irradiation, 64.4% degradation and 90.2% color removal were obtained. They concluded that the prepared perovskite catalysts demonstrated good performances (especially in decolorization) under both visible and UV light irradiation (Palas et al., 2017).

In 2016, Palas et al. examined heterogeneous photo-Fenton–like oxidation using walnut shell–based green catalysts for treatment of Procion Red MX-5B. Some parameters were tested like catalyst loading, the initial hydrogen peroxideconcentration (H_2O_2), pH, and UV light power. The optimum conditions for photo-Fenton–like oxidation was obtained as 0.25 gL^{-1} catalyst loading, 2 mM of the initial hydrogen peroxide concentration, pH = 3, and room temperature. Under these conditions, degradation and color removal performances were achieved as 49.8% and 88.3%, respectively (Palas et al., 2016).

In 2014, Blanco et al. published their study on the treatment of textile wastewater by sequencing batch reactor coupled to photo-Fenton processes. In this study, the optimum conditions were found as concentration of $Fe^{+2} = 216\ mgL^{-1}$; concentration of $H_2O_2 = 4950\ mgL^{-1}$; pH = 2.7, and T = 35 °C for the single photo-Fenton method with the removal of COD and TOC as 79% and 75%, respectively. This removal efficiencies were increased by coupled biological–photo-Fenton processes. Sequencing batch reactor was used for biological treatment with a hydraulic retention time of 1 day and 75% of TOC removal was obtained after 25 cycles. The optimum conditions were found for coupled photo-Fenton method concentration of $Fe^{+2} = 66.5\ mgL^{-1}$ and concentration of $H_2O_2 = 1518\ mgL^{-1}$ with the 97% and 95% removal of COD and TOC, respectively. It can be concluded that, using the combined system increased the removal efficiencies while decreasing the amounts of Fe^{+2} and H_2O_2 (Blanco et al., 2014).

The heterogeneous photocatalytic oxidation of Reactive Black B in the presence of an immobilized iron oxide supported with SiO_2 grain was studied by Shih et al. The photocatalytic oxidation was performed in an expanded-bed reactor. The decolorization was found to be faster (approximately 1.3 times faster) than that of degradation. This was explained by the azo group bonds being weaker than aromatic parts of the Reactive Black B. The percentage removal could approach 100% and 91.2% of color removal and degradation, respectively, in 5 h (Shih et al., 2013).

To treat Orange II containing wastewater, oxalate-enhanced mechanism of hydroxyl-Fe-pillared bentonite (H–Fe–P–B) was used with heterogeneous catalyst in UV-Fenton system by Chen and Zhu. They concluded that addition of oxalate increased the Fe leaching leading higher degradation efficiency of Orange II and lower energy consumption of treatment. The leached iron ions could be adsorbed back to the catalyst after the complete mineralization of organic intermediates (Chen and Zhu, 2011).

Photo-Fenton process in the presence of a heterogeneous photocatalyst $LiFe(WO_4)_2$ was investigated to treat methylene blue (MB) dye in wastewater by Ji et al. The effects of different parameters namely catalyst amount, $[H_2O_2]$, pH of the solution, and UV light intensity on the process efficiency were investigated. At low H_2O_2 concentration, the decolorization efficiency was low since enough hydroxyl radicals could not be generated, by the increase of hydrogen peroxide concentration leading to more hydroxyl radicals production, the efficiency increased. But the decolorization efficiency decreased when hydrogen peroxide increased further (>20 mM). They explained this situation by the production of the hydroperoxyl radicals. These radicals are less reactive when compared with hydroxyl radicals and they do not contribute to the mineralization of the organic substrate in the presence of excess H_2O_2 usage. The results showed that the process efficiency was significantly affected by the catalyst amount. One major disadvantages of Fenton reaction which is the tight range of pH was overcome by using this heterogeneous catalyst. They reported an increase in the efficiency as the pH value increased from 3 to 9. All of the results obtained showed that $LiFe(WO_4)_2$ could be effectively used in the decolorization of MB under UV light and hydrogen peroxide and decolorization rate could be accelerated by using UV radiation. They concluded that $LiFe(WO_4)_2$ proposed a wide pH range to be applied (Ji et al., 2011).

Photo-Fenton degradation of two azo dyes Reactive Orange 4 and Direct Red 23 was determined using a heterogeneous catalyst. As the catalyst Fe^{+2}—loaded Al_2O_3 was used. The degradation efficiencies of 25% and 15% were obtained for Direct Red 23 and Reactive Orange, respectively. The effects of catalyst amount, $[H_2O_2]$, and pH on photodegradation were investigated. The degradation increased with increase in UV power. They concluded that Direct Red 23 degradation was easier when compared with Reactive Orange 4 removal because of the stable triazine ring system in Reactive Orange 4. The heterogeneous catalyst was determined to be reusable and the leaching of Fe^{+2} was less than 10% in the pH range below 7 (Muthukumari et al., 2009).

2.3.2 Combination of Fenton process and sonolysis

In literature, there are many studies related on the sonolysis for the treatment of dyes; however, in most of them, it has been mentioned about the difficulties faced in removing dyes completely using sonolysis process single. Recent studies are present concerning on the use of sonolysis in combination with other AOPs to treat dye containing wastewaters.

Applying ultrasound process with Fenton's process uses the individual benefits of these two techniques to produce more hydroxyl radicals and may enhance the removal efficiency of organic contaminants. In order to remove the chemicals that have resistance to hydroxyl radicals, the cavitation serves an extra mechanism in terms of pyrolysis. Also in the case where the external mass transfer resistance is present, the turbulent conditions may help to prevent it.

Consequently, it can be concluded with this combination can be achieved and hence better and faster pollutant mineralization can be obtained (Adityosulindro et al., 2017; Chakma and Moholkar, 2014).

- more hydroxyl radical generation,
- improved mixing and better contact between the hydroxyl radicals and the pollutant,
- enhanced regeneration of ferrous ions.

In 2018, Çetinkaya et al. studied classic and modified Fenton processes on industrial textile wastewater. The efficiencies of classical Fenton and ultrasound Fenton processes were compared by investigating the effects of pH, amount of ferrous ion, H_2O_2, and reaction time. The best decolorization percentage and COD removal efficiency were found as 95% and 70%—80%, respectively, using 0.10 $g\ L^{-1}$ of Fe^{+2}, 2.20 $g\ L^{-1}$ of H_2O_2, and $pH = 3$ for 90 min in classic Fenton process. Similar experiments were carried out using under the ultrasonication, and the best decolorization percentage was obtained using 0.05 g L^{-1} of Fe^{+2} and 1.65gL^{-1} of H_2O_2 for 1 h at pH value of 3. They mentioned that when the Fenton process is coupled with ultrasound less amount of chemicals and shorter contact time was required with higher decolorization efficiency by comparing the performance with classic Fenton process. They concluded that the ultrasound coupled with Fenton oxidation may be utilized to treat textile wastewater in order to save time and operating cost (Çetinkaya et al., 2018).

The treatment of Rhodamine B using ultrasound combined with oxalic acid/iron foam Fenton-like process was evaluated by Li et al. The effects of various parameters such as dye concentration, oxalic acid concentration, and initial pH were tested. They found out that the rate and extent of color removal of the dye enhanced with increasing the oxalic acid concentration. However, it slightly decreased with increasing dye concentration. Good removal efficiencies were gained in the wide pH range (3–9). It was found that, the sonication significantly accelerated the Fenton-like reaction. The maximum removal of 92.16% was achieved at dye concentration of 5 mg L^{-1}; oxalic acid concentration of 25 mM at pH of 3. In general, they concluded that the combined methods give better removal efficiency when compared with the single processes based on Fenton type or ultrasound irradiation (Li et al., 2018).

The removal of Direct Red 81 (reactive azo dye) by Sono-Fenton method was investigated by Harichandran and Prasad. The combination of Fenton process with sonolysis showed a synergistic effect. This effect enhanced the degradation of Direct Red 81. The effect of the initial dye concentration, pH, and catalyst amount on the performance of the combined process was tested. The dye decolorization obeyed apparent first-order kinetics. The optimized parameters were pH = 3.0, Fe^{+2} concenration of 0.2 gL^{-1}, H_2O_2 concenration of 5.1×10^{-3} molL^{-1}, and ultrasonic frequency of 120 kHz, 60 W. Operating at these conditions, 99% of decolorization was obtained. This chapter shows that Fenton and Sono-Fenton methods can be used to remove color of Direct Red 81 dye in wastewaters (Harichandran and Prasad, 2016).

2.3.3 Combination of fenton process, ultraviolet light, and sonolysis

The application of ultrasound and UV along with Fenton reagent is known as sono-photo-Fenton oxidation. In this hybrid system, hydroxyl radicals and hydrogen atoms are produced by the sonocation process. It is pointed out that significant loss of hydrogen and hydroxyl radical species may be faced because of the recombination. But, introducing the UV light converts the H_2O_2 formed by recombination of hydroxyl radicals. By the help of sonolysis and photolysis, the intermediate complex generated due to the reaction between Fe^{+3} and H_2O_2 during the Fenton process can be reduced to Fe^{+2} (Sina and Mohsen, 2017; Matavos-Aramyan and Moussavi, 2017; Ribeiro et al., 2015; Pradhan and Gogate, 2010).

For photocatalytic applications, penetration of UV light may be low due to the fine suspension of photocatalyst, and hence the applicability of a large-scale photocatalytic process is still an open question. But, due to its good degradation performance and relatively low reaction times, the use of photochemical methods as a posttreatment or as a combined procedure remains as a promising solution (Barrera-Salgado et al., 2016; Fadzil et al., 2012).

Barrera-Salgado et al. investigated the combination of Fenton method with the ultrasound and UV light to enhance in the hydroxyl radical production making the treatment process more efficient. They applied the processes on a model pollutant acid green 50 textile dye which is a triarylmethane anionic dye by testing the parameters like iron (catalyst) amount, annealing temperature, concentration of dye, and pH. By the experimental results gained they pointed out that in the case when the heterogeneous Fenton process is combined with ultrasound and UV light irradiation, the process efficiencies were better than that obtained by the single process. They attributed the effect of the ultrasonic irradiation to the enhancement in the mass transfer of the substrates to the catalyst surface by cavitation generation of extra oxidizing agent such as hyroxyl radicals and increase in the activity of the catalyst surface area is related to the chemical activation of the catalyst surface (Birkin and Silva, 1996; Zhang et al., 2013). They concluded that such combined application is a promising AOP for the treatment of effluent with AG50 textile dye 3 (Barrera-Salgado et al., 2016).

2.3.4 Combination of fenton process with ultraviolet light and ozonation

Fadzil et al. examined various types of AOPs for the treatment of an azo dye, methyl orange. Their aim was to determine the performance of ozone-assisted homogeneous and heterogeneous photo catalytic systems in decolorization of methyl orange. In the homogeneous case, they applied the hybrid system of $UV/Fe^{+3}/H_2O_2$ and in the heterogeneous case UV/TiO_2 in the presence of ozone at near neutral pH. The presence of ozone in both catalytic systems was considerably effective in methyl orange removal efficiency. They concluded that, the best performance was gained in the homogeneous hybrid catalytic method (Fadzil et al., 2012).

2.3.5 Combination of ozonation and ultraviolet light

Application of combined ozonation process both in the presence of catalyst or noncatalyst, for the treatment of colored wastewater has been studied by many researchers and found to be more effective for the color removal (Dong et al., 2007; Hsing et al., 2007). As it is well known, dye containing effluents (textile industry effluents, etc.) have the characteristics of high coloration and the presence of toxic or carcinogenic pollutants. Hence, to minimize the environmental impact of these effluents, photochemical and ozonation processes were applied in the combined form in literature.

Moraes et al. studied the degradation of textile effluent by using photocatalytic and ozonation process alone and combined to investigate the reduction of total organic content, color, and toxicity of textile effluent. If the processes applied were compared, they mentioned that the ozonation process was the less efficient. They pointed out that the ozonation promoted only partial decolorization (60%) and negligible TOC removal. The photocatalytic method showed a fast color removal and an effective TOC reduction. But, the best results were obtained for the combined case by the total color removal and considerably high TOC removal (Moraes et al., 2000).

2.3.6 Combination of ozonation and sonolyis

Ince and Tezcanlı tried to solve the main problem of insufficient mineralization of pollutants faced in ozonation process by combining the process with ultrasonic irradiation. They investigated the performance of combining ultrasonic irradiation with ozonation in wastewaters containing reactive dye, CI Reactive Black 5. They concluded that the combined action of ultrasound and ozonation induced a synergistic effect on both color removal and degradation. They explained this synergy with the mechanical effects of ultrasound enhancing the mass transfer of ozone in solution and to the additional decomposition pathways for ozone by the radical chain reactions occurring during thermolysis of water and ozone. They concluded that the performance of ozonation is significantly increased by simultaneous irradiation of the wastewater with ultrasound (Ince and Tezcanlı, 2001).

2.3.7 Combination of sonolysis and ultraviolet light

In the sonolysis process for the cases where the hyroxyl radicals recombine with each other and form H_2O_2 and cause a loss of oxidation potential, the process may be used in combination with UV irradiation to prevent the resplitting of the generated H_2O_2 that regenerates the hydroxyl radicals (Chakma and Moholkar, 2014).

Chakma and Moholkar have investigated the combination of UV-Fenton oxidation and sonolysis in the treatment of bisphenol A. They found out that the techniques of sonolysis, Fenton reaction, and sono–Fenton reaction had produced a good combined effect, while the combination of Fenton reaction and UV had no synergy. They pointed out that procedure and the method used in the combination of sono–Fenton-like reaction and UV had been found to have negative synergy. The highest TOC removal efficiency was gained with Fenton oxidation combined with sonolyis (Chakma and Moholkar, 2014).

Four different configurations of two AOPs, photocatalytic oxidation, and sonolysis were studied by evaluating the removal of an azo dye, naphthol blue black by Stock et al. These are just sonolysis, just photocatalytic oxidation, simultaneous sonolysis, and photocatalysis and finally sequential sonolysis and photocatalysis. The intermediates formed during the oxidation of the azo dye were mineralized fast in the simultaneous process indicating its requirement for the environmental treatments. In both simultaneous and sequential case, a positive effect on the degradation rate was determined. Sonolysis process seemed to be effective in degradation rate, while the photocatalytic process was more effective for promoting the degradation. They concluded the main advantage of sonication on photocatalytic oxidation is that during simultaneous sonolysis and photocatalysis, the photocatalyst surface was to be constantly refreshed and the mass transfer of the both reactants and products to and from the surface of the catalyst improved as the slurry was mixed because the sonocation was preventing the agglomeration of the catalyst (Stock et al., 2000).

3. Decolorization and degradation of dyes by combination of advanced oxidation processes and other physicochemical processes

The physicochemical techniques such as coagulation, flocculation, adsorption, and precipitation membrane separation processes have been used intensively for the treatment of dye containing wastewaters. Individually, besides having advantages, all the treatment processes have certain disadvantages. Consequently, nowadays, application of physicochemical methods combined with the AOPs has also come out to be promising alternatives.

3.1 Individual physicochemical processes

The physicochemical techniques involve either physical process or chemical process or usually the combination of them. These techniques may be applied as pretreatment, final treatment, as well as specific treatment in wastewater treatment applications. They are applied for the removal of heavy metals, oils and greases, dyes, suspended matter organic inorganic components, toxic pollutants or high salt concentrations, etc.

Among the physicochemical processes, the coagulation–flocculation, electrocoagulation, adsorption with activated carbon, and processes with membrane are preffered to be applied in combination with the AOPs (Aguilar et al., 2011).

3.1.1 Coagulation

Coagulation technology is recently preffered as the first choice of physicochemical treatment technologies since it is an economic method. The coagulation method uses a chemical agent to convert pollutants that are dispersed in water into a destabilized state and accumulates them into a mixture that is easier to separate to treat contaminants (Zheng et al., 2013). It is generally utilized as a pretreatment in dye treatment (Li and Yang, 2018).

3.1.2 Electrocoagulation

Electrocoagulation is an attractive alternative for chemical coagulation which uses electrochemical treatment to treat dye containing effluents. The method serves a high degradation performance and presents a reliable, simple, and economical way for wastewater treatment without any requirement for extra chemicals (Khandegar and Saroha, 2013; Song et al., 2008). In the electrocoagulation process, usually aluminum or iron electrodes are used as coagulant because the adsorption capacities of the metal hydroxides such as $Al(OH)_3$, $Fe(OH)_2$, and $Fe(OH)_3$ are very high (Kabdaşlı et al., 2012). Electrocoagulation is related on generation of metal hydroxide flocs by dissolving anodes with the applied current (Ozyonar and Karagozoğlu, 2012). Electrocoagulation stabilizes suspended solids and colloidal materials. When the particles are properly contacted with the electrodes in electrical applications, the particles are neutralized. Large flocs are formed by the combination of different particles (Mollah et al., 2001). Electrocoagulation seems to be a better alternative than the classical coagulation process, since it can remove the smallest colloidal particles and form considerably small amount of sludge (Elimelech and O'Melia, 1990; Vik et al., 1984).

3.1.3 Adsorption

The adsorption process is the mass transfer process where the two phases are attaching each other by forming an interface. Adsorbent dosage, temperature, contact time, mixing speed, particle size, initial pH of solution, and adsorbate concentration are the main factors affecting the adsorption process efficiency (Patel and Vashi, 2010; Igwegbe et al., 2015).

3.1.4 Membrane separation

Membrane separation is a method which depends on three basic principles including adsorption, sieving, and electrostatic phenomenon diffusion (Nqombolo et al., 2018; Padaki et al., 2015). Membrane separation can be applied to the effluents that are hard to be treated by conventional processes, and the change of water characteristics will not make a considerable effect on the treatment performance. Various membrane processes with different separation processes have been investigated. These methods are mainly microfiltration, ultrafiltration, nanofiltration, and reverse osmosis (Nqombolo et al., 2018).

3.1.5 Ion exchange

Ion exchange can be defined as reversible interchange of ions between a solid (ion exchange material) and a liquid. There is no permanent change in the solid structure. In the process, by the exchange of cations or anions between the pollutants and the exchange medium, the ions are removed from the effluent. Ion exchange materials may be resins made from synthetic organic materials having ionic functional groups to which exchangeable ions are attached. They also may be inorganic and natural polymeric materials. Ion exchange process has been a widely preferred method since it is environmentally friendly, easily applied, and resistant to high temperatures (Kaur and Jindal, 2018).

3.2 Limitation of physicochemical processes

The physicochemical processes do not always present satisfactory performance because of the characteristics of the dye containing effluents. In addition, traditional physical–chemical processes have various limitations such as they are comparatively expensive and may lead to secondary pollution, etc.

Below some limitations of the physicochemical processes in the treatment of dye containing wastewaters are summarized.

3.2.1 Coagulation

The method has a low treatment rate of dissolved substances and has unsatisfactory results with azo, reactive, acid, and basic compounds. The process forms huge amount of sludge, and it is hard to remove completely pathogens and toxic trace materials in effluent. Hence, it is usually preferred as a pretreatment method. The removal is highly dependent on pH (Li and Yang, 2018; Robinson et al., 2001).

3.2.2 Adsorption

It is a pH-dependent process considering just the phase change of pollutants. Unsatisfactory result for disperse, sulfur, and vat dyes are generally obtained. It has problem of formation and disposal of the sludge. Generally, the adsorbents used are expensive, and hence this situation necessitates the adsorbents regeneration (Hai et al., 2007).

3.2.3 Membrane filtration

Concentrated sludge production is one of the main disadvantages. There is a requirement for membrane replacement but unfortunately the membrane module is usually very expensive. Generally, polymer-based membranes such as polysulfone and polyethersulfone are utilized membrane material, but since these polymers are hydrophobic, they have fouling tendency (Nqombolo et al., 2018).

3.2.4 Ion exchange

Suspended solids (>10 ppm), oil, and grease may clog the exchange resin. Consequently, before application of ion exchange, suitable pretreatment methods are necessary to remove suspended solids from wastewater. Oxidants in water, the pH of the water may damage the ion exchange resin. Also, appropriate ion exchanger resins are not present for all the metals. The costs (both capital and operating) are reported to be high (Ahmed et al., 1998). If the water to be treated contains bacteria, bacterial contamination can be faced. In the regeneration step, wastewater that needs extra treatment and disposal is formed. Inorganic ion exchangers have low chemical and mechanical strength, while organic ones are sensitive to strong dose of radiation (Kaur and Jindal, 2018).

3.3 Combination of advanced oxidation processes and other physicochemical processes

Recently, applications of physicochemical systems with combination of the AOPs have been an interesting and innovative approach for the treatment of dye containing wastewaters. By the help of combined applications, the strengths of individual methods can be enhanced and the limitations that are faced in the individual applications can be handled.

Among the physicochemical processes, the coagulation–flocculation, electrocoagulation, adsorption with activated carbon, and processes with membrane are preferred to be applied in combination with the AOPs (Aguilar et al., 2011).

Typical examples of combinations between AOPs and conventional physicochemical processes are as follows:

- Coagulation-based combinations
  - Fenton Reagent–Coagulation
  - Photocatalytic Oxidation–Coagulation
  - Photocatalytic Oxidation–Fenton Reagent–Coagulation
  - Ozonation–Coagulation
- Adsorption-based combinations
  - Photocatalytic Oxidation–Adsorption
  - Ozonation–Adsorption
- Membrane separation-based combinations
  - Photocatalytic Oxidation–Membrane Separation
  - Ozonation–Membrane Separation
  - Wet Air Oxidation–Membrane Separation
  - Wet Air Oxidation–Sonication–Membrane Separation
  - Photocatalytic Oxidation–Fenton Reagent–Membrane Separation
  - Photocatalytic Oxidation–Fenton Reagent–Coagulation–Membrane Separation

3.3.1 Coagulation-based combinations

In literature, coagulation-based combinations with AOPs applied in consecutive or in simultaneous manner have been studied to treat colored wastewater.

3.3.2 Fenton reagent oxidation–coagulation

Simultaneous usage of coagulation and Fenton reagent reaction has shown advanced performance over their individual applications (Kim et al., 2004). Application of coagulation and Fenton reagent process consecutively has been determined to decrease the settling time of the floc, improve the color removal efficiency, and to maintain the iron concentration in the determined limits (Lin and Lo, 1997).

The combination of coagulation and Fenton oxidation of disperse and reactive dyes was studied by Kim et al. The performances of Fe^{+3} coagulation and Fenton reaction of colored wastewaters were examined and compared in terms of COD and color removals. It was found that the hybrid process, Fenton oxidation with Fe^{+3} coagulation, had removed 90% of COD and 99% of dye successfully (Kim et al., 2004).

GilPavas et al. studied on the treatment of industrial wastewater by application of coagulation and flocculation sequential with an AOP. The advanced methods they tested were Fenton and photo-Fenton processes. The coagulation–flocculation as pretreatment removed 98% of turbidity and 48% of COD. The effluent of coagulation–flocculation process was subsequently subjected to either Fenton or photo-Fenton process. By the combined process including Fenton reagent 74% of COD was removed, whereas with photo-Fenton process the removal increased to 87%. They concluded that the results they gained showed that the application of coagulation–flocculation sequential with the AOPs for the industrial textile wastewater treatment was promising (GilPavas et al., 2017).

Zhang et al. tested the toxicity and genotoxicity of dye containing wastewater and performed the Fenton–coagulation method. The results showed that the dye effluent had high toxicity and genotoxicity. After application of Fenton-coagulation process, the dye effluent displayed no considerable acute toxicity and genotoxicity with a COD value less than 50 mgL^{-1}. They concluded that Fenton–coagulation process could lower the acute toxicity and genotoxicity as well as the chemical oxidation demand of the effluent (Zhang et al., 2014).

3.3.3 Photocatalytic oxidation–coagulation

The combination of coagulation and photocatalytic degradation with nanosized MgO was investigated for both Acid-Red 73 and real textile wastewater dye degradation by Jorfi and his friends. The operating parameters effecting the performance of the process such as pH and coagulant and photocatalyst amount were tested in synthetic wastewater that was prepared by Acid-Red 73. Complete degradation of Acid-Red 73 was achieved with MgO nanoparticles of 0.8 gL^{-1}, pH = 5, and reaction time of 1 h. Then these optimized parameters were applied for the treatment of real textile industry wastewater. In the case of real textile wastewater, with this sequence COD and TOC removals of 98.3% and 86.9% were obtained, respectively, in 5 h of reaction time (Jorfi et al., 2016).

Lam et al. realized a study on the hybrid proces including coagulation and photocatalytic oxidation of malachite green containing textile wastewater over ZnO micro/nanoflowers. The ZnO micro/nanoflowers were synthesized following reflux route. The effect of operating parameters such as the type and the amount of coagulant, pH, photocatalyst amount, and light power was investigated. The $CaCO_3$ was found to be the best coagulant because by this coagulant, high amount of precipitates were formed and the cationic dye molecules were adsorbed successfully. By this coagulant, 88.3% of malachite green was removed at pH = 9. Total removal of the dye was achieved with 0.5 gL^{-1} of ZnO micro/nanoflowers and 105 W light power. The kinetic data followed Langmuir–Hinshelwood model. In addition to these, 100% of malachite green dye and 80.0% of COD removal were observed over sequencing coagulation–photocatalytic process (Lam et al., 2018).

3.3.4 Photocatalytic oxidation–fenton reagent–coagulation

The photo-Fenton oxidation subsequent to coagulation was mentioned to give promising results, by reducing load on the AOP.

A multistage application including electrocoagulation, solar photo-Fenton, and biological oxidation methods for industrial wastewater treatment was investigated by Manenti and his friends. Each process was integrated in batch mode. The performances of the methods were assessed with respect to biodegradability index–dissolved organic carbon and COD. The optimum operating conditions for each process was determined. Partial degradation was achieved by the electrocoagulation method, with a 36% removal in COD. Eventhough, the electrocoagulation method ensured a considerable organic matter removal, the effluent was unsuitable for discharge into waterbodies For this reason, photo-Fenton process was tested as the subsequent step. 70% biodegradability index was achieved by performing the photo-Fenton at 100 mg Fe^{2+} per liter, pH = 2.8, 12 mM H_2O_2. Finally, with the biological process step a considerable decrease in COD was achieved. They concluded that treatment system consisting of electrocoagulation, solar photo-Fenton, and biological oxidation methods is promising technique that allows sufficient treatment performance with respect to the environmental limitations (Manenti et al., 2014).

3.3.5 Ozonation–coagulation

Application of coagulation followed by ozonation was verified to be better to their individual applications.

The treatment of chlorinated effluent coming from the production of azoic dyes was investigated by the combination of ozonation and $Ca(OH)_2$ coagulation by Sarasa et al. Mainly aniline derivatives and azo compounds and some organochloride compounds were present in the wastewater. After the application of ozonation, the concentration of the compounds present was decreased to acceptable levels but some by-products such as nitrobenzene derivatives and organic acid compounds were formed. Hence, the coagulation was applied to eliminate these remaining compounds. COD was decreased by 25% after ozonation, while it decreased by 50% after the complete removal. No change was observed in TOC in the ozonation step, but a 42% removal was achieved after chemical coagulation. In addition, the color of the wastewater was successfully removed (Sarasa et al., 1998).

Treatment of textile waste effluents by ozonation in a multiple reactor system and chemical coagulation was performed by Sheng and Chi. It was concluded that decolorization could be obtained in a very short time. In combination with coagulation, the COD of those wastewater could be consistently decreased by up to 70% or higher. Ozonation process is able to decompose complex structured molecules into simpler ones. These simple ones can be easily biodegraded with a further technique. Hence, combination of ozonation, coagulation, and the activated sludge methods could provide a promising application (Sheng and Chi, 1993).

Song et al. investigated the performance of ozone-promoted electrocoagulation method in the color removal of aqueous solution of CI Reactive Blue 19 utilizing iron electrodes. They tested the effects of different parameters such as pH, concentration of dye, current density, salt concentration, temperature, flowrate of ozone, and distance between electrodes on the color removal performance. All of the operating factors except for the initial dye concentration had both positive on the efficiency of the process. They concluded that the yield of the electrocoagulation method could be enhanced by using a subsequent ozonation process. The hybrid electrocoagulation and ozonation was found to be an efficient alternative to treat effluents having high content reactive dyes (Song et al., 2008).

3.3.6 Adsorption-based combinations

As it is well known, adsorption process is a process considering sole the phase change in the contaminants. Consequently, further treatment is always required for the disposal of the sludge. Hence, it can be concluded that combination of adsorption process with various methods seems to be promising.

3.3.7 Photocatalytic oxidation—adsorption

As it is well known, photocatalytic treatment has some limitations in practical applications. For example, using TiO_2 has the restriction since TiO_2 has a wide bandgap. In addition, photo-induced electron—hole pairs have short lifetime. Another is that photocatalytic applications are convenient for the degradation of wastewaters containing low concentration dyes. Hence, frequently in the treatment of high-strength wastewaters successful results cannot be achieved. Consequently, it is inevitable to prepare materials that have high adsorption ability and good photocatalytic performance. Having both these features will be an effective solution to enhance pollutant degradation (Nguyen and Juang, 2019).

The removal of cationic dyes from aqueous solution by hybrid application of adsorption and photocatalysis in the presence of two types of platinum-doped titanate nanoparticles (titanate nanosheets and titanate nanotubes) was studied by Nguyen and Juang. All prepared materials showed good performance in the removal of MB through the synergy of photocatalytic oxidation and adsorption. But, such a consequence was not valid for dyes that are anionic. For cationic and anionic dyes, titanate nanotubes demonstrated higher photocatalytic activity than titanate nanosheets. They concluded that Pt-doped titanate nanocomposites seemed to be effective catalysts to treat wastewaters especially containing cationic dyes (Nguyen and Juang, 2019).

Treatment of crystal violet dye by simultaneous adsorption and photocatalysis in a continuous flow microreactor in the presence of ZnO/zeolite pellets was investigated by Sacco et al. Treated water had a significant reduction of toxicity. A preliminary mathematical model was developed. They compared the performances of adsorption individual and simultaneous application of adsorption and photocatalysis. The results showed that UV irradiation can obtain a steady-state crystal violet concentration value corresponding to an equilibrium condition between adsorption and photocatalysis. Crystal violet adsorption constants were determined from experimental data performed at dark conditions in a batch mode (Sacco et al., 2018).

Photocatalytic oxidation of MB dye was tested in the presence of the titanate nanotubes. The nanotubes were prepared with different sodium amounts to be tested in the degradation in terms of adsorption and mineralization by Sandavol et al. Higher adsorption capacity was obtained when the sodium amount was higher. In this case, the MB dye adsorption followed the cation exchange mechanism. This high adsorption capacity had been explained by their structure, the sodium amount, crystalline phase, and specific surface area (Lee et al., 2008). But a high color removal was not accompanied by the complete degradation of the dye. This was related to a screen effect. The effect prevents the formation and diffusion of the hydroxyl radicals on the surface of the catalyst (Sandavol et al., 2017).

3.3.8 Ozonation–adsorption

Combined system having ozonation, activated carbon, and biological erated filter was investigated at labscale by Zou. In the experiments, the removal characteristics of organics, dye, and genotoxicity were evaluated. Zou concluded that ozonation changed the molecular structure of organic pollutants while destroying the chromophores and increased the biodegradability. The amount of ozone applied per mg of COD_0 is 0.25 mg, 40 min of ozonation contact time, and 30 min of hydraulic retention time were the optimized conditions for the hybrid application of ozonation, adsorption, and biological erated filter process. Under these conditions, the results obtained showed that the combined process could ensure the water reuse (Zou, 2015).

The study of Drimarene Red CL-3B decolorization by ozonation in a semibatch reactor followed by granular activated carbon adsorption was realized by Konsowa et al. In the tests, the effect of dye and ozone concentration, pH, and ozone-air flow rate on the rate of color removal was investigated. They mentioned that the rate of color removal increased with increasing ozone amount, ozone-air flow rate, and pH but decreased with increasing dye concentration. The method initiated with ozonation for efficient color removal, followed by adsorption method to achieve color and TOC removal. The adsorption method was found to be promising in removal of ozonation residual TOC (Konsowa et al., 2010).

3.3.9 Membrane separation-based combinations

Membrane separation gives various benefits for treatment of wastewater. These benefits are good process efficiency, consume lower energy, and have simple application. But, usage of membrane technology in the treatment of highly contaminated wastewater is limited because of the unavoidable fouling of the membranes. The hydrophobic nature of membrane materials and the various contaminants in wastewater may cause the membrane fouling. AOPs are widely used and suggested to be used in combination to eliminate the membrane fouling (Wang et al., 2019).

3.3.10 Photocatalytic oxidation—membrane separation

Using membrane technology in pilot-scale applications generally limited the fouling in the membrane since the fouling reduces both the performance of the separation and the lifetime of the membrane. By the presence of hydroxyl radicals, the photo-Fenton—like oxidation may remove different kinds of pollutants and combination of the process with membrane technology may enhance the process efficiency by improving the antifouling property (Wang et al., 2019).

The combined processes including membrane and photocatalytic methods have been studied also to overcome the drawback of catalyst separation from treated effluent. In addition, the membrane will serve as barrier for the species to be degraded. The membrane photocatalytic reactor systems use the catalytic features of the photoreactor and the separation features of the membrane in order to separate the treated water (Doruk et al., 2016).

The hybrid membrane photocatalytic reactor was designed by Doruk et al. by submerging a flat polyethersulfone low-pressure membrane module into a ZnO or TiO_2 slurry photocatalytic system to treat textile wastewater. They realized the study in two steps. In the first step, the performance of various parameters like the catalyst type and amount, concentration of wastewater, light type, and semiconductor powder type was investigated using a batch system without a membrane module. After the optimization step, the long-term tests were performed in the hybrid membrane photocatalytic reactor.

They concluded that approximately 100% color and 30%–55% COD removals were achieved at a hydraulic retention time of 6 h. The results presented that the treated water could be used as process water in the industry (Doruk et al., 2016).

Buscio et al. investigated the combination of photocatalysis with microfiltration to remove CI. Disperse Red 73 to overcome the main problem of photocatalytic method which is the recovery of the catalyst. The process was optimized by testing the effects of various parameters such as catalyst loading initial concentration of dye and pH. With the photocatalytic treatment degradation between 60% and 90% was achieved. Application of membrane treatment provided an uncolored permeate with considerably low concentration of residual organic matter. They concluded that the study they performed showed the applicability of the combined use of photocatalytic oxidation and membrane filtration for the treatment of colored wastewaters (Buscio et al., 2015).

3.3.11 Ozonation–membrane separation

Membrane and ozonation hybrid system was tested in the treatment of simulated textile wastewater. The effects of various parameters such as pressure, feed concentration, and membrane fouling on permeate flux were examined. The membrane filtration method formed a permeate with approximately 99% of the color removal, while 85% of salt by mass and 85% of the original water were reusable (Wu et al., 1998).

3.3.12 Wet air oxidation–membrane separation

A hybrid process containing membrane and WAO had been used in treatment of the disperse dye bath waste by Dhale and Mahajani. To demonstrate the process effluent, the synthetic wastewater was prepared with disperse dye CI 79. With nanofiltration membrane higher than 99% color and 97% COD removals were achieved. The concentrate was then treated by WAO process in the temperature range of 160–225 °C and oxygen partial pressure of 0.69–1.38 MPa in the presence of homogeneous copper sulfate The non catalytic WAO resulted in 75% COD removal, whereas using catalyst showed about 90% reduction in COD (Dhale and Mahajani, 2000).

The treatment of colored wastewater concentrate coming from the membrane separation was studied by a series of WAO experiments in an autoclave reactor. After membrane separation subsequently WAO was applied. The main goal was to be able to perform the tests at moderate conditions, and hence two new oxidation methods were investigated. In the first case it was aimed to replace oxygen partially or totally with hydrogen peroxide which is a stronger oxidant and the other was using a catalyst. Each method was found to attain a higher initial reaction rate of TOC and color removal at lower temperatures and pressures if compared with the traditional WAO. By using the two methods combined, approximately 80% of TOC and 90% of color removal were obtained within 30 min at 110 °C and 50 kPa (Lei et al., 1998).

3.3.13 Wet air oxidation–sonication–membrane separation

A hybrid system comprising membrane separation followed by sonication and WAO was tested to treat the dye (procion turquoize blue CI25) containing wastewater for reuse and discharge. The tests were first carried out with the reactive dye solution. A nanofiltration membrane (MPT 30) was first applied then the concentrate was subjected to WAO process. After parameters were optimized, the tests were continued with a real dye effluent. The effluent was found to be resistant to WAO at the determined conditions. Sonication of the concentrate obtained after membrane filtration, in the presence of $CuSO_4$, made the waste stream suitable for the WAO. Sonication followed by WAO was found to be more suitable at neutral conditions as compared with alkaline conditions (Dhale and Mahajani, 1999).

3.3.14 Photocatalytic oxidation–Fenton reagent–membrane separation

Wang et al. synthesized the $CuFe_2O_4$ particles and doped them in the PVDF@$CuFe_2O_4$ membranes. They tested them in the degradation of MB containing wastewater. The optimum conditions for the photo-Fenton–like oxidation were determined as $CuFe_2O_4$ concentration of 1.0%, pH value as 3.0, and the hydrogen peroxide concentration of 400 μL. With the photo-Fenton cleaning process, the membrane showed variable antifouling property to various types of foulants. These foulants are some organic dyes, nature organic matter, and protein. If the photo-Fenton and membrane process are used in combination, the separation efficiency of the membrane enhanced considerably. They concluded that the integration of membrane technology by the Fenton-like process seems to be an attractive methodology for the treatment of wastewater (Wang et al., 2019).

3.3.15 Photocatalytic oxidation–Fenton reagent–coagulation–membrane separation

Application of a hybrid system consisting of UV-Fenton, coagulation, and membrane filtration to textile dye-bath effluents was realized by Ince and Tezcanlı. The effluent was subjected to Fenton and Fenton-like processes (Fe^{+2}/H_2O_2 and Fe^{+3}/H_2O_2), in the presence and absence of UV light. At the optimized conditions, for 20 min of UV irradiation, complete decolorization and 79% total organic carbon removal were achieved. The increased dissolved solid concentration of the treated solution proved the requirement of a suitable filtration to ensure the reusabilty of the treated effluent (Ince and Tezcanlı, 1999).

4. Conclusions

Many manufacturing industries including textile, pharmaceutical, food, paints, plastic etc., are often using dyes. Colored effluents are highly toxic to the aquatic life and mutagenic to humans. The effluents containing dyes have become a serious problem requiring special care. In the last years, environmental regulations have become strict, needing an improved quality of treated effluent. The presence of even trace amounts of dyes in effluent is not desired.

AOPs such as ozonation, Fenton reagent reaction, and WAO, are promising methods to degrade and mineralize dyes. Today, a wide range of physicochemical treatment techniques such as precipitation, coagulation–flocculation, flotation, and membrane filtration have been investigated for the treatment of dye containing wastewater.

Usually, a single process cannot produce effluent that meets regulatory requirements. In order to reach these requirements, a combined approach may serve as a solution to overcome such drawbacks.

It is evident from the literature survey, using a combinative approach is certainly a positive step (energy efficient and economical) to achieve quick mineralization.

A great number of hybrid processes are being developed, with the combination among AOPs and the combination of physicochemical treatments with the AOPs. The hybrid applications in the treatment of dye containing wastewater have proven to be very effective. The use of hybrid processes represents improvement in the performance, including both decolarization and COD removal. In addition to the increase in the efficiency, the combined systems make the operation simpler in a shorter time.

References

Adityosulindro, S., Barthe, L., González-Labrada, K., Haza, U.J.J., Delmas, H., Julcour, C., 2017. Sonolysis and sono-Fenton oxidation for removal of ibuprofen in (waste)water. Ultrason. Sonochem. 39, 889–896.

Aguilar, H.S.R., Hernandez, R., Mendoza, R., Valencia, M., 2011. Physicochemical treatment (coagulation-flocculation-Fenton) of mature leachates from Tuxtla Gutierrez, Chiapas landfill. Sustain. Environ. Res. 21 (5), 313–319.

Ahmed, S., Chughtai, S., Keane, M.A., 1998. The removal of cadmium and lead from aqueous solution by ion exchange with Na–Y zeolite. Separ. Purif. Technol. 13, 57–64.

Asghar, A., Raman, A.A.A., Daud, W.M.A.W., 2015. Advanced oxidation processes for in-situ production of hydrogen peroxide/hydroxyl radical for textile wastewater treatment: a review. J. Clean. Prod. 87, 826–838.

Bagal, M.V., Gogate, P.R., 2014. Wastewater treatment using hybrid treatment schemes based on cavitation and Fenton chemistry: a review. Ultrason. Sonochem. 21, 1–14.

Barrera-Salgado, K.E., Ramírez-Robledo, G., Alvarez-Gallegos, A., Pineda-Arellano, C.A., SSierra-Espinosa, F.Z., Hernadez-Perez, J.A., Silva-Martinez, S., 2016. Fenton process coupled to ultrasound and UV light irradiation for the oxidation of a model pollutant. J. Chem. 7, 2016.

Birkin, P.R., Martinez, S.S., 1996. A study of the effect of ultrasound on mass transport to a microelectrode. J. Electroanal. Chem. 416 (1–2), 127–138.

Blanco, J., Torrades, F., Morón, M., Brouta-Agnésa, M., García-Montaño, J., 2014. Photo-Fenton and sequencing batch reactor coupled to photo-Fenton processes for textile wastewater reclamation: feasibility of reuse in dyeing processes. Chem. Eng. J. 240, 469–475.

Bokare, A.D., Choi, W., 2014. Review of iron-free Fenton-like systems for activating H_2O_2 in advanced oxidation processes. J. Hazard Mater. 275, 121–135.

Brillas, E., Sirés, I., Oturan, M.A., 2009. Electro-Fenton process and related electrochemical technologies based on Fenton's reaction chemistry. Chem. Rev. 109, 6570–6631.

Buscio, V., Brosillon, S., Mendret, J., Crespi, M., Gutierrez-Bouzan, C., 2015. Photocatalytic membrane reactor for the removal of C.I. Disperse red 73. Materials 8, 3633–3647.

Cetinkaya, S.G., Morcali, M.H., Akarsu, S., Ziba, C.A., Dolaz, M., 2018. Comparison of classic Fenton with ultrasound Fenton processes on industrial textile wastewater. Sust. Environ. Res. 28 (4), 165–170.

Chakma, S., Das, L., Moholkar, V.S., 2015. Dye decolorization with hybrid advanced oxidation processes comprising sonolysis/Fenton-like/photo-ferrioxalate systems: A mechanistic investigation. Sep. Purif. Technol. 156, 596–607.

Chakma, S., Moholkar, V.S., 2013. Physical mechanism of sono-Fenton process. AIChE J. 59, 4303–4313.

Chakma, S., Moholkar, V.S., 2014. Investigations in synergism of hybrid advanced oxidation processes with combinations of sonolysis + Fenton process + UV for degradation of bisphenol A. Ind. Eng. Chem. Res. 53, 6855–6865.

Chakma, S., Moholkar, V.S., 2011. Mechanistic features of ultrasonic desorption of aromatic pollutants. Chem. Eng. J. 175, 356–367.

Chen, J., Zhu, L., 2011. Oxalate enhanced mechanism of hydroxyl-Fe Pillared bentonite during the degradation of orange II by UV Fenton process. J. Hazard Mater. 185, 1477–1481.

Dhale, A.D., Mahajani, V.V., 1999. Reactive dye house wastewater treatment. Use of hybrid technology: membrane sonication followed by wet oxidation. Ind. Eng. Chem. Res. 38, 2058–2064.

Dhale, A.D., Mahajani, V.V., 2000. Studies in treatment of disperse dye waste:Membrane–wet oxidation process. Waste Manag. 20, 85.

Dong, Y., He, K., Zhao, B., Yin, Y., Yin, L., Zhang, A., 2007. Catalytic ozonation of azo dye active brilliant red X-3B in water with natural mineral brucite. Catal. Commun. 8, 1599–1603.

Doruk, N., Yatmaz, H.C., Dizge, N., 2016. Degradation efficiency of textile and wood processing industry wastewater by photocatalytic process using in situ ultrafiltration membrane. Clean 44 (3), 224–231.

Elimelech, M., O'Melia, C.R., 1990. Kinetics of deposition of colloidal particles in porous media. Environ. Sci. Technol. 24, 1528–1536.

Fadzil, N.A.M., Zainal, Z., Abdullah, A.H., 2012. Ozone-assisted decolorization of methyl orange via homogeneous and heterogeneous photocatalysis. Int. J. Electrochem. Sci. 7, 11993–12003.

GilPavas, E., Dobrosz-Gómez, I., Gómez-García, M.A., 2017. Coagulation-flocculation sequential with Fenton or Photo-Fenton processes as an alternative for the industrial textile wastewater treatment. J. Environ. Manag. 191, 189–197.

Gogate, P.R., Pandit, A.B., 2004. A review of imperative technologies for wastewater treatment I:oxidation technologies at ambient conditions. Adv. Environ. Res. 8, 501–551.

Habib, A., İsmail, I.M.I., Mahmood, a.J., Ullah, R., 2012. Photocatalytic decolorization of brilliant golden yellow in TiO_2 and ZnO suspensions. J. Saudi Chem. Soc. 16 (4), 423–429.

Hai, F.I., Yamamoto, K., Fukushi, K., 2007. Hybrid treatment systems for dye wastewater. Crit. Rev. Environ. Sci. Technol. 37, 315–377.

Harichandran, G., Prasad, S., 2016. SonoFenton degradation of an azo dye, Direct red. Ultrason. Sonochem. 29, 178–185.

Hassan, D.U.B., Aziz, A.R.A., Daud, W.M.A.W., 2012. On the limitation of Fenton oxidation operational parameters: a review. Int. J. Chem. React. Eng. 10.

He, Y., Wang, X., Xu, J., Yan, J., Ge, Q., Gu, X., Jian, L., 2013. Application of integrated ozone biological aerated filters and membrane filtration in water reuse of textile effluents. Bioresour. Technol. 133, 150–157.

Hsing, H.-J., Chiang, P.-C., Chang, E.-E., Chen, M.-Y., 2007. The decolorization and mineralization of acid orange 6 azo dye in aqueous solution by advanced oxidation processes: a comparative study. J. Hazard Mater. 141, 8–16.

Hua, L., Ma, H., Zhang, L., 2013. Degradation process analysis of the azo dyes by catalytic wet air oxidation with catalyst CuO/Al_2O_3. Chemosphere 90, 143–149.

Igwegbe, C.A., Onyechi, P.C., Onukwuli, O.D., Nwokedi, I.C., 2015. Adsorptive treatment of textile wastewater using activated carbon produced from Mucuna pruriens seed shells. World J. Eng. Technol. 4 (1), 21.

Ince, N.H., Tezcanı, G., 2001. Reactive dyestuff degradation by combined sonolysis and ozonation. Dyes Pigments 49, 145.

Ince, N.H., Tezcanlı, G., 1999. Treatability of textile dye-bath effluents by advanced oxidation: preparation for reuse. Water Sci. Technol. 40 (1), 183.

Ji, F., Li, C., Zhang, J., Deng, L., 2011. Heterogeneous photo-fenton decolorization of methylene blue over LiFe(WO4)2 catalyst. J. Hazard Mater. 186, 1979–1984.

Jing, G., Luan, M., Chen, T., 2016. Progress of catalytic wet air oxidation technology. Arab. J. Chem. 9, 1208–1213.

Jorfi, S., Barzegar, G., Ahmadi, M., Soltani, R.D.C., Haghighifard, N.J., Takdastan, A., Saeedi, R., Abtahi, M., 2016. Enhanced coagulation-photocatalytic treatment of Acid red 73 dye and real textile wastewater using UVA/synthesized MgO nanoparticles. J. Environ. Manag. 177, 111–118.

Kabdaşlı, I., Arslan-Alaton, I., Ölmez-Hancı, T., Tünay, O., 2012. Electrocoagulation applications for industrial wastewaters: a critical review. Environ. Technol. Rev. 1 (1), 2–45.

Kaur, K., Jindal, R., 2018. Synergistic effect of organic-inorganic hybrid nanocomposite ion exchanger on photocatalytic degradation of Rhodamine-B dye and heavy metal ion removal from industrial effluents. J. Environ. Chem. Eng. 6, 7091–7101.

Keav, S., Martin, A., Barbier Jr., J., Duprez, D., 2010. Deactivation and reactivation of noble metal catalysts tested in the Catalytic Wet Air Oxidation of phenol. Catal. Today 151, 143.

Khandegar, V., Saroha, A.K., 2013. Electrocoagulation for the treatment of textile industry effluent–a review. J. Environ. Manag. 128, 949–963.

Kim, T.-H., Park, C., Yang, J., Kim, S., 2004. Comparison of disperse and reactive dye removals by chemical coagulation and Fenton oxidation. J. Hazard Mater. B 112, 95.

Konsowa, A.H., Ossman, M.E., Chen, Y., Crittenden, J.C., 2010. Decolorization of industrial wastewater by ozonation followed by adsorption on activated carbon. J. Hazard Mater. 176, 181–185.

Krishnakumar, B., Swaminathan, M., 2011. Influence of operational parameters on photocatalytic degradation of a genotoxic azo dye Acid Violet 7 in aqueous ZnO suspensions. Spectrochim. Acta A. Mol. Biomol. Spectrosc. 81, 739–744.

Krishnan, S., Rawindran, H., Sinnathambi, C.M., Lim, J.W., 2017. Comparison of various advanced oxidation processes used in remediation of industrial wastewater laden with recalcitrant pollutants. In: IOP Conference Series: Materials Science and Engineering, vol. 206. IOP Publishing (1).

Lam, S., Low, X.D., Wong, K., Sin, J., 2018. Sequencing coagulation−photodegradation treatment of Malachite Green dye and textile wastewater through ZnO micro/nanoflowers. Chem. Eng. Commun. 205 (8), 1143−1156.

Lee, C.-K., Wang, C.-C., Juang, L.-C., Lyu, M.-D., Hung, S.-H., Liu, S.-S., 2008. Effects of sodium content on the microstructures and basic dye cation exchange of titanate nanotubes. Colloid. Surface. Physicochem. Eng. Aspect. 317, 164−173.

Lei, L., Hu, X., Yue, P.-L., 1998. Improved wet oxidation for the treatment of dyeing wastewater concentrate from membrane separation process. Water Res. 32−9, 2753.

Li, G., Qiu, S., Ma, F., Ji, Y., Jiang, X., 2018. Degradation of RhB by a sono-Fenton-like process with an iron-foam in the presence of oxalic acid. Anal. Methods 10, 3976−3983.

Li, Z., Yang, P., 2018. Review on physicochemical, chemical, and biological processes for pharmaceutical wastewater. IOP Conf. Ser. Earth Environ. Sci. 113, 012185.

Lin, S.H., Lo, C.C., 1997. Fenton process for treatment of desizing wastewater. Water Res. 31 (8), 2050−2056.

Manenti, D.R., Aparecido, N., Módenes, A.N., Petrick, A., Soares, P.A., Espinoza-Quiñones, F.R., Rui, A.R., Boaventura, R.A.R., Bergamasco, R., Vilar, V.J.P., 2014. Assessment of a multistage system based on electrocoagulation, solar photo-Fenton and biological oxidation processes for real textile wastewater treatment. Chem. Eng. J. 252, 120−130.

Matavos-Aramyan, S., Moussavi, M., 2017. Advances in Fenton and Fenton based oxidation processes for industrial effluent contaminants control-A review. Int. J. Environ. Sci. Nat. Res. 2 (4).

Mollah, M.Y.A., Schennach, R., Parga, J.R., Cocke, D.L., 2001. Electrocoagulation (EC)—science and applications. J. Hazard Mater. 84 (1), 29−41.

Moraes, S.G., Freire, R.S., Duran, N., 2000. Degradation and toxicity reduction of textile effluent combined photocatalytic and ozonation processes. Chemosphere 40, 369−373.

Muthukumari, B., Selvam, K., Muthuvel, I., Swaminathan, M., 2009. Photoassisted hetero-Fenton mineralisation of azo dyes by Fe(II)-Al_2O_3 catalyst. Chem. Eng. J. 153, 9−15.

Nguyen, C.H., Juang, R., 2019. Efficient removal of cationic dyes from water by a combined adsorption-photocatalysis process using platinum-doped titanate nanomaterials. J. Taiwan Inst. Chem. Eng. 99, 166−179.

Nqombolo, A., Mpupa, A., Moutloali, R.M., Nomngongo, P.N., 2018. In: Yonar, T. (Ed.), Wastewater and Water Quality Chapter 2: Wastewater Treatment Using Membrane Technology. IntechOpen. https://doi.org/10.5772/intechopen.76624.

Oancea, P., Meltzer, V., 2013. Photo-Fenton process for the degradation of Tartrazine (E102) in aqueous medium. J. Taiwan Inst. Chem. Eng. 44 (6), 990−994.

Oğuz, E., Keskinler, B., Çelik, Z., 2005. Ozonation of aqueous Bomaplex Red CR-L dye in a semi-batch reactor. Dyes Pigments 64, 101−108.

Orak, C., Atalay, S., Ersöz, G., 2017. Photocatalytic and photo-Fenton-like degradation of methylparaben on monolith-supported perovskite-type catalysts. Separ. Sci. Technol. 52 (7), 1310−1320.

Ozyonar, F., Karagozoglu, B., 2012. Systematic assessment of electrocoagulation for the treatment of marble processing wastewater. Int. J. Environ. Sci. Technol. 9 (4), 637−646.

Padaki, M., Murali, R.S., Abdullah, M.S., Misdan, N., Moslehyani, A., Kassim, M.A., Hilal, N., Ismail, A.F., 2015. Membrane technology enhancement in oil—water separation. A review. Desalination 357, 197—207.

Palas, B., Ersöz, G., Atalay, S., 2016. Heterogeneous photo Fenton-like oxidation of Procion Red MX-5B using walnut shell based green catalysts. J. Photochem. Photobiol. Chem. 324, 165—174.

Palas, B., Ersöz, G., Atalay, S., 2017. Photo Fenton-like oxidation of Tartrazine under visible and UV light irradiation in the presence of $LaCuO_3$ perovskite catalyst. Process Saf. Environ. Protect. 111, 270—282.

Park, H., Park, Y., Kimb, W., Choi, W., 2013. Surface modification of TiO_2 photocatalyst for environmental applications. J. Photochem. Photobiol. C Photochem. Rev. 15, 1—20.

Patel, H., Vashi, R.T., 2010. Treatment of textile wastewater by adsorption and coagulation. J. Chem. 7 (4), 1468—1476.

Pradhan, A.A., Gogate, P.R., 2010. Degradation of p-nitrophenol using acoustic cavitation and Fenton chemistry. J. Hazard Mater. 173 (1—3), 517—522.

Pushpalatha, M., Krishna, M.B., 2017. Int. J. Adv. Res. Ideas Innovat. Technol. 3 (1), 439—451.

Ribeiro, A.R., Nunes, O.C., Pereira, M.F., Silva, A.M., 2015. An overview on the advanced oxidation processes applied for the treatment of water pollutants defined in the recently launched. Directive 2013/39/EU. Environ. Int. 75, 33—51.

Robinson, T., McMullan, G., Marchant, R., Nigam, P., 2001. Remediation of dyes in textile effluent: a critical review on current treatment technologies with a proposed alternative. Bioresour. Technol. 77, 247—255.

Sacco, O., Matarangolo, M., Vaiano, V., Libralato, G., Guiga, M., Lofrano, G., Carotenuto, M., 2018. Crystal violet and toxicity removal by adsorption and simultaneous photocatalysis in a continuous flow micro-reactor. Sci. Total Environ. 644, 430—438.

Sandoval, A., Hernández-Ventura, C., Klimova, T.E., 2017. Titanate nanotubes for removal of methylene blue dye by combined adsorption and Photocatalysis. Fuel 198, 22—30.

Sarasa, J., Roche, M.P., Ormad, M.P., Gimeno, E., Puig, A., Ovelleiro, J.L., 1998. Treatment of a wastewater resulting from dyes manufacturing with ozone and chemical coagulation. Water Res. 32, 2721.

Sevimli, M., Sarıkaya, H., 2002. Ozone treatment of textile effluents and dyes: effect of applied ozone dose, pH and dye concentration. J. Chem. Technol. Biotechnol. 77, 842—850.

Sheng, H.L., Chi, M.L., 1993. Treatment of textile waste effluents by ozonation and chemical coagulation. Water Res. 27, 1743—1748.

Sheriff, T.S., Cope, S., Ekwegh, M., 2007. Calmagite dye oxidation using in situ generated hydrogen peroxide catalysed by manganese(II) ions. Dalton Trans. 5119.

Shih, Y., Ho, C., Huang, Y., 2013. Photo-fenton oxidation of azo dye reactive black B using an immobilized iron oxide as heterogeneous catalyst. Water Environ. Res. 85.

Sina, M.A., Mohsen, M., 2017. Advances in Fenton and Fenton based oxidation processes for industrial effluent contaminants control-A review. Int. J. Environ. Sci. Nat. Res. 2 (4), 555594.

Solmaz, S.K.A., Birgül, A., Üstün, G.E., Yonar, T., 2006. Colour and COD removal from textile effluent by coagulation and advanced oxidation processes. Color. Technol. 122, 102—109.

Song, S., Yao, J., He, Z., Qiu, J., Chen, J., 2008. Effect of operational parameters on the decolorization of C.I. Reactive Blue 19 in aqueous solution by ozone-enhanced electrocoagulation. J. Hazard Mater. 152, 204—210.

Sreeja, P.H., Sosamony, K.J., 2016. A comparative study of homogeneous and heterogeneous photo-Fenton process for textile wastewater treatment. Procedia Technol. 24, 217—223.

Stock, N.L., Peller, J., Vinodgopal, K., Kamat, P.V., 2000. Combinative sonolysis and photocatalysis for textile dye degradation. Environ. Sci. Technol. 34 (9), 1747–1750.

Su, C.X., Low, L.W., Teng, T.T., Wong, Y.S., 2016. Combination and hybridisation of treatments in dye wastewater treatment: a review. J. Environ. Chem. Eng. 4, 3618–3631.

Tekin, G., Ersöz, G., Atalay, S., 2018. Visible light assisted Fenton oxidation of tartrazine using bismuth oxyhalides as novel photocatalysts. J. Environ. Manag. 228, 441–450.

Torrades, F., Montano, J., Garcia-Hortal, J.A., Domenech, X., Peral, J., 2004. Decolorization and mineralization of commercial reactive dyes under solar light assisted photo-Fenton conditions. Sol. Energy 77, 573–581.

Tyagi, V.L., Lo, S.L., 2011. Application of physico-chemical pretreatment methods to enhance the sludge disintegration and subsequent anaerobic digestion: an up to date review. Rev. Environ. Sci. Biotechnol. 10, 215–242.

Vik, E.I., Carlsoon, D.A., Eikum, A.S., Gjessing, E.T., 1984. Electrocoagulation of potable water. Water Res. 18, 1355–1360.

Wang, T., Wang, Z., Wang, P., Tang, Y., 2019. An integration of photo-Fenton and membrane process for water treatment by a PVDF@$CuFe_2O_4$ catalytic membrane. J. Membr. Sci. 572, 419–427.

Wu, J., Eiteman, M.A., Law, S.E., 1998. Evaluation of membrane filtration and ozonation processes for treatment of reactive-dye wastewater. J. Environ. Eng. 124 (3), 272.

Zhang, H., Gao, H.C., Zhang, C., Chen, L., 2013. Decolorization of crystal violet by ultrasound/heterogeneous Fenton process. Water Sci. Technol. 68 (11), 2515–2520.

Zhang, J., Chen, S., Zhang, Y., Quan, X., Zhao, H., Zhang, Y., 2014. Reduction of acute toxicity and genotoxicity of dye effluent using Fenton-coagulation process. J. Hazard Mater. 274 (15), 198–204.

Zheng, H., Sun, Y., Zhu, C., Guo, J., Zhao, C., Liao, Y., Guan, Q., 2013. UV-initiated polymerization of hydrophobically associating cationic flocculants: synthesis, characterization, and dewatering properties. Chem. Eng. J. 234, 318–326.

Zou, X., 2015. Combination of ozonation, activated carbon, and biological aerated filter for advanced treatment of dyeing wastewater for reuse. Environ. Sci. Pollut. Res. 22, 8174–8181.

Chapter 8

Wastewater Treatment Plants as emerging source of antibiotic resistance

Neha Sharma[a], Sanjay K. Sharma[b]
[a]*Department of Microbiology, Maharishi Markandeshwar Medical College and Hospital, Maharishi Markandeshwar University, Solan, Himachal Pradesh, India;* [b]*Green Chemistry and Sustainability Research Group, Department of Chemistry, JECRC University, Jaipur, Rajasthan, India*

1. Introduction

Water is the only element known to man that exists naturally in all three states of matter. It has a very significant role in chemistry as an excellent solvent that can dissolve many ionic and polar substances, being considered as an effective medium for carrying nutrients to plants as well as to animals (Sharma et al., 2012).

Water may be contaminated with different types of impurities including organic, inorganic, and biological waste originating from anthropogenic as well as industrial activities. Management of water resources is therefore a herculean task. The current selection of water and wastewater management options, particularly in the developing world, has the opportunity to learn from 200 years of experience. So, the decisions being made today will impact on the future generations (Chanan et al., 2013).

In the last decades, exponential increase in global population, economic development, and agricultural activities has increased water demand substantially; therefore, mitigation of incoming pollutants and toxicants has urged an important role of green economy (Atalay et al., 2020).

One such important classes of emerging pollutants are antibiotics which have found their way into industrial effluents and are rendered to exhibit recalcitrance and resistance. The prolific use of antibiotics in the last decades for disease containment has been a focal theme of recent research investigations. The discovery of penicillin by Alexander Fleming, in 1929, is often described as one of the breakthroughs of modern medicine in the 20th century. Ever since the introduction of antibiotics into medicine dating back to

Green Chemistry and Water Remediation: Research and Applications.
https://doi.org/10.1016/B978-0-12-817742-6.00008-6

1940s and until 1970s, they have occupied a central position in contemporary medicine and health care with over 160 new antibiotics and semisynthetic derivatives molecules commercialized (Davies, 2006). Their role not only remains confined in treating serious bacterial infections but also circumventing infections in patients who have undergone surgery and as a defense mechanism in immunocompromized patients by overcoming opportunistic infections. Besides, their promiscuous role in human health and antibiotics has been considered of immense utility in livestock production (Ekpeghere, 2017) for Animal Feeding Operations and as a scaled-up tool for over 1000 animals in Concentrated Animal Feeding Operations, prevention of disease in livestock (Gelband et al., 2015), growth promoters and enhancers at herd level (Woolhouse et al., 2015), as well as aquaculture. Aquaculture has been considered an emerging food sector globally due to intensive farming practices (Henrikkson et al., 2018; Cabello et al., 2013) and broiler feedlots (He et al., 2014); the exact amount of antibiotics used in such facilities is cumbersome to quantify, but it stands much higher than that being used by health care facilities (Chee-Sanford et al., 2009). A recent report suggests that antibiotic consumption in livestock reached 63,151 tons in 2010 and is predicted to increase by another 67% by 2030. Approximately 80% of antibiotics sold in the United States are for veterinary use (as a public health concern), excessive usage of antibiotics followed by its unprecedented release into the environment attributed to incorrect disposal methods, and excretion by humans and animals is the getting through of antibiotics, their metabolites, and transformation products into the hospital and municipal sewage. This is more aptly referred to as Antibiotic Pollution (AP) (Boy-Roura et al., 2018). Therefore, the prevalence of antibiotics across biotic and abiotic resources has emerged as one of the major global challenges which has led to the inception of Antibiotic Resistance (AR) (Barancheshme and Munir, 2019). AR is depicted as any reduction in susceptibility of bacterial strain for an antibiotic as contrasted with susceptible wild-type strains (Andersson and Hughes, 2012). This extremely dangerous phenomenon, which arrests effective treatment of bacterial infections, may lead to fatal epidemic outbreaks with high mortality (Pazda et al., 2019) as antibiotics remain a common strategy for circumventing bacterial infections. In addition to their persistence in the environment and exerting toxic effects, they have also been held responsible for causing selective pressure. Selective pressure is being exerted by antibiotics which in turn are dependent upon the type of antibiotic and bacterial species in concern (Kolár et al., 2001). This leads to elimination of microorganisms (sensitive to antibiotics) which favors survival of resistant cells, which allows them to overcome the adverse effects of antibiotics. The emergence of microbial pathogens being able to resist antibiotics and acquire resistance is a pertinent public health care issue (CDCP, 2013; Davies et al., 2013), for the first time observed *in vitro* followed by clinical medicine (Abraham and Chain, 1940; Barber, 1947). Antibiotic

Resistance Bacteria (ARB) which harbors Antibiotic Resistance Genes (ARGs) are being placed under the category of *emerging pollutants* (Karkman et al., 2018; Icgen and Yilmaz, 2014) as a potential threat to food safety in addition to public health (Xu et al., 2017). A recent report published by WHO in September 2017 has released first-ever list of AR "*Priority Pathogens*" which includes categorization into three classes (critical, high, and medium priority). It is reported that drug-resistant tuberculosis kills around 250,000 people each year (WHO, 2017). AR claims 25,000 lives in Europe and 23,000 in the United States every year (Sachdeva et al., 2017). ARB-like Methicillin-Resistant *Staphylococcus aureus* (MRSA) is causative agent of skin infections, pneumonia, endocarditis, and Toxic Shock Syndrome. Other prominent fatal ARB commonly referred to as *superbugs* includes Vancomycin-resistant *Enterococcus sp*, Multi-Drug Resistant (MDR) *Mycobacterium tuberculosis*, and Carbapenem-resistant Enterobacteriaceae gut bacteria (Barancheshme and Munir, 2019).

1.1 Evolution, emergence, and dissemination of antibiotic resistance

Prevalence and persistence of resistant bacterial nosocomial infections renders health care facilities among high-interest environments to study the *evolution, emergence*, and *dissemination* of AR (Kraemer et al., 2019). The microbial communities harboring ARG in health care facilities are believed to have originated from human microbiomes (Pitout et al., 2005) and those inhabiting water and airflow systems (Kizny Gordon et al., 2017; Breathnach et al., 2012). Inevitably, prolonged use of antibiotics in health care systems leads to *de novo* resistance evolution, common example being during long-term treatment of chronic infections (Folkesson et al., 2012). Pathogens harboring newly evolved ARG can subsequently disseminate between patients through an epidemic route (López-Causap et al., 2013).

A logarithmic increase in ARB inhabiting different reservoirs has been observed globally (Segura et al., 2009). Based on the data obtained by the European Union and European Economic Activity countries, ARB caused 671,689 infections in 2015 and contributed to >33,000 deaths each year (Cassini et al., 2019). Socioeconomic findings according to European Center for Disease Prevention and Control (ECDC) in 2009 revealed hospitalization cost of patients infected by selected MDR bacteria was estimated at least EUR 1.5 billion each year (ECDC/EMEA Technical Report, 2009), while in the United States in 2014, the same was estimated to be approximately $ 2.2 billion annually (Thorpe et al., 2018). Similar situation is prevalent with a more intensive outbreak in low- and middle-income economies, with a high incidence of infectious diseases (Bougnom and Piddock, 2017). Noteworthy is the fact that WHO and ECDC have declared AR to be an emerging challenge for contemporary medicine and public health concern which needs to be addressed immediately, given its exponential spike (Pazda et al., 2019; Pal C et al., 2017; Versporten et al., 2014) as the paucity of viable treatment options exists (Lawe-Davis and Bennet, 2017). Attributed to above-cited facts, distinctly AR is poised to be a daunting task which necessarily seeks the involvement of regional management regimes for clinical and agricultural use of antibiotics with effective treatment policies leading to abatement of AR spread, nationally (Pruden et al., 2013). The rate of antibiotic discovery has witnessed a sharp decline over a period of time due to technical and socioeconomic challenges, leading up to an "Antibiotic Crisis" (AC) (Livermore, 2011) (Fig. 8.1).

2. Wastewater Treatment Plants: potential reservoirs of AR

Wastewater Treatment Plants (WWTPs), by a highly complex and specific mode of functioning, are being considered as powerhouses of AR, ARB, and

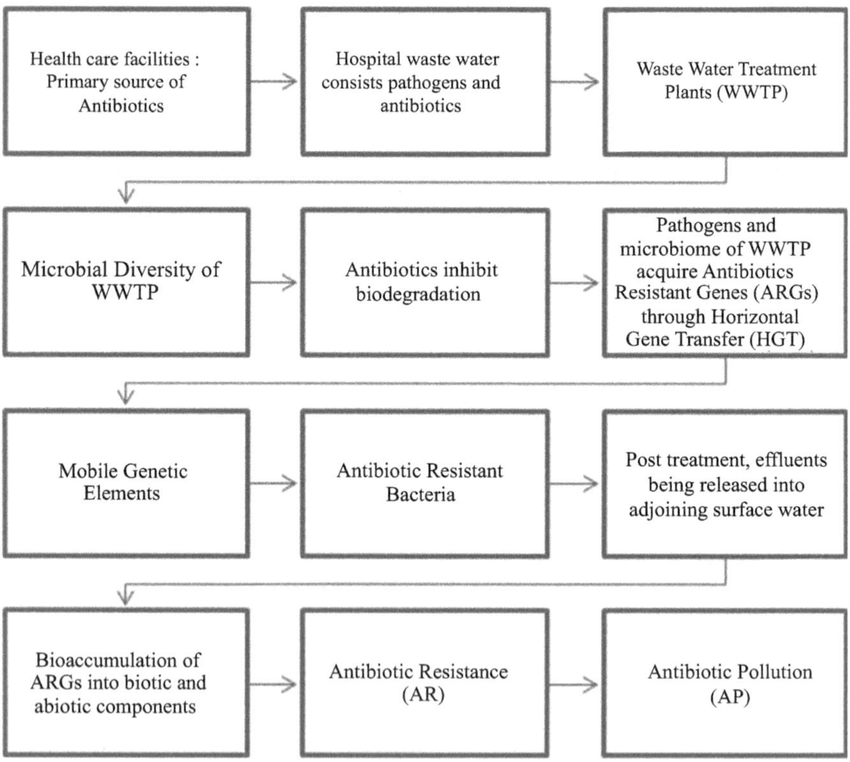

FIG. 8.1 Schematic representation of corelation between health care facilities and Antibiotic Pollution.

ARG (Rizzo et al., 2013) which leads to Antibiotics Spread Mechanism. A typical WWTP is essentially a closed system which deploys *three-tier end of pipe treatment strategy* to remove toxicants, pollutants, and pathogens by conglomerating different physicochemical, physicomechanical, chemical, and microbiological processes (Sharma et al., 2017). In a Common Effluent Treatment Plant (CETP), wastewater from divergent origins viz. municipal, industrial, and hospitals are allowed to enter through a common drain to undergo treatment (Sharma et al., 2019). Municipal wastewater, primarily being composed of organic matter both in dissolved and particulate fractions (Becerra-Castro et al., 2015) and high microbial load, provides a nutrient-rich environment for growth of autochthonous microflora and it becomes steadily seeded with incoming ARBs, ARGs associated with clinical pathogens (Gao et al., 2012; Pruden et al., 2012; Riquelme et al., 2013; Devarajan et al., 2015; Di Cesare et al., 2016; Tseng et al., 2009; Huerta et al., 2013; Von Wintersdorff et al., 2016), and antibiotics. Many ARGs including tet(M), tet(O), tet(S), tet(Q), tet(W), and mec(A) have been identified in microbial communities of hospital wastewaters attributable to unprecedented consumption of human antibiotics in hospitals (Zhang et al., 2009). Moreover, a range of antibiotics is present at sub Minimum Inhibitory Concentration (MIC) concentrations (Ju et al., 2019; Yang et al., 2014). Effluents released by hospitals are enriched for ARB such as *Escherichia coli* carrying Extended-Spectrum Beta-Lactamase (ESBL), the resistant opportunistic pathogen *Pseudomonas aeruginosa* and Vancomycin-Resistant *Enterococci* (VRE) (Szczepanowski et al., 2009; Reinthaler et al., 2003). ARB and ARG abundances are highly reduced in the water fraction of sewage (Chen et al., 2013). However, the fate of these emerging pollutants remains poorly understood in applied technologies of conventional WWTP systems (Czekalski et al., 2015; Pruden et al., 2006). Despite their central role in AC, WWTPs are currently not specifically tasked with the removal of ARGs in effluents (Bürgmann et al., 2018). In a study conducted at WWTP, Beijing (China) revealed that there was no significant difference in the concentration of antibiotics in effluent, both upstream and downstream of WWTP (Xu et al., 2015). Unparalleled to this study, another study conducted in Spain highlighted the fact that downstream, the concentration of Ofloxacin, Azithromycin, Trimethoprim, and Metronidazole was increased by manifolds indicating selective pressure for induction of AR (Rodriguez-Mozaz et al., 2015). A study conducted at WWTP in Patancheru, near Hyderabad, India, operated by Patancheru Environment Technology Limited (PETL) which treats effluents being generated by 90 regional bulk drug manufacturers, reported presence of Fluoroquinolone antibiotics, Entrofloxacin, Ofloxacin, Norfloxacin, Lomefloxacin, and Ciprofloxacin being the most common ones, at concentrations as high as 30 mg/L (Larsson et al., 2007). Unsurprisingly, such high contaminations of antibiotics have led to a significant increase of AR in environments influenced by industrial-scale antibiotics contamination (Johhning et al., 2013). Although beta-lactams, especially penicillins, are the most commonly used antimicrobial agents, their detection in effluent and biosolids systems is relatively low (Bailón-Pérez et al., 2008; Benito- Peña et al., 2006; Brown et al., 2006; Craig, 2003), the fact attributed to proven instability of the beta-lactam ring and its sensitivity and susceptibility to hydrolysis. Penicillin V indicated a maximum concentration as 13,800 ng/L in influent and 2000 ng/L in purified wastewater (Watkinson et al., 2007), whereas the highest measured concentration of Amoxicillin was 6940 ng/L (Watkinson et al., 2009) and 1670 ng/L (Minh et al., 2009) in influent and effluent, respectively. Next, Cloxacillin was determined in quantities of up to 4600 ng/L in raw sewage and 700 ng/L in wastewater after the purification process (Gulkowska et al., 2008). The highest concentration of Cefalexin in influent was 64,000 ng/L (Watkinson et al., 2009), while up to 5070 ng/L of this compound was detected in the effluent (Gulkowska et al., 2008). Furthermore, the highest concentration of Cefaclor in raw wastewater was 6150 and 1800 ng/L in treated wastewater (Watkinson et al., 2009). Macrolides, which are the main metabolites of Erythromycin, Erythromycin-H_2O do not undergo metabolism in the human body and are consequently excreted mainly via bile and faeces and detectable as unchanged parent compound (Nnadozie et al., 2017). The highest concentration of erythromycin-H_2O detected in raw sewage has reportedly been 10,025 ng/L (Kasprzyk-Hordern et al., 2009) and 4330 ng/L in treated wastewater (Minh et al., 2009). The highest concentration of erythromycin was found to be 830 and 620 ng/L in influent and effluent, respectively (Hirsch et al., 1999). Concentration of Tylosin was detectable as 1150 ng/L (Yang and Carlson, 2004) in raw sewage and 3400 ng/L (Watkinson et al., 2009) in wastewater post treatment. Clarithromycin was detected in influent and effluent at concentrations of 1433 ng/L (Lin et al., 2009) and 996 ng/L (Spongberg and Witter, 2008), respectively. Nevertheless, recent reports have confirmed the presence of these antibiotics in WWTP systems leveraging for new insights for sustained and viable treatment options (Figs. 8.2a and 8.2b).

2.1 Ecotoxicological implications of antibiotic resistance genes: a bioaccumulative pathway

Concentration of ARGs in the effluents exceeds than in natural rivers which have been proved by correlation of ARG occurrence in sediments, lakes, rivers, and soils (Sharma et al., 2016). Biosolids which remain buried in landfills have found their usage as fertility enhancers for agriinputs. Their role in bioaccumulating ARB and ARGs has been well established (Munir et al., 2011). A study on a WWTP operating on activated sludge and chlorination processes, respectively, in their treatment process exhibited concentration of total Tetracycline resistance genes was 6.4×10^5 copies/mL in biosolids and

FIG. 8.2A Wastewater: Potential hotspots of Antibiotic Pollution and Antibiotic Resistance.

FIG. 8.2B Partially treated wastewater: Utility vehicle for irrigation and agriinputs.

6.4×10^3 copies/mL in the effluent (Al-Jassim et al., 2015). In a similar study conducted, the concentrations of tet(Q) and tet(G) in biosolid and effluent were reportedly 2.2×10^9 and 3.4×10^4 copies/mL, respectively. Also, the copies of resistance genes in accordance with the number of bacterial 16S rRNA genes reported at different sites of a natural river (Cache La Poudre River) ranged from 10^{-7} to 10^{-3} for ARGs [sul(1), sul(2), tet(W), and tet(O)] (Pei et al., 2006). Norfloxacin, Ofloxacin, Ciprofloxacin, and Doxycycline present in biosolids were quantified in the sludge of a Swedish WWTP. Similarly, notable concentrations of Ciprofloxacin and Norfloxacin have also been reported in sewage sludge samples from Switzerland (Clarke and Smith, 2011).

The ecotoxicological manifestation of antibiotics needs to be stringently monitored on routine basis (Eamens et al., 2008). The same is being carried out by environmental risk assessment including a two-stage process; estimation of *Expected Introductory Concentration (EIC)* entering the environment and Predicted *Environmental Concentration* (PEC). Antibiotics being detected in their partially degraded moieties in wastewater provided insights that partial degradation does not co-relates with the removal of antibiotics and hence provided an impetus about cross-linkage between microbial ecology and human health (Kim and Aga, 2007). Owing to their ecotoxicological route, screening, characterization determination of antibiotics in different environmental samples is being prioritized (Martin et al., 2015). In places where no treatment options are available, urban wastewaters are directly the primary contributors of antimicrobial compounds present in detectable concentration (Rodriguez-Mozaz et al., 2015) (Fig. 8.2c).

A low-level AR can occur via natural selection, contrary to higher levels attributed to anthropogenic pathway. Antibiotics in high concentration along with their associated ARB and ARG in WWTP effluent stream gain entry into the environment through the release of treated effluents into adjoining surface waters (rivers, lakes, seas, or oceans) (Allen et al., 2011), reuse of wastewater, agriculture inputs which utilize reclaimed water for irrigational purposes (Christou et al., 2017) which might be responsible for cross-transmittance of

FIG. 8.2C A natural wetland: Potential reservoir for containment of Antibiotic Resistance.

resistance between pathogens like *E. coli*, *Klebsiella pneumoniae*, and *S. aureus* which are the major culprits of Hospital Acquired Infections (HAIs) and microflora of WWTP (Barancheshme and Munir, 2019). Few reports suggest that there is evidence of the increased abundance of AR determinants at the outlet of WWTP (Neudorf et al., 2017; Bengtsson-Palme et al., 2016; Alexander et al., 2015). Another hotspot for the ARG is a constructed wetland, where bioaccumulated aqueous ARGs are an imperative source which embarks subsequent release of microbes from sediment into the water. It remains obligatory to understand that variability in the concentration of ARGs is dependent synergistically both upon operational and environmental factors of wetland and not on pollutant source (Barancheshme and Munir, 2019).

At the herd level, antibiotics being used as growth accelerators are amalgamated with microbiomes of farm animals resulting in potential contamination with ARBs (Pruden et al., 2006). This is followed by a collection of manure from Animal Feeding Operation in waste lagoons where it undergoes biodegradation (Chee-Sanford et al., 2009) and the biodegraded product is used or agricultural purposes where the second Point of Contact is soil microbiome. Pig manure, with its abundant and diverse ARGs and sheer volume, is a major source of ARGs (Zhu et al., 2013) (Fig. 8.2d).

In open-water aquaculture, antibiotics are incorporated into pens for disease containment and growth inducers, culminating into emergence, dispersal, and transmission of ARBs and ARGs in sediments and open waters (Shah et al., 2014; Buschmann et al., 2012). Closed systems are believed to contribute to the transfer of resistance genes between systems as waste is fed to fish in aquaculture (Watts et al., 2017).

FIG. 8.2D Common drain releasing wastewater into the environment.

These ARGs can traverse through water cycle via treated or partially treated effluents being dumped into aquatic environments (Rizzo et al., 2013) that are supposedly prominent reservoirs for the occurrence and spread of ARG. Wastewaters are spiked by antimicrobial compounds in a continuum (Rodriguez-Mozaz et al., 2015). Many perennial rivers have been found to possess high concentrations of antibiotics, ARB, and ARGs in their sediments as inflicted by urban wastewater. Plausibly, their pristine origins have no antibiotic or antibiotic-resistant contaminations (Pei et al., 2006) (Fig. 8.2e).

3. Mechanism of ARG acquisition by ARB: crosswalk between microbiome and resistome

The emergence of multidrug resistance in bacterial human pathogens is one of the most serious challenges for health care sector, globally. Pathogens that earlier were sensitive to antibiotics are becoming increasingly resistant. Rerouting to biogeochemical pathways, antibiotics have played a crucial role in microbial ecosystems, leveraging the fact that AR is an ancient phenomenon and as evidence, ARG was persistent in the environment pre-antibiotic era (Bhullar et al., 2012; D'Costa et al., 2011; Martínez et al., 2008). Most of the ARGs carried by pathogens have their origins in environments other than the clinical world in disparate environments (Wellington et al., 2013; Forsberg et al., 2012) of human-independent extent like pristine, nonhuman influenced environments, ranging from permafrost soils (Perron et al., 2015) to caves. This feature is concomitant with the "natural" resistance state of diversified microbial ecosystems (Lee and Lee, 2011). Bacterial diversity of caves has

FIG. 8.2E Natural aquaculture pond contaminated with wastewater.

validated a significant presence of 65% of gram-negative and 70% of gram-positive isolates which were found to possess multiple ARGs indicating the same to be an inclusive environmental bioprocess rather than an exhibition of selective pressures (Bhullar et al., 2012).

Hotspots of AR where genetic evolution takes place which culminates into the acquisition of AR are aptly referred to as *genetic reactors,* the most common ones being (Baquero et al., 2008)

- Human and animal microbiota
- Health care facilities
- Wastewater
- Soil or ground water

It is apprehensive that antibiotics are used for competitive ecological interactions across divergent habitats like rhizosphere (Haas and Keel, 2003) to water columns (Long and Azam, 2001); their concentrations remain ineffectively low to eliminate competitive bacterial species. Antibiotics have shown to stimulate conjugation (Zhang et al., 2011).

3.1 Mobile Genetic Elements: a model for genetic acquisition of antibiotic resistance genes

Bioefficacy of bacteria to exchange genes across species barrier deems them *most fit* for harboring ARGs. This often is facilitated by Mobile Genetic Elements (MGE) such as (Marathe et al., 2013).

- Plasmids
- Transposons (integrative and conjugative elements)
- Integrons
- Genomic Islands

It has been found that multiple ARGs are located on MGEs which have found to exchange ARGs across diverse taxonomic groups (Johnson, 2015). Transposons and integrons facilitate the transfer of genetic information between bacterial cells or within the genome of a cell. Integrons have proved to be highly efficient, natural *gene capturing systems* that can integrate individual gene cassettes under the control of a strong endogenous promoter (Rowe-Magnus and Mazel, 2002) and their *int* gene has been inclusive of different gene panels substantiating their crucial role in AR transfer mechanisms (Gillings et al., 2015). It has been postulated that different genes encoding for specific antibiotics occupy the same position of chromosomes or MGE which leads to multiple resistances and cross selection (Xu et al., 2017). Therefore,

MGE plays a vital role in the emergence and spread of ARGs (Zhu et al., 2013). The transfer of ARGs (genetic acquisition and emergence) across diverse taxonomic groups encompassing native microbiome (of the specific genetic reactor), pathogenic strains, and existing ARBs is carried out by

- Vertical Gene Transfer (VGT), i.e., transmission of modified genetic information located on the bacterial chromosome to subsequent generations of daughter cells (Pazda et al., 2019)
- Horizontal Gene Transfer (HGT) (You et al., 2019)

HGT is one of the most important mechanisms leading to the dispersal, distribution, and dissemination of AR in both biotic and abiotic components of the environment. The transfer is mediated by donor bacteria, phages, free DNA, or even from the dead cells to living cells (Sharma et al., 2016; McKinney and Pruden, 2012). Resistance gene transfer mechanism by MGEs through HGT is inclusive of the following pathways:

- Conjugation (bacteria to bacteria transfer) Barancheshme and Munir, 2019).
- Transduction (bacteriophage mediated) (Springman et al., 2009).
- Transformation (uptake of free DNA from the environment by bacterial cells) (Mazaheri Nezhad Fard et al., 2011).
- Gene Transfer Agents (GTA) (bacteriophage like elements made by bacteria that carry random segments of DNA of host bacterium).

3.2 Heavy metal contamination and AR: a co-selection pathway

With industrialization gaining momentum by leaps and bounds, different industrial setups utilize diversified xenobiotic compounds in their unit operations and processes. As a part of the *industrial ecology* regime, effluents so generated are being treated by WWTP and CETPs followed by their subsequent release into adjoining surface waters or are being used for irrigation (Sharma et al., 2019). Owing to the recalcitrant attribute of xenobiotics including heavy metals, they are omnipresent in industrial effluents. Because of their resistance to microbial degradation, it has been conferred that heavy metals can act as a trigger for inducing *Heavy Metal Antibiotic Resistance* (HMAR) subsequently leading to the genetic acquisition of ARB. They are referred to as Heavy Metal Resistance Genes (HMRG). *E. coli* and *Salmonella* sp. have acquired resistance to multiple antibiotics and heavy metals (Icgen and Yilmaz, 2014). Co-selection corroborates the relation between the acquisition of HMRGs and ARGs (Di Cesare et al., 2016). It is a genetic mechanism mediated by the constant presence of MGE (Yu et al., 2016) where stress (antibiotics) acts as an inducer for selection of the associated resistance gene results in the persistence of additional resistance genes (Wellington et al.,

2013). It has been speculated that metals (Cu and Zn) (Poole, 2017) in high concentration in effluents exert pressure and lead to the *metal selection* of MGE that harbor both ARG and HMRG through genetic couplings (Li et al., 2017). On the contrary, in soil ecosystems, metal (Cu, Cr, Ni, Fe, and Pb) contamination in low concentrations has been associated with 11 abundant ARGs as evaluated by Quantitative Polymerase Chain Reaction (qPCR) (Knapp et al., 2011).

Baker-Austin et al. (2006) provided experimental insights that genes corresponding to resistant phenotypes (As, Cu, Zn, Mn, Co, Ag, Hg, Cd, and Ni) when present on same MGE favor co-selection of other resistant genes (Chloramphenicol, Ciprofloxacin, Coumermycin, Rifampicin, Tetracycline, and Trimethoprim) located on same MGE in *Enterobacteriaceae.* These findings also provided an impetus to our understanding of **cross-resistance** in addition to **cross-selection** resistance.

Some recent findings are of viewpoint, depending upon the mode of action of antibiotics on bacterial cells; resistance can be conferred by different mechanisms (Wright, 2010; Walsh, 2000).

- Enzymatic inactivation of antibiotics
- Transportation to the cell's exterior by an efflux pump
- Modification of antibiotics target
- *De novo* route
- Persistent in environment
- Alternative metabolic pathway

Acquisition of *de novo* AR in response to anthropogenic selective pressures and high consumption of antibiotics, resistance genes are presumed to increase radically in abundance within the microbial ecosystem (Kim et al., 2014; Kristiansson et al., 2011; Dethlefsen et al., 2008) (Figs. 8.3 and 8.4).

4. Microbial diversity of genetic reactors: "black box" of antibiotic-resistant bacteria harboring antibiotic resistance genes

Microbes are the most ubiquitous organisms which occupy different ecological niche and are the major performance indicators of biogeochemical processes such as nutrient cycling, decomposition, and primary productivity in a wide array of the environment (Gibbons and Gilbert, 2015). A unique and remarkable feature of *phenotypic heterogeneity* among diverse bacterial species or even lineages within a species is exhibited in response to different antibiotics. A change in gene expression or fluctuations in critical physiological traits has been observed (Meouche and Dunlop, 2018). High-end use of antibiotic-associated selective pressures can act detrimentally on microbial community composition by shrinking taxa diversity or microbial shift. An occurrence of *weak antibiotic selective pressure* favors Gram-negative bacteria more in

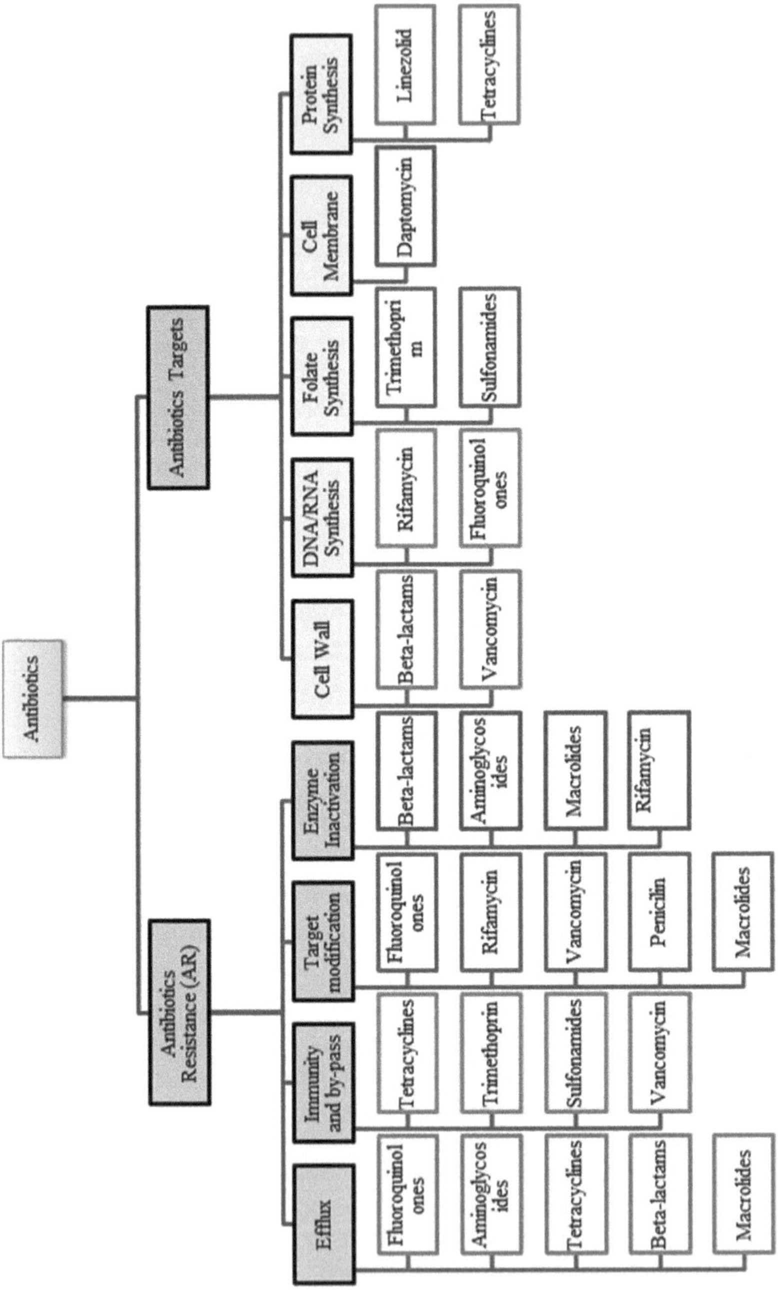

FIG. 8.3 Mode of action of antibiotics on a bacterial cell and their probable targets.

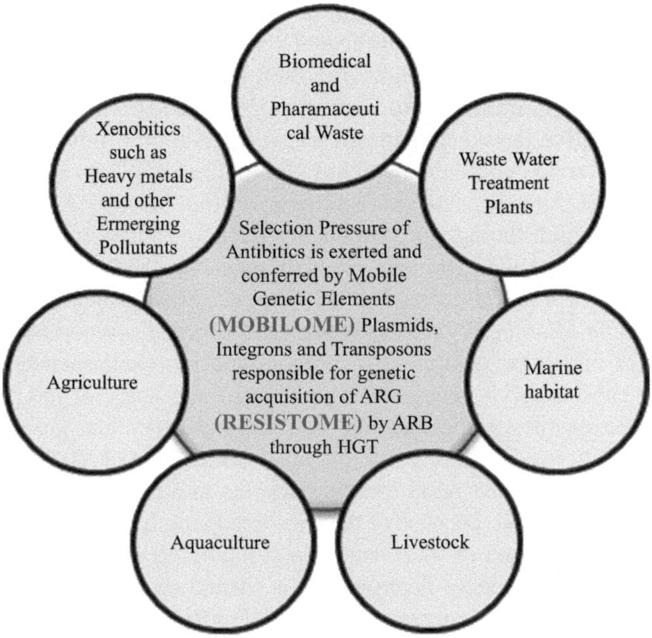

FIG. 8.4 Selection pressure exerted by antibiotics for the genetic acquisition of resistance by Mobile Genetic Elements through Horizontal Gene Transfer.

contrast to Gram-positive bacteria. This unique phenomenon is attributed to increased susceptibility to antibacterial compounds as the cell is devoid of an outer cell membrane (Delcour, 2009). This can lead to complete or partial elimination of key taxa involved in carrying out nutrient cycle in aquatic habitats under influence of *strong antibiotic selective pressure* (Ecker et al., 2019; Grenni et al., 2018). Subminimal composition of antibiotics has exhibited an increase in bacterial genetic diversity by activation of the SOS response, leading to an increased rate of mutation throughout the genome (Foster, 2007) by exerting a direct mutagenic effect on DNA (Andersson and Hughes, 2014).

Likewise, presence of antibiotics in soil has also altered microbial community structure, leading to biomass loss, limited metabolic activities like nitrification, denitrification, respiration, partial inhibition of enzymatic activity, including dehydrogenases, phosphatases, and ureases, which play pivotal role in maintaining soil health (Cyckon et al., 2019).

WWTPs have emerged as model systems providing an interface between humans and environment, housing large microbial genetic diversity. This is aimed at exemplifying genetic acquisition of AR by ARBs through HGT. Antibiotics and other co-selection factors in sewage generate an inductive selection pressure for ARGs and ARB associated with MGE in the WWTP

(Di Cesare et al., 2016). Identification of prospective ARB and genetic contexts (including MGEs) using multifaceted approaches, including conventional isolation strategy, has been recently reported (Paiva et al., 2017). Culturable bacterial diversity is important to understand the phenotypic traits of isolates and their respective resistance patterns and it accounts for only 1% of species inhabiting different ecological niches including WWTP. Environmental Resistome (ER) (influent, activated sludge and effluent) in WWTP can hence be explored by high-throughput sequencing techniques like 16S rDNA, community dynamics studies through metagenomics analysis, and 454 pyrosequencing to name a few (Luo et al., 2017; Ju et al., 2016) to identify *Viable but Non-Culturable Microbes* (VBNC). Wastewater samples are complex matrices composed of higher concentration of organic and inorganic matter, inhibitors like humic substances or detergents which retard efficiency of the process by making detection of specific bacterial species and functional genes difficult. This insensitivity may lead to erratic output (Volkmann et al., 2007). However, the information about the ARG-carrying species and the genetic contexts remains barely understood, because of the short read length generated by Illumina sequencing. Assemblages of short reads can certainly provide information about frequent repetitive sequences flanking ARGs carried on MGEs which hampers the effective assembly of genetic contexts of ARGs (Ashton et al., 2015). Identification of probable hosts bearing ARGs has been carried out based on the linking of functional genes with universal phylogenetic markers (16S rDNA) (Spencer et al., 2016) by EpicPCR (Hultman et al., 2018).

Municipal WWTP operating on anaerobic/anoxic/aerobic Membrane Bio-Reactor (MBR) methodology may lead to subsequent HGT and the spread and occurrence of ARGs and multiresistant microorganisms, the fact attributed to significant amounts ARGs, such as class 1 integrongene (intI1) (Du et al., 2015). Integrons possess resistance cassettes encoding different ARGs including aminoglycoside and sulfonamide resistance genes, as well as qacE11 efflux pump genes (Zhu et al., 2013) which can encode different ARGs under influence of a mutual promoter. This leads to the co-selection of ARGs, hence, selection pressure applied by one antibiotic may lead to selection for ARGs associated with diverse antibiotics within the gene cassette of the integron (Miller et al., 2013). Trimethoprim resistance mechanism is also the replacement of a trimethoprim-sensitive dihydrofolate reductase by cassette borne trimethoprim-resistant dihydrofolate reductase (Zhang et al., 2009).

The effluent of WWTP harbors ARGs and HMRGs which when released into aquatic environment results in the dissemination of ARGs across different genetic reactors through the food chain. The associated risks must be accounted to address potential toxicological implication effluents as effluents are reused as irrigation water (Becerra-Castro et al., 2015).

When a bacterial community is exposed to heavy metal selection pressure in WWTP, the phenomenon of co-selection is of high-order magnitude. The class 1 integrons are frequently associated with gene cassettes in which both HMRGs and ARG are coassociated. In a study, six ARGs (tetA, sul(2), blaTEM, blaCTX-M, ermB, and *qnrS*), two HMRGs (czcA and arsB), and the class I integron (int1) in different phases of three WWTPs were measured and dynamics of sul(2), HMRGs, and int1 correlated which provided an evidence a possible relation between heavy metal contamination, HMRGs and spread of ARGs (Di Cesare et al., 2016).

The discharge of antimicrobials to the river has resulted in co-selection of HMRG for metals (Pb, Sn, Ni, Ba, Al, Sr, Ag, Li) in the range of 50%–92% and acquisition of ARGs for Cephalosporin, Quinolone, Sulfonamide, and Aminoglycoside classes of antibiotics (Icgen and Yilmaz, 2014). Antibacterial efficacy of metals (Ti, Co, Ni, Cu, Zn, Zr, Mb, Sn, and Pb) has been explored for *S. aureus* and *E. coli.* Apart from Ti and Sn, none of the metals exhibited antibacterial properties (Yasuyuki et al., 2010) with Cu and Zn being the most common in WWTPs (Azizi et al., 2016; Jingjing Ruan, 2014) which are in contact with ARGs and HMRGs.

Zn or Cu selected for populations of *Betaproteobacteria* and *Flavobacteria* has been shown to induce MDR against Carbapenems and third-generation Cephalosporins (Becerra-Castro et al., 2015).

4.1 Probabilistic models for containing the spread of antimicrobial resistance

Different techniques for the abatement of antimicrobial resistance are into existence for quite some time now. Few emerging techniques based on platform technology like bionanotechnology have gained momentum in the recent past.

- ***Anaerobic and/or Aerobic Treatment Reactors***

Both aerobic and anaerobic treatment processes are low energy and environmentally benign strategies meant to minimize levels of Chemical Oxygen Demand and have also reportedly removed ARB and ARGs (Christgen et al., 2015). Based on WWTP studies conducted to evaluate the variation of five ARGs [tet(G), tet(W), tet(X), sul(1), and intI(1)] in influent and effluent, it was found that anaerobic anoxic process reduced the levels of ARGs significantly as compared with aerobic MBR process of each treatment unit on municipal WWTP has been studied to. The WWTP possessed the anaerobic/anoxic/aerobic MBR process. There was a significant positive correlation observed between the reduction of tet(W), intI(1), and sul(1) and the reduction of 16S rDNA in the wastewater treatment process (Du et al., 2015). Another low energy concept of Anaerobic–Aerobic Sequence (AAS) bioreactors has also been explored through metagenomics and it was reported more than 85% of ARGs in the influent were found to be removed (Christgen et al., 2015). Removal of antibiotics including Sulfamethazine, Sulfamethoxazole, Trimethoprim, and Lincomycin had been studied in five different WWTPs using aerobic/anaerobic treatment methods with low removal efficacy (Behera et al., 2011).

- ***Constructed Wetlands***

Constructed wetlands are small, simple, cost-effective, and eco-friendly semiaquatic microbial ecosystems, which operate on biodegradative efficacies of naturally attenuated microbial strains (Chen et al., 2016) used for removal of ARGs from agroindustrial effluents. Strong positive correlations between concentrations of intI1 and ARGs, indicating that MGE affects the dissemination of ARGs in CW, have also been substantiated (Fang et al., 2017). The removal of ARB and ARGs in raw domestic wastewater by differently constructed wetlands has been investigated and 8 antibiotics and 12 genes and 16S rRNA (bacteria) were studied in different matrices. The aqueous removal efficiencies of 8 antibiotics, 12 genes, and 16S rDNA sequence constructs in presence of plants exhibited removal capabilities in the range of 75.8% −98.6%, while those of total ARGs fluctuated between 63.9% and 84.0% (Chen et al., 2016).

- ***Disinfection***

Ozonation, chlorination, and UV radiation for subsequent removal of Cephalexin, Ciprofloxacin, Chloramphenicol, Erythromycin, Gentamicin, Rifampicin, Sulfadiazine, Tetracycline, and Vancomycin have been previously monitored along with its effect on inactivation of ARB and ARGs have been studied in a WWTP. Pang et al., 2016; Yuan et al., 2015 studied fates of nine different ARB and two series of ARGs [ere(A), ere(B), erm(A), erm(B), tet(A), tet(B),tet(M), and tet(O)] by real-time PCR in treated wastewater using chlorination and exhibited 60% and 20% removal of these genes, respectively, applying various doses of chlorine ARGs including sul(1), ARtet(X), tet(G), intI(1), and 16S rRNA were considered and a co-relation between inactivation ofARG and chlorine dosage, contact time was monitored (Sharma et al., 2016). Lab-scale analysis for a combinatorial treatment approach (chlorination, UV irradiation, and ozonation) was conducted to evaluate inactivation of sul(1), tet(G), and intI(1), gene copies of sul(1), tet(G), intI(1), and 16S rDNA (Zhuang et al., 2015).

- ***Nanomaterials***

Antimicrobial Nanotechnology has led to the development of nanomaterials seemingly promiscuous against AR, ARBs, and ARGs (Lam et al., 2016; Monteiro et al., 2009; Shahverdi et al., 2007). Their bioefficacy is exemplified by two probabilistic models exhibiting antimicrobial property (Aruguete et al., 2013).

1. *The combinatorial* approach is also known as *Trojan* horse (Behera et al., 2011) of functionalized nanomaterial with antibiotics upon gaining entry in ARB, releases toxic ions.
2. *Synergistic* approach in which combination of antibiotic and nanomaterials combat ARGs separately.

Functionalized nanomaterials comprise of *molecular antibiotics* that are prepared by a combination of liposomes, dendrimers, and an antibiotic which is embedded inside a water-soluble polymer polyvinyl pyrrolidone and silver nanoparticle capsules. In this chapter, silver nanoparticles coated with a water-soluble polymer called were used to combat nanomaterial-resistant organisms (Aruguete et al., 2013). Antibacterial efficacy of functionalized nanomaterials is attributed to the physicochemical properties of nanoparticle in concern and type of bacteria (Hajipour et al., 2012). AR removal efficiency mechanism of four ARGs including *tetA*, *sul2*, *ermB*, and *ampC* has been found using graphene oxide nanosheet. The removal efficiency was reported in the range of 2.88–3.11 logs at 300 μg/mL nanosheet solution showing the potential of grapheme oxide nanosheet as an innovative and effective adsorbent for treatment of ARGs (Yu et al., 2017).

Some contrasting facts have also been established in which they have been linked with the inception of AR within different biotic and biotic components (Lam et al., 2016; Miller et al., 2013) and have induced cytotoxicity, tissue ulceration, and reduction of cell capability in different organisms. Citrate-coated silver nanoparticles and iron and iron oxide nanoparticles which have been in use for targeted drug delivery have demonstrated genotoxicity and cytotoxicity both *in vitro* and *in vivo* (Srivastava et al., 2015).

5. Future implications

In a broader context, different strategies could hold promise in order to restrain further *surfacing* of ARGs within gene pool of "*microbe mediators*." This remains essential to address issues related to Sustainable Development Goals aimed to address sustained environment through *greener* route and conceptualizing best practices to keep AP at bay. This can be accomplished by

1. Controlled use of antibiotics.
2. Alternative medicine with phytochemical interventions.
3. Metagenomic analysis to extract the community DNA from WWTPs to explore untapped reservoir of Viable But Non Culturable (VBNC) microbes which would help us ascertain novel microbes harboring ARGs.
4. Metabolomics and transcriptomics to decipher genetic machinery of unknown ARBs.
5. Deciphering biological pathways of indigenous microbes, for bioremediation of effluents originating from health care facilities as a promising, cost-effective, viable, and sustained model for end of pipe treatment.

6. Conclusion

WWTPs are emerging reservoirs of AR, ARGs, HMRGs, and most importantly ARBs. Owing to multifaceted high consumption and excessive usage of antibiotics, their unprecedented release into the environment (Genetic Reactors) via anthropogenic route, the dearth of proper treatment methods, and lack of policies for disposal standards to name a few; acquisition and genetic evolution of ARGs have been declared a global threat by WHO and appropriately stated as "Antibiotic Crisis" or "Antibiotic Resistance Pollution." This regime has not only led to socioeconomic losses but has also rendered the emergence of resistance to newer antibiotics and nonantibiotic exposed bacterial species. Fundamentally speaking, this emergence, dispersal, dissemination, and transmission has led to cross-resistance and has co-selection which is proving detrimental not only in health care facilities but also ecosystem at large. Speaking in stricter sense about need-based solutions, high throughput, and advanced *platform technology*—based treatment practices with an amalgamation of Biotechnology, Nanotechnology, Biomedical engineering, Artificial Intelligence (AI), and Internet of Things (IoT) should be devised. This should be followed by adherence with stricter guidelines for the release of effluents/solid waste into the environment posttreatment. This certainly might prevent in aggravating this global threat further and lead to sustenance. The need of an hour is to formulate stringent policies for need-based consumption. This perhaps should prove a benchmark in addressing toxicological implications imposed across species by virtue of MGE which through HGT are the major culprits. Until recently, few developed and developing economies of the world are looking for traditional and alternate medicine, *botanical biotherapeutics*; *multicomponent drug mixture nutraceuticals* have recently gained momentum as a part of exploratory research.

6.1 Declaration of a competing interest

The authors declare they have no competing financial or personal interests.

References

Abraham, E.P., Chain, E., 1940. An enzyme from bacteria able to destroy penicillin. Nature 146, 837.

Al-Jassim, N., kram Ansari, M.I., Harb, M., Hong, P.-Y.Y., 2015. Removal of bacterial contaminants and antibiotic resistance genes by conventional wastewater treatment processes in Saudi Arabia: is the treated wastewater safe to reuse for agricultural irrigation? Water Res. 73, 277–290.

Alexander, J., Bollmann, A., Seitz, W., Schwartz, T., 2015. Microbiological characterization of aquatic microbiomes targeting taxonomical marker genes and antibiotic resistance genes of opportunistic bacteria. Sci. Total Environ. 512–513, 316–325. https://doi.org/10.1016/j.scitotenv.2015.01.046.

Allen, S.E., Boerlin, P., Janecko, N., Lumsden, J.S., Barker, I.K., Pear, D.L., Reid-Smith, R.J., Jardine, C., 2011. Antimicrobial resistance in generic *Escherichia coli* isolates from wild small mammals living in swine farm, residential, landfill, and natural environments in Southern Ontario, Canada. Appl. Environ. Microbiol. 77 (3), 882–888. https://doi.org/10.1128/AEM.01111-10.

Andersson, D.I., Hughes, D., 2012. Evolution of antibiotic resistance at non-lethal drug concentrations. Drug Resist. Updates 15 (3), 162–172. https://doi.org/10.1016/j.drup.2012.03.005.

Andersson, D.I., Hughes, D., 2014. Microbiological effects of sublethal levels of antibiotics. Nat. Rev. Microbiol. 2, 465–478.

Aruguete, D.M., Kim, B., Hochella Jr., M.F., Ma, Y., Cheng, Y., Hoegh, A., et al., 2013. Antimicrobial nanotechnology: its potential for the effective management of microbial drug resistance and implications for research needs in microbial nanotoxicology. Environ. Sci. Processes Impacts 15, 93–102. https://doi.org/10.1039/c2em30692a.

Ashton, P.M., Nair, S., Dallman, T., Rubino, S., Rabsch, W., Mwaigwisya, S., et al., 2015. MinION nanopore sequencing identifies the position and structure of a bacterial antibiotic resistance island. Nat. Biotechnol. 33 (3), 296–330.

Atlay, S., Ersoz, G., Sharma Sanjay, K., 2020. Green chemistry and its applications in water remediation. In: Sharma, S.K. (Ed.), Bioremediation A Sustainable Approach to Preserving Earth's Water. CRC Press, taylor and Francis Group, Boca Raton, Fl, pp. 1–18.

Azizi, S., Kamika, I., Tekere, M., 2016. Evaluation of heavy metal removal from wastewater in a modified packed bed biofilm reactor. PLoS One 11 (5), 1–13.

Bailón-Pérez, M.I., García-Campaña, A.M., Cruces-Blanco, C., del Olmo Iruela, M., 2008. Trace determination of β-lactam antibiotics in environmental aqueous samples using off-line and on-line preconcentration in capillary electrophoresis. J. Chromatogr. A 1185, 273–280. https://doi.org/10.1016/j.chroma.2007.12.088.

Baker-Austin, C., Wright, M.S., Stepanauskas, R., McArthur, J.V., 2006. Co-selection of antibiotic and metal resistance. Trends Microbiol. 14 (4), 176–182.

Baquero, F., Martínez, J.L., Cantón, R., 2008. Antibiotics and antibiotic resistance in water environments. Curr. Opin. Biotechnol. 19, 260–265. https://doi.org/10.1016/j.copbio.2008.05.006.

Barancheshme, F., Munir, M., 2019. Development of antibiotic resistance in wastewater treatment plants. Antimicrob. Resis. - A Glob. Threat. https://doi.org/10.5772/intechopen.81538.

Barber, M., 1947. Staphylococcal infection due to penicillin-resistant strains. Br. Med. J. 2, 863–865.

Becerra-Castro, C., Machado, R.A., Vaz-Moreira, I., Manaia, C.M., 2015. Assessment of copper and zinc salts as selectors of antibiotic resistance in Gram-negative bacteria. Sci. Total Environ. 530–531, 367–372.

Behera, S.K., Kim, H.W., Oh, J.E., Park, H.S., 2011. Occurrence and removal of antibiotics, hormones and several other pharmaceuticals in wastewater treatment plants of the largest industrial city of Korea. Sci. Total Environ. 409, 4351–4360. https://doi.org/10.1016/j.scitotenv.2011.07.015.

Bengtsson-Palme, J., Hammarén, R., Pal, C., Östman, M., Björlenius, B., Flach, C.F., Fick, J., Kristiansson, E., Tysklind, M., Larsson, D.G.J., 2016. Elucidating selection processes for antibiotic resistance in sewage treatment plants using metagenomics. Sci. Total Environ. 572, 697–712. https://doi.org/10.1016/j.scitotenv.2016.06.228.

Benito-Peña, E., Partal-Rodera, A.I., León-González, M.E., Moreno-Bondi, M.C., 2006. Evaluation of mixed mode solid phase extraction cartridges for the pre concentration of beta-lactam antibiotics in wastewater using liquid chromatography with UV-DAD detection. Anal. Chim. Acta 556, 415–422.

Bhullar, K., Waglechner, N., Pawlowski, A., Koteva, K., Banks, E.D., Johnston, M.D., Barton, H.A., Wright, G.D., 2012. Antibiotic resistance is prevalent in an isolated cave microbiome. PLoS One 7, e34953.

Bougnom, B.P., Piddock, L.J.V., 2017. Wastewater for urban agriculture: a significant factor in dissemination of antibiotic resistance. Environ. Sci. Technol. 51, 5863–5864. https://doi.org/10.1021/acs.est.7b01852.

Boy-Roura, M., Mas-Pla, J., Petrovic, M., Gros, M., Soler, D., Brusi, D., Menció, A., 2018. Towards the understanding of antibiotic occurrence and transport in groundwater: findings from the Baix Fluvià alluvial aquifer (NE Catalonia, Spain). Sci. Total Environ. 612, 1387–1406.

Breathnach, A.S., Cubbon, M.D., Karunaharan, R.N., Pope, C.F., Planche, T.D., 2012. Multidrug-resistant *Pseudomonas aeruginosa* outbreaks in two hospitals: association with contaminated hospital waste-water systems. J. Hosp. Infect. 82, 19–24.

Brown, K.D., Kulis, J., Thomson, B., Chapman, T.H., Mawhinney, D.B., 2006. Occurrence of antibiotics in hospital, residential, and dairy effluent, municipal wastewater, and the Rio Grande in New Mexico. Sci. Total Environ. 366, 772–783. https://doi.org/10.1016/j.scitotenv.2005.10.007.

Bürgmann, H., Frigon, D., Gaze, W.H., Manaia, C.M., Pruden, A., Singer, A.C., Smets, B.F., Zhang, T., 2018. Water and sanitation: an essential battlefront in the war on antimicrobial resistance. Fed. Eur. Microbiol. Soc. 94 fiy101.

Buschmann, A.H., Tomova, A., López, A., Maldonado, M.A., Henríquez, L.A., Ivanova, L., Moy, F., Godfrey, H.P., Cabello, F.C., 2012. Salmon aquaculture and antimicrobial resistance in the marine environment. PLoS One 7, e42724.

Cabello, F.C., Godfrey, H.P., Tomova, A., Ivanova, L., Dölz, H., Millanao, A., Buschmann, A.H., 2013. Antimicrobial use in aquaculture re-examined: its relevance to antimicrobial resistance and to animal and human health. Environ. Microbiol. 15, 1917–1942. https://doi.org/10.1111/1462-2920.12134.

Cassini, A., Högberg, L.D., Plachouras, D., Quattrocchi, A., Hoxha, A., Simonsen, G.S., Colomb-Cotinat, M., Kretzschmar, M.E., Devleesschauwer, B., Cecchini, M., Ouakrim, D.A., Oliveira, T.C., Struelens, M.J., Suetens, C., Monnet, D.L., 2019. Attributable deaths and disability-adjusted life-years caused by infections with antibiotic-resistant bacteria in the EU and the European Economic Area in 2015: a population-level modelling analysis. Lancet Infect. Dis. 19, 56–66. https://doi.org/10.1016/S1473-3099(18)30605-4.

Centers for Disease Control and Prevention, 2013. Antibiotic Resistance Threats in the United States. Centers for Disease Control and Prevention, Atlanta, GA, USA.

Chanan Amit, P., Vigneswaran, S., Kandasamy, J., Simmons, B., 2013. Waste water management journey- from Indus valley civilization to the twenty first century. In: Sharma, S.K., Sanghi, R. (Eds.), Waste Water Reuse and Management. Springer Dordrecht Heidelberg, New York, London, pp. 3–18.

Chee-Sanford, J.C., Mackie, R.I., Koike, S., Krapac, I.G., Lin, Y.F., Yannarell, A.C., Maxwell, S., Aminov, R.I., 2009. Fate and transport of antibiotic residues and antibiotic resistance genes following land application of manure waste. J. Environ. Qual. 38 (3), 1086–1108. https://doi.org/10.2134/jeq2008.0128.

Chen, H., Zhang, M., 2013. Occurrence and removal of antibiotic resistance genes in municipal wastewater and rural domestic sewage treatment systems in eastern China. Environ. Int. 55, 9–14.

Chen, J., Ying, G.G., Wei, X.D., Liu, Y.S., Liu, S.S., Hu, L.X., et al., 2016. Removal of antibiotics and antibiotic resistance genes from domestic sewage by constructed wetlands: effect of flow configuration and plant species. Sci. Total Environ. 571, 974–982. https://doi.org/10.1016/j.scitotenv.2016.07.085.

Christgen, B., Yang, Y., Ahammad, S.Z., Li, B., Rodriquez, D.C., Zhang, T., et al., 2015. Metagenomics shows that low-energy anaerobic-aerobic treatment reactors reduce antibiotic resistance gene levels from domestic wastewater. Environ. Sci. Technol. 49, 2577–2584. https://doi.org/10.1021/es505521w.

Christou, A., Agüera, A., Bayona, J.M., Cytryn, E., Fotopoulos, V., Lambropoulou, D., Manaia, C.M., Michael, C., Revitt, M., Schröder, P., Fatta-Kassinos, D., 2017. The potential implications of reclaimed wastewater reuse for irrigation on the agricultural environment: the knowns and unknowns of the fate of antibiotics and antibiotic resistant bacteria and resistance genes - a review. Water Res. 123, 448–467. https://doi.org/10.1016/j.watres.2017.07.004.

Clarke, B.O., Smith, S.R., 2011. Review of 'emerging' organic contaminants in biosolids and assessment of international research priorities for the agricultural use of biosolids. Environ. Int. 37 (1), 226–247.

Craig, W.A., 2003. Basic pharmacodynamics of antibacterials with clinical applications to the use of β-lactams, glycopeptides, and linezolid. Infect. Dis. Clin. 17, 479–501. https://doi.org/10.1016/S0891-5520(03.

Cycoń, M., Mrozik, A., Piotrowska-Seget, Z., 2019. Antibiotics in the soil environment—degradation and their impact on microbial activity and diversity. Front. Microbiol. Microbiotechnol. 10, 338. https://doi.org/10.3389/fmicb.2019.00338.

Czekalski, N., Sigdel, R., Birtel, J., Matthews, B., Bürgmann, H., 2015. Does human activity impact the natural antibiotic resistance background? Abundance of antibiotic resistance genes in 21 Swiss lakes. Environ. Int. 81, 45–55. https://doi.org/10.1016/j. envint.2015.04.005.

Davies, J., 2006. Where have all the antibiotics gone? Can. J. Infect Dis. Med. Microbiol. 17, 287–290.

Davies, S., Grant, J., Catchpole, M., 2013. The Drugs Don't Work. A Global Threat. Penguin Specials, London.

Delcour, A.H., 2009. Outer membrane permeability and antibiotic resistance. Biochim. Biophys. Acta Protein Proteonomics 1794, 808–816.

Dethlefsen, L., Huse, S., Sogin, M.L., Relman, D.A., 2008. The pervasive effects of an antibiotic on the human gut microbiota, as revealed by deep 16S rRNA sequencing. PLoS Biol. 6 (11), e280. https://doi.org/10.1371/journal.pbio.0060280.

Devarajan, N., Laffite, A., Graham, N.D., Meijer, M., Prabakar, K., Mubedi, J.I., et al., 2015. Accumulation of clinically relevant antibiotic-resistance genes, bacterial load, and metals in freshwater lake sediments in Central Europe. Environ. Sci. Technol. 49, 6528–6537. https://doi.org/10.1021/acs.est.5b01031.

Di Cesare, A., Eckert, E.M., D'Urso, S., Bertoni, R., Gillan, D.C., Wattiez, R., et al., 2016. Co-occurrence of integrase 1, antibiotic and heavy metal resistance genes in municipal wastewater treatment plants. Water Res. 94, 208–214. https://doi.org/10.1016/j.watres.2016.02.049.

Du, J., Geng, J., Ren, H., Ding, L., Xu, K., Zhang, Y., 2015. Variation of antibiotic resistance genes in municipal wastewater treatment plant with A2O-MBR system. Environ. Sci. Pollut. Control Ser. 22 (5), 3715–3726.

D'Costa, V.M., Osta, V.M., King, C.E., Kalan, L., Morar, M., Sung, W.W.L., Schwarz, C., Froese, D., Zazula, G., Calmels, F., et al., 2011. Antibiotic resistance is ancient. Nature 477, 457–461.

Eamens, G.J., Waldron, A.M., 2008. Salmonella uptake in sheep exposed to pastures. After biosolids application to agricultural land. Soil Res. 46, 302–308.

ECDC/EMEA Technical Report, 2009. The Bacterial Challenge: Time to React (EMEA Doc Ref. EMEA/576176/2009).

Eckert, E.M., Quero, G.M., Di Cesare, A., Manfredini, G., Mapelli, F., Borin, S., Fontaneto, D., Luna, G.M., Corno, G., 2019. Antibiotic disturbance aspects aquatic microbial community composition and food web interactions but not community resilience. Mol. Ecol. 28, 1170–1182.

Ekpeghere, K.I., Lee, J.W., Kim, H.Y., Shin, S.K., Oh, J.E., 2017. Determination and characterization of pharmaceuticals in sludge from municipal and livestock wastewater treatment plants. Chemosphere 168, 1211–1221.

Fang, H., Zhang, Q., Nie, X., Chen, B., Xiao, Y., Zhou, Q., et al., 2017. Occurrence and elimination of antibiotic resistance genes in a long-term operation integrated surface flow constructed wetland. Chemosphere 173, 99–106. https://doi.org/10.1016/j.chemosphere.2017.01.027.

Folkesson, A., Jelsbak, L., Yang, L., Johansen, H.K., Ciofu, O., Hoiby, N., Molin, S., 2012. Adaptation of *Pseudomonas aeruginosa* to the cystic fibrosis airway: an evolutionary perspective. Nat. Rev. Microbiol. 10, 841–851.

Forsberg, K.J., Reyes, A., Wang, B., Selleck, E.M., Sommer, M.O.A., Dantas, G., 2012. The shared antibiotic resistome of soil bacteria and human pathogens. Science 337, 1107–1111.

Foster, P., 2007. Stress-induced mutagenesis in bacteria. Crit. Rev. Biochem. Mol. Biol. 42 (5), 373–397. https://doi.org/10.1080/10409230701648494.

Gao, P., Munir, M., Xagoraraki, I., 2012. Correlation of tetracycline and sulfonamide antibiotics with corresponding resistance genes and resistant bacteria in a conventional municipal wastewater treatment plant. Sci. Total Environ. 421–422, 173–183.

Gelband, H., Miller-Petrie, M., Pant, S., Gandra, S., Levinson, J., Barter, D., White, A., Laxminarayan, R., 2015. Center for Disease Dynamics, Economics & Policy. State of the World's Antibiotics. CDDEP, Washington, D.C.

Gibbons, S.M., Gilbert, J.A., 2015. Microbial diversity-exploration of natural ecosystems and microbiomes. Curr. Opin. Genet. Dev. 35, 66–72.

Gillings, M.R., Gaze, W.H., Pruden, A., Smalla, K., Tiedje, J.M., Zhu, Y.G., 2015. Using the class 1 integron-integrase gene as a proxy for anthropogenic pollution. J. Int. Soc. Microb. Ecol. 9, 1269–1279.

Grenni, P., Ancona, V., Caracciolo, B.A., 2018. Ecological effects of antibiotics on natural ecosystems: a review. Microchem. J. 136, 25–39.

Gulkowska, A., Leung, H.W., So, M.K., Taniyasu, S., Yamashita, N., Yeung, L.W.Y., Richardson, B.J., Lei, A.P., Giesy, J.P., Lam, P.K.S., 2008. Removal of antibiotics from wastewater by sewage treatment facilities in Hong Kong and Shenzhen, China. Water Res. 42, 395–403. https://doi.org/10.1016/j. watres.2007.07.031.

Haas, D., Keel, C., 2003. Regulation of antibiotic production in root-colonizing *Pseudomonas* spp. and relevance for biological control of plant disease. Annu. Rev. Phytopathol. 41, 117–153.

Hajipour, M.J., Fromm, K.M., Ashkarran, A.A., Jimenez de Aberasturi, D., de Larramendi, I.R., Rojo, T., et al., 2012. Antibacterial properties of nanoparticles. Trends Biotechnol. 30, 499–511. https://doi.org/10.1016/j.tibtech.2012.06.004.

He, L.Y., Liu, Y.S., Su, H.C., Zhao, J.L., Liu, S.S., Chen, J., et al., 2014. Dissemination of antibiotic resistance genes in representative broiler feedlots environments: identification of indicator ARGs and correlations with environmental variables. Environ. Sci. Technol. 48, 13120–13129. https://doi.org/10.1021/es5041267.

Henriksson, P.J.G., Rico, A., Troell, M., Klinger, D.H., Buschmann, A.H., Saksida, S., Chadag, M.V., Zhang, W., 2018. Unpacking factors influencing antimicrobial use in global aquaculture and their implication for management: a review from a systems perspective. Sustain. Sci. 13, 1105–1120. https://doi.org/10.1007/s11625-017-0511-8.

Hirsch, R., Ternes, T., Haberer, K., Kratz, K.L., 1999. Occurrence of antibiotics in the aquatic environment. Sci. Total Environ. 225, 109–118. https://doi.org/10.1016/S0048-9697. (98) 00337-4.

Huerta, B., Marti, E., Gros, M., López, P., Pompêo, M., Armengol, J., et al., 2013. Exploring the Links between antibiotic occurrence, antibiotic resistance, and bacterial communities in water supply reservoirs. Sci. Total Environ. 456–457, 161–170. https://doi.org/10.1016/j.scitotenv.2013.03.071.

Hultman, J., Tamminen, M., Pärnänen, K., Cairns, J., Karkman, A., Virta, M., 2018. Host range of antibiotic resistance genes in wastewater treatment plant influent and effluent. Fed. Eur. Microbiol. Soc. 94 (4).

Icgen, B., Yilmaz, F., 2014. Co-occurrence of antibiotic and heavy metal resistance in Kizilirmak River isolates. Bull. Environ. Contam. Toxicol. 93 (6), 735–743.

Jingjing Ruan, Y.X., 2014. Stress of five heavy metals on the resistance of isolates from swine wastewater to four antibiotics. J. Waste Water Treat. Anal. 05 (03).

Johnning, A., Moore, E.R.B., Svensson-Stadler, L., Shouche, Y.S., Joakim Larsson, D.G., Kristiansson, E., 2013. Acquired genetic mechanisms of a multiresistant bacterium isolated from a treatment plant receiving wastewater from antibiotic production. Appl. Environ. Microbiol. 79, 7256–7263.

Johnson, A.P., 2015. Surveillance of antibiotic resistance. Philos. Trans. R. Soc. Lond. B Biol. Sci. 370 (1670), 20140080. https://doi.org/10.1098/rstb.2014.0080.

Ju, F., Li, B., Ma, L., Wang, Y., Huang, D., Zhang, T., 2016. Antibiotic resistance genes and human bacterial pathogens: co-occurrence, removal, and enrichment in municipal sewage sludge digesters. Water Res. 91, 1–10.

Ju, F., Beck, K., Yin, X., Maccagnan, A., McArdell, C.S., Singer, H.P., Johnson, D.R., Zhang, T., Bürgmann, H., 2019. Wastewater treatment plant resistomes are shaped by bacterial composition, genetic exchange, and upregulated expression in the effluent microbiomes. J. Int. Soc. Microb. Ecol. 13, 346–360.

Karkman, A., Do, T.T., Walsh, F., Virta, M.P.J., 2018. Antibiotic-resistance genes in waste water. Trends Microbiol. 26, 220–228. https://doi.org/10.1093/femsec/fiw014.

Kasprzyk-Hordern, B., Dinsdale, R.M., Guwy, A.J., 2009. The removal of pharmaceuticals, personal care products, endocrine disruptors and illicit drugs during wastewater treatment and its impact on the quality of receiving waters. Water Res. 43, 363–380. https://doi.org/10.1016/j.watres.2008.10.047.

Kim, S., Aga, D.S., 2007. Potential ecological and human health impacts of antibiotics and antibiotic- resistant bacteria from wastewater treatment plants. J. Toxicol. Environ. Health, Part B 10 (8), 559–573.

Kim, S., Lieberman, T.D., Kishony, R., 2014. Alternating antibiotic treatments constrain evolutionary paths to multidrug resistance. Proc. Natl. Acad. Sci. U.S.A. 111, 14494–14499.

Kizny Gordon, A.E., Mathers, A.J., Cheong, E.Y.L., Gottlieb, T., Kotay, S., Walker, A.S., Peto, T.E.A., Crook, D.W., Stoesser, N., 2017. The hospital water environment as a reservoir for carbapenem-resistant organisms causing hospital-acquired infections—a systematic review of the literature. Clin. Infect. Dis. 64, 1435–1444.

Knapp, C.W., McCluskey, S.M., Singh, B.K., Campbell, C.D., Hudson, G., Graham, D.W., 2011. Antibiotic resistance gene abundances correlate with metal and geochemical conditions in archived Scottish soils. PLoS One 6 (11), e27300.

Kolár, M., Urbánek, K., Látal, T., 2001. Antibiotic selective pressure and development of bacterial resistance. J. Antimicrob. Agents 17 (5), 357–363.

Kraemer, S.A., Arthi Ramachandran, A., Perron, G.G., 2019. Antibiotic pollution in the environment: from microbial ecology to public policy. Microorganisms 7, 180. https://doi.org/10.3390/microorganisms.

Kristiansson, E., Fick, J., Janzon, A., Grabic, R., Rutgersson, C., Weijdegård, B., Söderström, H., Joakim Larsson, D.G., 2011. Pyrosequencing of antibiotic-contaminated river sediments reveals high levels of resistance and gene transfer elements. PLoS One 6, e17038.

Lam, S.J., O'Brien-Simpson, N.M., Pantarat, N., Sulistio, A., Wong, E.H., Chen, Y.Y., et al., 2016. Combating multidrug-resistant gram-negative bacteria with structurally nanoengineered antimicrobial peptide polymers. Nat. Microbiol. 1, 16162. https://doi.org/10.1038/nmicrobiol.2016.162.

Larsson, D.G.J., de Pedro, C., Paxeus, N., 2007. Effluent from drug manufactures contains extremely high levels of pharmaceuticals. J. Hazard Mater. 148, 751–755.

Lawe-Davies, O., Bennett, S., 2017. WHO - List of Bacteria for Which New Antibiotics Are Urgently Needed. WHO Department of Communications.

Lee, S.H., Lee, J.H., 2011. Molecular characterization of TEM-type beta-lactamases identified in cold-seep sediments of edison seamount (south of lihir island, Papua New Guinea). In: Handbook of Molecular Microbial Ecology II: Metagenomics in Different Habitats. Wiley-Blackwell, Hoboken, NJ, USA, pp. 545–552.

Li, L.G., Xia, Y., Zhang, T., 2017. Co-occurrence of antibiotic and metal resistance genes revealed in complete genome collection. Int. Soc. Microb. Ecol. J. 11 (3), 651–662.

Lin, A.Y.C., Yu, T.H., Lateef, S.K., 2009. Removal of pharmaceuticals in secondary wastewater treatment processes in Taiwan. J. Hazard Mater. 167, 1163–1169. https://doi.org/10.1016/j.jhazmat.2009.01.108.

Livermore, D.M., 2011. Discovery research: the scientific challenge of finding new antibiotics. J. Antimicrob. Chemother. 66, 1941–1944.

Long, R.A., Azam, F., 2001. Antagonistic interactions among marine Pelagic bacteria. Appl. Environ. Microbiol. 67 (110), 4975–4983. https://doi.org/10.1128/AEM.67.11.4975–4983.

López-Causapé, C., Rojo-Molinero, E., Mulet, X., Cabot, G., Moyà, B., Figuerola, J., Togores, B., Pérez, J.L., Oliver, A., 2013. Clonal dissemination, emergence of mutator lineages and antibiotic resistance evolution in *Pseudomonas aeruginosa* cystic fibrosis chronic lung infection. PLoS One 8, e71001.

Luo, G., Li, B., Li, L.G., Zhang, T., Angelidaki, I., 2017. Antibiotic resistance genes and correlations with microbial community and metal resistance genes in fullscale biogas reactors as revealed by metagenomic analysis. Environ. Sci. Technol. 51 (7), 4069–4080.

Marathe, N.P., Regina, V.R., Walujkar, S.A., Charan, S.S., Edward, R.B., Moore, D.G., Larsson, J., Shouche, Y.S., 2013. A treatment plant receiving waste water from multiple bulk drug manufacturers is a reservoir for highly multi-drug resistant integron-bearing bacteria. PLoS One 8 (10), e77310.

Martin, J., Santos, J.L., Aparicio, I., Alonso, E., 2015. Pharmaceutically active compounds in sludge stabilization treatments: anaerobic and aerobic digestion, wastewater stabilization ponds and composting. Sci. Total Environ. 503–504, 97–104.

Martínez, J.L., 2008. Antibiotics and antibiotic resistance genes in natural environments. Science 321, 365–367.

Mazaheri Nezhad Fard, R., Barton, M.D., Heuzenroeder, M.W., 2011. Bacteriophage-mediated transduction of antibiotic resistance in enterococci. Lett. Appl. Microbiol. 52, 559–564.

McKinney, C.W., Pruden, A., 2012. Ultraviolet disinfection of antibiotic resistant bacteria and their antibiotic resistance genes in water and wastewater. Environ. Sci. Technol. 46 (24), 13393–13400. https://doi.org/10.1021/es303652q.

Meouche, E., Dunlop, M.J., 2018. Heterogeneity in efflux pump expression predisposes antibiotic-resistant cells to mutation. Science 9 (6415), 686–690. https://doi.org/10.1126/science.aar7981, 362.

Miller, J.H., Novak, J.T., Knocke, W.R., Young, K., Hong, Y., Vikesland, P.J., et al., 2013. Effect of silver nanoparticles and antibiotics on antibiotic resistance genes in anaerobic digestion. Water Environ. Res. 85, 411–421. https://doi.org/10.2175/106143012X13373575831394.

Minh, T.B., Leung, H.W., Loi, I.H., Chan, W.H., So, M.K., Mao, J.Q., Choi, D., Lam, J.C.W., Zheng, G., Martin, M., Lee, J.H.W., Lam, P.K.S., Richardson, B.J., 2009. Antibiotics in the Hong Kong metropolitan area: ubiquitous distribution and fate in Victoria Harbour. Mar. Pollut. Bull. 58 (7), 1052–1062. https://doi.org/10.1016/j. marpolbul.2009.02.004.

Monteiro, D.R., Gorup, L.F., Takamiya, A.S., Ruvollo-Filho, A.C., de Camargo, E.R., Barbosa, D.B., 2009. The growing importance of materials that prevent microbial adhesion: antimicrobial effect of medical devices containing silver. Int. J. Antimicrob. Agents 34, 103–110. https://doi.org/10.1016/j.ijantimicag.2009.01.017.

Munir, M., Wong, K., Xagoraraki, I., 2011. Release of antibiotic resistant bacteria and genes in the effluent and biosolids of five wastewater utilities in Michigan. Water Res. 45, 681–693. https://doi.org/10.1016/j.watres.2010.08.033.

Neudorf, K.D., Huang, Y.N., Ragush, C.M., Yost, C.K., Jamieson, R.C., Truelstrup Hansen, L., 2017. Antibiotic resistance genes in municipal wastewater treatment systems and receiving waters in Arctic Canada. Sci. Total Environ. 598, 1085–1094. https://doi.org/10.1016/j.scitotenv.2017.04.151.

Nnadozie, C.F., Kumari, S., Bux, F., 2017. Status of pathogens, antibiotic resistance genes and antibiotic residues in wastewater treatment systems. Rev. Environ. Sci. Biotechnol. 16, 491–515. https://doi.org/10.1007/s11157-017-9438-x.

Paiva, M.C., Reis, M.P., Costa, P.S., Dias, M.F., Bleicher, L., Scholte, L.L.S., et al., 2017. Identification of new bacteria harboring qnrS and aac(6′)-Ib/cr and mutations possibly involved in fluoroquinolone resistance in raw sewage and activated sludge samples from a full-scale WWTP. Water Res. 110, 27–37.

Pal, C., Asiani, K., Arya, S., Rensing, C., Stekel, D.V., Larsson Joakim, D.G., Hobman, J.L., 2017. Metal resistance and its association with antibiotic resistance. In: Advances in Microbial Physiology, first ed. Elsevier Ltd., pp. 261–313

Pang, Y., Huang, J., Xi, J., Hu, H., Zhu, Y., 2016. Effect of ultraviolet irradiation and chlorination on ampicillin-resistant *Escherichia coli* and its ampicillin resistance gene. Front. Environ. Sci. Eng. 10, 522–530. https://doi.org/10.1007/s11783-015-0779-9.

Pazda, M., Kumirska, J., Stepnowski, P., Mulkiewicz, E., 2019. Antibiotic resistance genes identified in wastewater treatment plant systems – a review. Sci. Total Environ. 697, 134023.

Pei, R., Kim, S.C., Carlson, K.H., Pruden, A., 2006. Effect of river landscape on the sediment concentrations of antibiotics and corresponding Antibiotic Resistance Genes (ARG). Water Res. 40, 2427–2435. https://doi.org/10.1016/j.watres.2006.04.017.

Perron, G.G., Whyte, L., Turnbaugh, P.J., Goordial, J., Hanage, W.P., Dantas, G., Desai, M.M., 2015. Functional characterization of bacteria isolated from ancient arctic soil exposes diverse resistance mechanisms to modern antibiotics. PLoS One 10, e0069533.

Pitout, J.D.D., Nordmann, P., Laupland, K.B., Poire, L., 2005. Emergence of Enterobacteriaceae producing extended-spectrum beta lactamases (ESBLs) in the community. J. Antimicrob. Chemother. 56, 52–59.

Poole, K., 2017. At the nexus of antibiotics and metals: the impact of Cu and Zn on antibiotic activity and resistance. Trends Microbiol. 25 (10), 820–832.

Pruden, A., Pei, R., Storteboom, H., Carlson, K.H., 2006. Antibiotic resistance genes as emerging contaminants: studies in northern Colorado. Environ. Sci. Technol. 40, 7445–7450. https://doi.org/10.1021/es060413l.

Pruden, A., Arabi, M., Storteboom, H.N., 2012. Correlation between upstream human activities and riverine antibiotic resistance genes. Environ. Sci. Technol. 46, 11541–11549. https://doi.org/10.1021/es302657r.

Pruden, A., Larsson, D.G.J., Amézquita, A., Collignon, P., Brandt, K.K., Graham, D.W., et al., 2013. Management options for reducing the release of antibiotics and antibiotic resistance genes to the environment. Environ. Health Perspect. 121, 878–885. https://doi.org/10.1289/ehp.1206446.

Reinthaler, F.F., Posch, J., Feierl, G., Wüst, G., Haas, D., Ruckenbauer, G., Mascher, F., Marth, E., 2003. Antibiotic resistance of *E. Coli* in sewage and sludge. Water Res. 37, 1685–1690.

Riquelme, B., Maria, V., Novak, J.T., Vikesland, P.J., Pruden, A., 2013. Effect of wastewater colloids on membrane removal of antibiotic resistance genes. Water Res. 47, 130–140. https://doi.org/10.1016/j.watres.2012.09.044.

Rizzo, L., Manaia, C., Merlin, C., Schwartz, T., Dagot, C., Ploy, M.C., Michael, I., Fatta-Kassinos, D., 2013. Urban wastewater treatment plants as hotspots for antibiotic resistant bacteria and genes spread into the environment: a review. Sci. Total Environ. 447, 345–360. https://doi.org/10.1016/j.scitotenv.2013.01.032.

Rodriguez-Mozaz, S., Chamorro, S., Marti, E., Huerta, B., Gros, M., Sànchez- Melsió, A., et al., 2015. Occurrence of antibiotics and antibiotic resistance genes in hospital and urban wastewaters and their impact on the receiving river. Water Res. 69, 234–242. https://doi.org/10.1016/j.watres.2014.11.021.

Rowe-Magnus, D.A., Mazel, D., 2002. The role of integrons in antibiotic resistance gene capture. Int. J. Med. Microbiol. 292, 115–125.

Sachdeva, S., Palur, R.V., Sudhakar, K.U., Rathinavelan, T., 2017. *E. Coli* group 1 capsular polysaccharide exportation nano machinary as a plausible antivirulence target in the perspective of emerging antimicrobial resistance. Front. Microbiol. 8, 70. https://doi.org/10.3389/fmicb.2017.00070.

Segura, P.A., François, M., Gagnon, C., Sauvé, S., 2009. Review of the occurrence of anti infectives in contaminated wastewaters and natural and drinking waters. Environ. Health Perspect. 117, 675–684. https://doi.org/10.1289/ehp.11776.

Shah, S.Q.A., Cabello, F.C., L'Abée-Lund, T.M., Tomova, A., Godfrey, H.P., Buschmann, A.H., Sørum, H., 2014. Antimicrobial resistance and antimicrobial resistance genes in marine bacteria from salmon aquaculture and non-aquaculture sites. Environ. Microbiol. 16, 1310–1320.

Shahverdi, A.R., Fakhimi, A., Shahverdi, H.R., Minaian, S., 2007. Synthesis and effect of silver nanoparticles on the antibacterial activity of different antibiotics against *Staphylococcus aureus* and *Escherichia coli*. Nanomedicine 3, 168–171. https://doi.org/10.1016/j.nano.2007.02.001.

Sharma, V.K., Johnson, N., Cizmas, L., McDonald, T.J., Kim, H., 2016. A review of the influence of treatment strategies on antibiotic resistant bacteria and antibiotic resistance genes. Chemosphere 150, 702–714. https://doi.org/10.1016/j. chemosphere.2015.12.084.

Sharma, N., Bhatnagar, P., Chatterjee, S., John, P.J., Soni, I.P., 2017. Bio nanotechnological intervention: a sustainable alternative to treat dye bearing waste waters. Indian J. Pharmaceut. Biol. Res. 5 (1), 17–24.

Sharma, N., Bhatnagar, P., Chatterjee, S., 2019. Degradation of direct red 28 by *alcaligenes* sp. TEX S6 isolated from aeration tank of common effluent treatment plant (CETP), Pali, Rajasthan. Nat. Environ. Pollut. Technol. 18 (1), 9–20.

Sharma Sanjay, K., Sanghi, R., Mudhoo, A., 2012. Green practices to save our precious ' water resource'. In: Sharma, S.K., Sanghi, R. (Eds.), Advances in Water Treatment and Pollution Prevention. Springer Dordrecht Heidelberg, New York, London, pp. 1–36.

Spencer, S.J., Tamminen, M.V., Preheim, S.P., Guo, M.T., Briggs, A.W., Brito, I.L., et al., 2016. Massively parallel sequencing of single cells by epicPCR link functional genes with phylogenetic markers. J. Int. Soc. Microb. Ecol. 10 (2), 427–436.

Spongberg, A.L., Witter, J.D., 2008. Pharmaceutical compounds in the wastewater process stream in Northwest Ohio. Sci. Total Environ. 397, 148–157. https://doi.org/10.1016/j.scitotenv.2008.02.042.

Springman, A.C., Lacher, D.W., Wu, G., Milton, N., Whittam, T.S., Davies, H.D., Manning, S.D., 2009. Selection, recombination, and virulence gene diversity among group B streptococcal genotypes. J. Bacteriol. 191, 5419–5427.

Srivastava, V., Gusain, D., Sharma, Y.C., 2015. Critical review on the toxicity of some widely used engineered nanoparticles. Ind. Eng. Chem. Res. 54, 6209–6233. https://doi.org/10.1021/acs.iecr.5b01610.

Szczepanowski, R., Linke, B., Krahn, I., Gartemann, K.H., Gützkow, T., Eichler, W., Pühler, A., Schlüter, A., 2009. Detection of 140 clinically relevant antibiotic-resistance genes in the plasmid metagenome of waste water treatment plant bacteria showing reduced susceptibility to selected antibiotics. Microbiology 55, 2306–2319.

Thorpe, K.E., Joski, P., Johnston, K.J., 2018. Antibiotic-resistant Infection Treatment Costs Have Doubled since 2002, Now Exceeding $2 Billion Annually. 2018., Pharmaceutical and Medical Technology. https://doi.org/10.1377/hlthaff. 2017.115.

Tseng, Y.S., Wu, D.C., Chang, C.Y., Kuo, C.H., Yang, Y.C., Jan, C.M., et al., 2009. Amoxicillin resistance with b-lactamase production in *Helicobacter pylori*. Eur. J. Clin. Invest. 39, 807–812. https://doi.org/10.1111/j.1365-2362.2009. 02166.x.

Versporten, A., Bolokhovets, G., Ghazaryan, L., Abilova, V., Pyshnik, G., Spasojevic, T., Korinteli, I., Raka, L., Kambaralieva, B., Cizmovic, L., Carp, A., Radonjic, V., Maqsudova, N., Celik, H.D., Payerl-Pal, M., Pedersen, H.B., Sautenkova, N., Goossens, H., 2014. Antibiotic use in eastern Europe: a cross-national database study in coordination with the WHO Regional Office for Europe. Lancet Infect. Dis. 14, 381–387. https://doi.org/10.1016/S1473-3099(14)70071-4.

Volkmann, H., Schwartz, T., Kirchen, S., Carmen, S.C., Silke, K., Obst, U., 2007. Evaluation of inhibition and cross-reaction effects on real-time PCR applied to the total DNA of wastewater samples for the quantification of bacterial antibiotic resistance genes and taxon-specific targets. Mol. Cell. Probes 21, 125–133. https://doi.org/10.1016/j.mcp.2006.08.009.

Von Wintersdorff, C.J., Penders, J., Van Niekerk, J.M., Mills, N.D., Majumder, S., van Alphen, L.B., et al., 2016. Dissemination of antimicrobial resistance in microbial ecosystems through horizontal gene transfer. Front. Microbiol. 7, 173. https://doi.org/10.3389/fmicb.2016.00173.

Walsh, C., 2000. Molecular mechanisms that confer antibacterial drug resistance. Nature 406, 775–781.

Watkinson, A.J., Murby, E.J., Costanzo, S.D., 2007. Removal of antibiotics in conventional and advanced wastewater treatment: implications for environmental discharge and wastewater recycling. Water Res. 41, 4164–4176. https://doi.org/10.1016/j.watres.2007.04.005.

Watkinson, A.J., Murby, E.J., Kolpin, D.W., Costanzo, S.D., 2009. The occurrence of antibiotics in an urban watershed: from wastewater to drinking water. Sci. Total Environ. 407, 2711–2723. https://doi.org/10.1016/j.scitotenv.2008.11.059.

Watts, J.E.M., Schreier, H.J., Lanska, L., Hale, M.S., 2017. The rising tide of antimicrobial resistance in aquaculture: sources, sinks and solutions. Mar. Drugs 15 (6), 158. https://doi.org/10.3390/md15060158.

Wellington, E.M., Boxall, A.B., Cross, P., Feil, E.J., Gaze, W.H., Hawkey, P.M., et al., 2013. The role of the natural environment in the emergence of antibiotic resistance in Gram-negative bacteria. Lancet Infect. Dis. 13 (2), 155–165.

WHO, 2017. Not enough new antibiotics in the pipeline. UN report warns. https://news.un.org/en/story/2017/09/565592-not-enough-new-antibiotics-pipeline-un-report-warns.

Woolhouse, M., Ward, M., Bunnik, B.V., Farrar, J., 2015. Antimicrobial resistance in humans livestock and the wider environment. Phil. Trans. R. Soc. B 370, 0083. https://doi.org/10.1098/rstb.2014.008.

Wright, G.D., 2010. Q&A: antibiotic resistance: where does it come from and what can we do about it? BMC Biol. 8. https://doi.org/10.1186/1741-7007-8-123. http://www.biomedcentral.com/1741-7007/8/123.

Xu, J., Xu, Y., Wang, H., Guo, C., Qiu, H., He, Y., Zhang, Y., Li, X., Meng, W., 2015. Occurrence of antibiotics and antibiotic resistance genes in a sewage treatment plant and its effluent-receiving river. Chemosphere 119, 1379–1385. https://doi.org/10.1016/j.chemosphere.2014.02.040.

Xu, Y.B., Hou, M.Y., Li, Y.F., Huang, L., Ruan, J.J., Zheng, L., Qiao, Q.X., Du, Q.P., 2017. Distribution of tetracycline resistance genes and AmpC β-lactamase genes in representative non-urban sewage plants and correlations with treatment processes and heavy metals. Chemosphere 170 (100), 274–281.

Yang, S., Carlson, K.H., 2004. Solid-phase extraction-high-performance liquid chromatography-ion trap mass spectrometry for analysis of trace concentrations of macrolide antibiotics in natural and wastewater matrices. J. Chromatogr. A 1038, 141–155. https://doi.org/10.1016/j.chroma.2004.02.084.

Yang, Y., Li, B., Zou, S., Fang, H.H.P., Zhang, T., 2014. Fate of antibiotic resistance genes in sewage treatment plant revealed by metagenomics approach. Water Res. 62, 97–106.

Yasuyuki, M., Kunihiro, K., Kurissery, S., Kanavillil, N., Sato, Y., Kikuchi, Y., 2010. Antibacterial properties of nine pure metals: a laboratory study using *Staphylococcus aureus* and *Escherichia coli*. Biofouling 26 (7), 851–858.

You, C., Yu, X., Lei, L., Li, A.D., Yang, Y., Zhang, T., 2019. Mobile antibiotic resistome in waste water treatment plants revealed by Nanopore metagenomic sequencing. Microbiome 7, 44. https://doi.org/10.1186/s40168-019-0663-0.

Yu, W., Zhan, S., Shen, Z., Zhou, Q., Yang, D., 2016. Efficient removal mechanism for antibiotic resistance genes from aquatic environments by graphene oxide nanosheet. Chem. Eng. J. 313, 836–846. https://doi.org/10.1016/j.cej.2016.10.107.

Yu, Z., Gunn, L., Wall, P., Fanning, S., 2017. Antimicrobial resistance and its association with tolerance to heavy metals in agriculture production. Food Microbiol. 64, 23–32.

Yuan, Q.B., Guo, M.T., Yang, J., 2015. Fate of antibiotic resistant bacteria and genes during wastewater chlorination: implication for antibiotic resistance control. PLoS One 10, e0119403. https://doi.org/10.1371/journal.pone.0119403.

Zhang, X.X., Zhang, T., Fang, H.H., 2009. Antibiotic resistance genes in water environment. Appl. Microbiol. Biotechnol. 82, 397–414. https://doi.org/10.1007/s00253-008-1829-z.

Zhang, T., Zhang, X.X., Ye, L., 2011. Plasmid metagenome reveals high levels of antibiotic resistance genes and mobile genetic elements in activated sludge. PLoS One 6, 1–7. https://doi.org/10.1371/journal.pone.0026041.

Zhu, Y.G., Johnson, T.A., Su, J.Q., Qiao, M., Guo, G.X., Stedtfeld, R.D., et al., 2013. Diverse and abundant antibiotic resistance genes in Chinese swine farms. Proc. Natl. Acad. Sci. U.S.A 110, 3435–3440. https://doi.org/10.1073/pnas.1222743110.

Zhuang, Y., Ren, H., Geng, J., Zhang, Y., Zhang, Y., Ding, L., Xu, K., 2015. Inactivation of antibiotic resistance genes in municipal wastewater by chlorination, ultraviolet, and ozonation disinfection. Environ. Sci. Pollut. Res. Int. 22, 7037–7044. https://doi.org/10.1007/s11356-014-3919-z.

Chapter 9

Green chemistry and its applications in hospital wastewater and its treatment

Reetu Rani[a,b], Suman Singh[a,b]
[a]*CSIR-Central Scientific Instruments Organisation, Chandigarh, India;* [b]*Academy of Scientific and Innovative Research (AcSIR), Ghaziabad, Uttar Pradesh, India*

1. Introduction

Hospitals are a substantial consumer of water depending upon different activities such as sanitary, medical purpose, laundry, and HVAC systems. It is observed that generally 400–1200 L of water is consumed on average by per person per day depending on the size and type of hospital (Verlicchi et al., 2010; Carraro et al., 2016; El-Ogri et al., 2016). Consequently, hospitals generate huge quantities of wastewater every year which contains various harmful contaminants such as pathogens, pharmaceuticals and antibiotics residues, radioisotopes, heavy metals, endocrine disruptors, etc. Pharmaceutical and antibiotics residues persist in the environment for long period of time and have a tendency to accumulate in the environment which results in the production of antibiotic-resistant bacteria (Verlicchi et al., 2010; Verlicchi et al., 2012). It is well reported that almost 15%–20% of medicines consumed in medicals are bioaccumulative. Organically bound halogens in pharmaceuticals and their residues cause AOX (adsorbable organic halides) emissions. Pathogens present in hospital wastewater (HWW) make it source of various infectious diseases. All these contaminants make HWW more toxic and complex than urban effluents. About 4 to 150 times higher concentration of micropollutants is reported in HWW compared with urban effluents (Kosma et al., 2010). It is well stated that wastewater discharged from a hospital with 1000 beds is as polluted as wastewater of a town with population of 10,000 (Boillot et al., 2008).

Green Chemistry and Water Remediation: Research and Applications.
https://doi.org/10.1016/B978-0-12-817742-6.00009-8

1.1 Micropollutants in hospital wastewater and environmental risk

1.1.1 Pharmaceutically active compounds

Pharmaceuticals along with their active residues are present everywhere in the environment. Pharmaceuticals from different classes such as antivirals, antibiotics, anesthetics, and antidepressants are commonly observed as micropollutants. These compounds are released from different sources such as industries, households, and hospitals. Out of these, hospitals are the major source of pharmaceutical pollution in the environment (Kosma et al., 2010). Wastewater generated from hospitals are usually discharged into wastewater treatment plants (WWTPs) without prior treatments. But most of WWTPs are not designed to remove these micropollutants from the wastewater and therefore a large number of these pharmaceuticals are observed in WWTPs effluent. In general, the concentration of different pharmaceuticals in WWTPs effluent is reported in the range of ngL^{-1} to μgL^{-1}. In a study, about 56 active pharmaceutical ingredients were found in treated effluent of wastewater plants in United States (Kostich et al., 2014). In another study, ecotoxicity and environmental hazard of treated WWTPs effluent was performed and it was observed that about 197 pharmaceuticals were identified in the effluent out of which 15 were very hazardous compounds (Frédéric and Yves, 2014). The common micropollutants reported in HWW (Carraro et al., 2016; Aydin et al., 2019; Balakrishna et al., 2017) are discussed below and shown in Table 9.1.

1.1.1.1 Antibiotics

Antibiotics are the major micropollutants observed in HWW. These are commonly used to treat bacterial infections both in humans and animals. About 1–2 lakh tonnes of antibiotic consumption is estimated worldwide and out of total antibiotics consumption, about 26% of antibiotics are used in hospitals. Consumption of antibiotics is usually higher in winters compared with summers. A study showed that antibiotic's load in winter was about 303.2 g/day in hospital effluent which was much higher than antibiotic's load found in summers, i.e., 3.46 g/day (Aydin et al., 2019). Much of the common antibiotics used for treatment of patients get excreted unmetabolized and through urine and feces of treated patients antibiotics and their metabolites reach in WWTPs. These micropollutants remain untreated in WWTP effluent and therefore found in surface and drinking waters. Penicillins, sulfamethoxazole, clarithromycin, and ciprofloxacin antibiotics are commonly observed in HWW. Along with antibiotics, presence of their metabolites such as metronidazole-OH, N-acetyl sulfamethoxazole, and erythromycin-H_2O is also reported in wastewaters by different researchers. The major concern of presence of antibiotics in wastewater is the development of antibiotic-resistant bacteria and producing implications in human health (Aydin et al., 2019; Diwan et al., 2009; Pauwels and Verstraete, 2006). Along with this, presence of antibiotics in waterbodies poses a potential risk for aquatic species (Aydin et al., 2019; Kim et al., 2007).

TABLE 9.1 List of some common micropollutants observed in HWW.

Category	Common micropollutants
Antiinflammatories and analgesics	Acetaminophen, codeine, ibuprofen, diclofenac, paracetamol, naproxen, ketoprofen, dipyrone, mefenamic acid, mevastatin, salicylic acid derivatives
Antibiotics	Macrolides (erythromycin, azithromycin, roxithromycin), tetracyclines (chlortetracycline, tetracycline, oxytetracycline), quinolones (ciprofloxacin, norfloxacin, ofloxacin), trimethoprim, ronidazole, penicillin, cefazolin, chloramphenicol, clindamycin, lincomycin, spiramycin, oleandomycin
Disinfectants	Glutaraldehyde, triclosan
Anesthetics	Propofol
Heavy metals	Platinum, mercury, lead, gadolinium, chromium, cadmium
Iodized contrast media (ICM)	Iopamidol, iopromide
β-blockers	Atenolol, betaxolol, carazolol, timolol, pindolol, metoprolol, propranolol, sotalol, nadolol
Steroids	Estriol, estrone, 17-β-estradiol
Antihistamines	Ranitidine, cimetidine
Psychiatric drug, antidepressants, anticonvulsants	Carbamazepine, diazepam, lorazepam, paroxetine, fluoxetine
Antineoplastic drugs	Tamoxifen
Lipid regulators	Mevastatin, pravastatin, clofibric acid, bezafibrate, atorvastatin
Musk fragrances	Musk ketone, musk moskene

1.1.1.2 Endocrine disruptors

Endocrine disruptors are the chemicals that adversely affect the endocrine system (thyroid, pituitary, pancreas, adrenal glands, testes, and ovaries) of human beings and animals. These chemicals upshot hormonal system of humans and cause various adverse effects on reproductive and neurological health. These can be estrogenic, androgenic, or thyroidal depending upon their target area. Chemicals such as estrone, estriol, and 17-β-estradiol are the common estrogens detected in hospital effluents making it estrogenically active. Other than these, some chemicals such as bisphenol A and triclosan are also observed in HWW. These compounds are also classified as endocrine disruptors as they can mimic or interfere with the endocrine system (Johnson et al., 2005; Joss et al., 2004).

1.1.1.3 Other pharmaceuticals and personal care products

Other common pharmaceuticals that are frequently observed in HWW are antiinflammatories and analgesics, disinfectants, anesthetics, β-blockers, antineoplastic drugs, and lipid regulators. Among these diclofenac, ketoprofen, acetaminophen, ibuprofen, sulfamethoxazole, propranolol, carbamazepine, ranitidine, and atenolol are the most frequently recovered pharmaceuticals in hospital effluent. Carbamazepine is an antiepileptic drug that acts as an endocrine disruptor and is very difficult to remove using common removal techniques. Along with these, ingredients of personal care products such as cosmetics, sunscreens, deodorant compounds, etc., are also recovered from hospital effluent. Compounds such as antibacterials (triclocarban, triclosan), UV-blocking agents (3-(4-methyl benzylidene) camphor, octocrylene), and musk ketones come under this category (Balakrishna et al., 2017). Cytostatic agents present in HWW pose serious health threats to human beings due to their carcinogenic, mutagenic, and embryotoxic properties. A number of studies have confirmed potential ecotoxicity of different pharmaceuticals and personal care products (Jjemba, 2006; Hernando et al., 2004).

1.1.2 Contrast agents

Contrast agents are the compounds given to patients to augment the contrast between the organs to be investigated and the surrounding tissue during use of imaging technology. Different contrast agents such as iodinated contrast agents (diatrizoate, iohexol, iobitridol, iopromide, etc.), barium and gadolinium contrast agents are used for different applications in medicals. Out of these iodinated contrast agents are used in larger amounts (Lusic and Grinstaff, 2012; Pasternak and Williamson, 2012). These compounds are not metabolized in the body and get excreted as it is and accumulated into the environment. Different studies confirmed their presence in surface waters and drinking water (Mendoza et al., 2015; Borowska et al., 2015). Contrast agents are not of much ecotoxicological concern but their metabolites are harmful as they carry free amino groups that can be responsible for mutations. Various studies show presence of these contrast agents in HWWs (Ternes et al., 2003; Schrap et al., 2003).

1.1.3 Heavy metals

A group of metals with a relatively higher density are classified as heavy metals. These are nonbiodegradable and therefore possess the ability to accumulate in the environment. Heavy metals are toxic even if present at ppb levels and most of them are reported with carcinogenic properties. Long-term exposure to these causes adverse health effects on humans. Platinum, gadolinium, and mercury are the heavy metals present majorly in HWW along with minor concentrations of cadmium, copper, lead, nickel, zinc, and iron (Danchaivijitr et al., 2005; Nyamukamba et al., 2019). Platinum is released from the drugs used for the treatment of cancer patients. Gadolinium is employed in Magnetic Resonance Imaging applications.

1.2 Microbial contaminants in hospital wastewater and environmental risks

The microbial composition of HWW is similar to UWW in terms of fecal origin bacteria such as *E. Coli, Salmonella*, etc. Concentration of fecal bacteria is low in HWW because of presence of antibiotics and disinfectants (Emmanuel et al., 2005). But it is reported HWW is highly contaminated due to presence of hospital-generated pathogenic bacteria such as *Staphylococcus aureus*, *Pseudomonas, etc.* Load of antibiotic resistant bacteria and viruses is also high in HWW compared with UWW. Pathogens such as hepatitis B virus, norovirus, *clostridium difficile, acinobacter*, *candida* sp, *and Acinetobactor baumannii* are some hospital-acquired infection pathogens reported in HWW (Carraro et al., 2016; Weinstein and Hota, 2004; Liu et al., 2010). The problem of pathogenic microbes in water causes, outbreaks from time to time. Apart from this, antibiotic-resistant strains of bacteria are of major health concern as they can cause chronic effects on human health.

2. Characteristics of hospital wastewater

A huge increase in production of HWW has been observed in recent years due to expansion of medical services (Amouei et al., 2010). Generally, HWW is considered similar to UWW in characteristics and therefore discharged into WWTPs without treatment. But HWW is very different from UWW both qualitatively and quantitatively (El-Ogri et al., 2016; Kosma et al., 2010). Values for the parameters such as biological oxygen demand (BOD), chemical oxygen demand (COD), and suspended solids (SSs) are found much higher for HWW than UWW and, consequently, a separate treatment for HWW is suggested by different researchers (Verlicchi et al., 2010; Kosma et al., 2010; Vieno et al., 2007). The effluent generated in hospitals is contributed from both medical and nonmedical activities. Activities like research, diagnosis, first aids, operations, etc., fall under medical activities and activities like laundry, kitchen, and toilets come under nonmedical activities. The effluent released from normal wards, kitchen, and laundry is similar to domestic discharge in characteristics. But effluent discharged from most of the medical activities is considered highly toxic due to presence of hazardous substances such as pharmaceuticals and personal care product residues, pathogens, acids, alkalis, infectious, and radioactive agents (Amouei et al., 2010). The wastewater generated by different services employed in hospitals is very toxic to environment when discharged without treatment. Characteristics of hospital effluent greatly vary from hospital to hospital depending upon the type of patients and services in particular hospital. Consequently, characterization of HWW is of crucial importance to track its environmental risks. Various researches have been performed in different geographical areas to characterize physical, chemical, and biological parameters of HWW (Mesdaghinia et al., 2009; Kumarathilaka et al., 2015; Prayitno et al., 2013). Under physicochemical characteristics, different parameters such as pH, electrical conductivity, temperature, BOD, COD, total solids (TS), total dissolved solids and total suspended solids (TSS), heavy metals, nitrogen, and phosphates are monitored. Main physical, chemical, and microbiological characteristics of HWW are given in Table 9.2. pH of wastewater is recommended in the range of 6−9 otherwise highly acidic or basic wastewater can hinder the treatment process using biological methods. Other than pH, TSS is another important parameter to determine quality of wastewater. TSS of domestic wastewater is normally reported in the range of 120−400 mg/L. The organic content of wastewater is measured in terms of BOD_5 and COD. The average values of BOD_5 and COD for domestic wastewater is reported in range 110−350 and 250−800 mg/L, respectively (Mesdaghinia et al., 2009). Different studies by various researches showed that HWW possess a high range of concentrations for various parameters compared with domestic wastewater (Emmanuel et al., 2005). Microbial composition of HWW is similar to UWW in terms of coliforms but different hospital-acquired infection pathogens and antibiotic-resistant bacteria are found in large numbers.

In some of the developed countries, hospital effluent is generally discharged into municipal sewage systems after partial treatment following the guidelines set by national or international agencies. But in most of developing countries, legal guidelines are not well established to manage HWW. Therefore, hospital effluent is usually released into sewage systems without prior treatments. But the conventional WWTPs are designed to remove macropollutants and HWW being more complex as contains polar micropollutants such as pharmaceuticals and personal care products and is difficult to treat in WWTPs.

There are many reports giving evidence of occurrence of different antibiotics and personal care products in treated wastewater effluents of WWTP (Rodriguez-Mozaz et al., 2015; Ben et al., 2017; Santos et al., 2013). This treated water is used for various nonpotable applications in different regions such as for gardening, irrigation purpose. Consecutively, these contaminants reach the surface water and then leach into groundwater (Benotti et al., 2008; Huerta-Fontela et al., 2011). Along with this, the sludge of WWTPs is used as a fertilizer and therefore acts as a potential source of these micropollutants in the environment. Prevalence of these micropollutants in treated HWW is of great concern for environment and human health and therefore there is need to design eco-friendly treatment technologies with efficient removal potential for nonbiodegradable micropollutants from HWW before discharge into environment. To save our environment from the potential risks of toxic wastewater generated from hospitals, green chemistry principles can be employed as possible solutions. Different possible solutions for sustainable treatment and reclamation of hospital effluent are discussed in the next section.

3. Green chemistry approach for sustainable treatment and management of hospital wastewater

Green chemistry or clean chemistry approach involves the design, development, and use of new products or techniques with minimum waste production and negligible toxicity to the environment. In another terms, we can say that it is the key to sustainable development (Anastas and Eghbali, 2010; Warner John et al., 2004). There are twelve principles of green chemistry that set clear guidelines to achieve goal of sustainable development. The principles include prevention of waste production, utilization of all precursors to yield products, synthesis processes with less toxic precursors and at ambient conditions of temperature and pressure for energy efficiency, use of catalysts, design of safer chemicals, solvents, and auxiliaries and their degradation pathways, promote practices of renewable resources, cut down the use of derivatives, real-time monitoring and control of hazardous pollutants, and use of greener technology for prevention of accidents (Anastas and Eghbali, 2010; Horváth and Anastas, 2007). Fig. 9.1 shows all the principles of green chemistry.

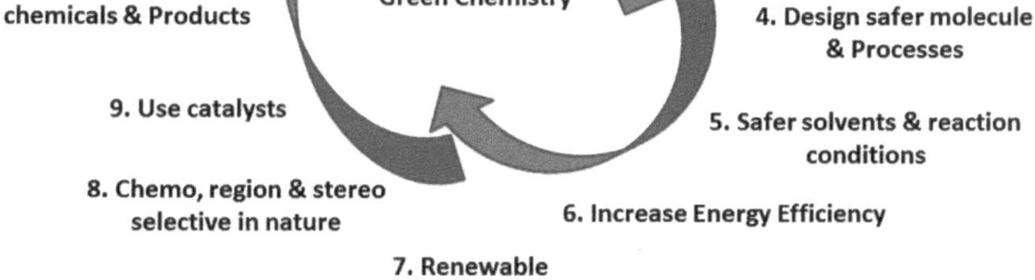

FIG. 9.1 Green chemistry principles.

Following all these principles, one can combat environmental pollution without harming environment and mankind. These green chemistry principles are already adapted in some developed counties but there is need to apply these principles worldwide to manage HWW effectively. In this context, some physicochemical methods (advanced oxidation processes (AOPs), adsorption, electrocoagulation, etc.) and biological treatment methods which are based on green chemistry principles can be exploited for possible solution to treatment of harmful HWW. Recently, nanotechnology has emerged as one of the potential methods of wastewater treatment. Current research in nanotechnology enabled to develop technically and economically feasible substitutions to traditional wastewater treatment. In this technology, materials of nanoscale dimensions are employed to build new structures and components. The properties of nanomaterials are quite different and advanced than micromaterials. To date, various nanomaterials have been explored for photocatalysis, ozonation, adsorption, and filtration of different pollutants from wastewater. In this regard, it is observed that nano-sized photocatalyst such as nano-TiO_2 and MoS_2 nanosheets shows higher photocatalytic activity as compared with bulk materials. Nanomaterials with higher surface area such as MWCNTs, metal oxide nanoparticles, possess efficient potential to replace different macromaterials used for adsorption or oxidation of contaminants (Bora and Dutta, 2014). Various form of nanotechnology can be collaborated with different conventional physicochemical and biological methods of HWW treatment to achieve goal of efficient removal of micropollutants in sustainable means. Different physicochemical and biological methods employed for HWW treatment and reclamation are discussed below along with latest advances and developments made in nanotechnology for improving the effectiveness of present remediation methods of HWW.

3.1 Physicochemical methods

Different conventional physicochemical techniques such as AOPs, adsorption, catalytic degradation, oxidation, and filtration showed great potential for treatment of HWW. Low cost, eco-friendly nature, and easy operation make these methods attractive solution for the reclamation of wastewater generated from hospitals. Merits and demerits of these methods are discussed below with possible new modifications for better removal efficiency.

3.1.1 Advanced oxidation processes

As the name suggests in AOPs, oxidation of organic matter present in HWW is carried out by various high energy oxidants such as H_2O_2, ozone and Fenton reagent, UV, etc., by generating reactive oxygen species (ROS). The main goal of AOPs is the oxidation of organic matter to produce harmless inorganic matter. AOPs are carried out basically in two steps; first is the generation of

TABLE 9.2 Main physical, chemical, and microbiological characteristics of HWW.

Physical parameter	Range of concentrations
Conductivity, μS/cm	300–1000
pH	6–9
Redox potential, mV	850–950
Fat and oil, mg/L	50–210
Chlorides, mg/L	80–400
Total N, mg N/L	60–98
NH_4, mgNH_4/L	10–68
Nitrite, mg NO_2/L	0.1–0.58
Nitrate, mg NO_3/L	1–2
Phosphate, mg P-PO_4/L	6–19
Suspended solids, mg/L	120–400
Chemical parameters	
COD, mg/L	1350–2480
Dissolved COD, mg/L	380–700
DOC, mg/L	120–130
TOC, mg/L	31–180
AOX, μg/L	550–10000
BOD_5/COD (biodegradability index)	0.3–0.4
Microorganisms, MPN/100 mL	
E. coli	10^3–10^6
Enterococci	10^3–10^6
Fecal coliform	10^3–10^4
EC_{50} (*Daphnia*), TU	9.8–117
Total surfactants, mg/L	4–8
Total disinfectants, mg/L	2–200
Specific disinfectants[a]	
BAC_C12, μg/L	34
BAC_C12–18, μg/L	49

Continued

TABLE 9.2 Main physical, chemical, and microbiological characteristics of HWW.—cont'd

Physical parameter	Range of concentrations
DDAC-C10, μg/L	102
Antibiotics, μg/L	30–200
Antiinflammatories, μg/L	5–1500
Lipid regulators, μg/L	1–10
Cytostatic agents, μg/L	5–50
ICM, μg/L	0.2–2600
Beta-blockers, μg/L	0.4–25

[a]*Disinfectants: quaternary ammonia disinfectant: BAC_C12–18: benzalkonium chloride; DDAC-C10: dimethyldidecylammonium chloride.*
Adapted with permission from reference Verlicchi, P., Al Aukidy, M., Zambello, E., 2015. What have we learned from worldwide experiences on the management and treatment of hospital effluent? an overview and a discussion on perspectives. Sci. Total Environ. 514, 467–491.

ROS or hydroxyl radicals and in second step use of these radicals to oxidize the organic matter. These oxidation processes can be further categorized as following depending upon the type of oxidation conditions used like ozonation, photocatalysis, sonolysis, UV photolysis, Fenton, and wet air oxidation as shown in Fig. 9.2 (Huber et al., 2003; Mishra et al., 2017; Machado et al., 2007). Table 9.3 shows the list of different AOPs used for the treatment of HWW. These processes are potential candidate to treat HWW as they could completely eliminate micropollutants from HWW or could convert the pollutants into less harmful products. AOPs are greener and sustainable options for decontamination of HWW as these do not produce any harmful sludge or transfer any pollutants from one phase to another (Jjemba, 2006).

3.1.1.1 Ozonation

In this method, ozone (O_3) is used for oxidation of organic matter. Ozone is a strong oxidant as well as it possess properties of disinfectant. Ozonation is a promising method to remove micropollutants and other organic–inorganic contaminants, degrading pathogenic microorganisms and to decolorize the wastewater (Jjemba, 2006; Huber et al., 2003; Hansen et al., 2016; Rame et al., 2018). However, ozonation kinetics of pharmaceuticals is very slow. In this regard, catalytic ozonation in which nanocatalysts such as carbon nanotubes, metal oxides, etc., are employed for the degradation of organic contaminants, is a better option. Ozonation can be used in combination with other physicochemical processes to completely eliminate the micropollutants from wastewater. Therefore, ozone is also used with H_2O_2 and metal oxide catalysts

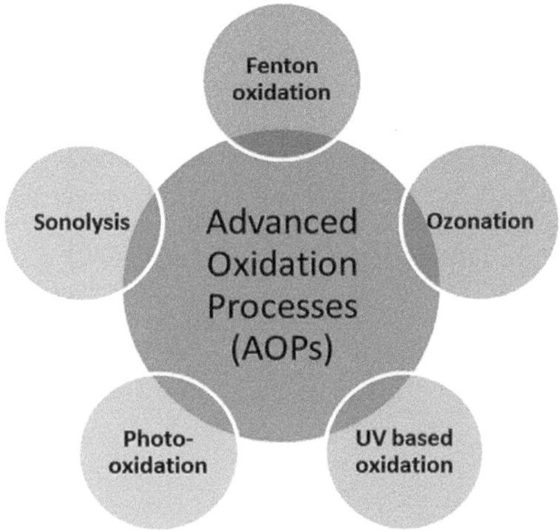

FIG. 9.2 Different advanced oxidation processes (AOPs).

(Fe_2O_3, MnO_2, etc.) to enhance the oxidation rate or ozone transfer rate. Lee et al. studied the micropollutant removal efficiency of ozone in HWW. The authors concluded that around 38 micropollutants out of 56 detected in HWW possess ozone reactive moieties and therefore efficiently eliminated at a lower dose of O_3. But some of the pollutants such as contrast agents were removed inefficiently (Lee et al., 2014). In another study, Kovalova et al. performed posttreatment of HWW using O_3, O_3/H_2O_2, activated carbon, and UV/TiO_2 in a pilot-scale study. The elimination efficiency of these methods was tested for 56 micropollutants. Through ozonation, the majority of the pollutants get eliminated except diatrizoate (ICM) which were efficiently removed by UV/TiO_2 (Kovalova et al., 2013). Recently Malik and coworkers performed nanocatalyzed ozonation of pharmaceutical wastewater using O_3/Fe^{2+}, O_3/nZVI (nano zero valent iron), and observed enhanced biodegradability of wastewater. Along with this removal of color, toxicity and COD was also achieved (Malik et al., 2019). It is clear from the above studies that ozonation is potential method to eliminate most of the micropollutants but contrast agents, surfactants, are resistant to oxidative degradation by ozone in alone; therefore, it is suggested to use this technique in combination with other physicochemical remediation methods.

3.1.1.2 UV photolysis

In the UV photolysis method, UV light is applied to degrade organic matter and microorganisms present in wastewater. Energy of UV light is utilized to mineralize organic matter by breaking bonds present in pollutants. The different UV lamps are employed to generate UV radiations of different wavelengths depending upon the nature of pollutants. As there is no use of

TABLE 9.3 List of different AOPs with merits and demerits.

Process	Principle	Merits	Demerits	Reference
Ozonation	Use of ozone to degrade micropollutants and microbes	• O_3 simultaneously used as oxidant and disinfectant	• Ozone decomposition in water is complex • Low oxidation efficiency of ozone for some chemicals	Huber et al. (2003), Hansen et al. (2016), Rame et al. (2018)
UV photolysis	Use of UV light to degrade microbes and UV absorbing organic matter. UV light causes homolytic cleavage of organic molecules to generate radicals	• Short duration of time • Eco-friendly as no residuals left	• Expensive method • Partial degradation of organic matter	Machado et al. (2007), Avisar et al. (2010)
Fenton oxidation	Use of fenton reagent (H_2O_2 and Fe^{2+}) to generate free radicals for degradation of organic pollutants	• Degrade organic as well inorganic pollutants • Easy and economical method	• Effective only in highly acidic conditions (pH 2–5) • Production of iron sludge	Velichkova et al. (2017), Diao et al. (2017)
Sonolysis	Use of ultrasonic sound waves (frequency 20 KHz –10 MHz) to generate cavitation bubbles which degrade organic pollutants and generate OH free radicals	• Eco-friendly process	• Higher energy consumption and low mineralization efficiency	Naddeo et al. (2012), Fraiese et al. (2019)
Photocatalysis	Use of photocatalyst to break down organic pollutants	• Eco-friendly process	• Low photo thermal stability of photocatalyst	Kanakaraju et al. (2014), Chatzitakis et al. (2008)

chemicals in this method, problem of residues is not encountered. Therefore, it is one of the eco-friendly technology used to combat the concentration of organic pollutants in water and wastewater. A number of organic compounds such as endocrine disrupters, pharmaceuticals, pesticides, and steroids are degraded by application of this technology by various researchers (Kovalova et al., 2013; Kim et al., 2009; Sanches et al., 2010; Avisar et al., 2010). UV technology is also used in combination with ozone technology to group the benefits of both. Different researchers reported effective wastewater treatment by a combination of both the technologies (Gong et al., 2008; Dudziak and Burdzik, 2016).

3.1.1.3 Fenton oxidation

In the Fenton oxidation method, generation of hydroxyl radicals is accomplished using Fenton reagent, i.e., ferrous salts and H_2O_2. This method is easy to use and comparatively cheaper and there is no need for input energy. Since nonhazardous chemicals are used for the process, this process is also considered under green approaches. These advantages make the Fenton process a sustainable solution to HWW treatment (Velichkova et al., 2017). In Fenton process, a series of chain reactions occur between Fe^{2+} and H_2O_2 under low pH and dark conditions to break bonds of H_2O_2 to generate hydroxyl radical (OH). As generated free radical is then used to oxidize the organic matter present in wastewater. The mechanism of Fenton oxidation is presented below.

$$Fe^{2+} + H_2O_2 \rightarrow Fe^{3+} + OH^{\bullet} + OH \tag{9.1}$$

$$OH^{\bullet} + Fe^{2+} \rightarrow OH + Fe^{3+} \tag{9.2}$$

$$RH + OH^{\bullet} \rightarrow R^{\bullet} + H_2O \tag{9.3}$$

Fenton oxidation is pH-dependent process and to generate maximum free radicals the pH of the solution must be in the acidic range. Oxidation of a number of hazardous organic pollutants including antibiotics such as -lactams, trimethoprim, quinolones, and tetracyclines has been performed using Fenton oxidation method (Umadevi, 2015; Affam and Chaudhuri, 2014; Benitez et al., 2011; Diao et al., 2017).

3.1.1.4 Sonolysis

In this method, ultrasonic sound waves in a frequency range from 20 KHz to 10 MHz are employed to degrade hazardous pollutants present in wastewater. Cavitation bubbles are generated as a result of ultrasonic waves. An explosion of these cavitation bubbles breaks down the organic pollutants. In this process, no secondary pollutants are generated making it a sludge-free sustainable process. But compared with other AOPs, its mineralization efficiency is lower; therefore, it is used in combination with other AOPs such as photocatalysis, ozonation, and Fenton process (Naddeo et al., 2012; Fraiese et al., 2019).

3.1.1.5 Electrochemical oxidation

In this AOP method, hydroxyl radicals are electrogenerated. Oxidation and reduction of pollutants is carried out with the aid of electrons generated at the electrode surface. Electrode of different shapes (plates, bars, fibres, and porous) and materials (platinum, glassy carbon, gold, boron-doped diamond (BDD) electrode, etc.) are explored for electrochemical degradation of different organic pollutants present in wastewater (Sopaj et al., 2015). The choice of electrode material greatly depends upon the type of pollutants and the type of sample. Other than this the electrode material should be inert, cost-effective, and stable (Mhemdi et al., 2013; Ganzenko et al., 2014). Electrochemical oxidation of pharmaceutical present in HWW was carried out using BDD and platinum electrodes and reported pollutants abatement up to 100% after 120 min (Ouarda et al., 2019). To enhance the efficiency of electrochemical oxidation of pollutants, different nanomaterials such as graphene oxide, MWCNTs, CNTs, metal nanoparticle, etc., can be used to modify the electrode surface. The modified surface shows enhanced catalytic activity and results in complete mineralization of organic pollutants present in HWW (Pogacean et al., 2015; Motoc et al., 2012).

3.1.1.6 Photo-Fenton process

In this process, Fenton reaction mixtures are irradiated with UV and visible lights to increase the rate of the reaction of the reacting mixture. During this procedure, iron salts perform as photo catalyst and H_2O_2 as oxidant. Free radicals are generated to greater extent compared to normal Fenton process. The rate of reaction is very fast which depends on different factors such as pH of the solution, the concentration of iron salts, and H_2O_2 added (Elmolla and Chaudhuri, 2009). Although photo-Fenton method is more efficient compared with simple Fenton oxidation, narrow pH range and high cost sludge processing limit the application of this process in wastewater treatment (Mishra et al., 2017; Trovó et al., 2011). Reaction mechanism of photo-Fenton process is given below in Eq. (9.4).

$$\mathrm{Fe(OH)}^{+2} + h\nu \rightarrow \mathrm{Fe}^{+3} + \mathrm{OH}^{\bullet} \tag{9.4}$$

3.1.1.7 Photocatalysis

In the photocatalysis method, a light harnessing molecule is used (photocatalyst) which uses UV rays from sunlight to break down bonds of organic pollutants. This technique has enormous potential to degrade organic pollutants from HWW. Due to low price, high efficiency, and nontoxic properties, photocatalysis has become widespread wastewater treatment method. Different photocatalysts such as TiO_2, zinc oxide (ZnO), titanium perovskites, and graphitic carbon nitride are explored by various researchers to degrade different micropollutants present in wastewater. Out of these TiO_2 is used vastly because of its high photostability, nontoxicity, and good photoreactivity, and chemical and biological inert behavior. Different researchers reported treatment of HWW pollutants such as antibiotics of different classes, antiepileptic, antiinflammatory, and analgesics drugs using photocatalysis (Kanakaraju et al., 2014; Achilleos et al., 2010). Though the conventional photocatalysis method still has many drawbacks in terms of low quantum efficiency, high energy intake and low removal efficiency of micropollutants. Therefore, new photocatalyst materials are required with improved properties. In this regard, nano-TiO_2, nano-ZnO, nano-WO_3, and g-C_3N_4 have emerged as potential photocatalysts due to their unique properties such as presence of abundant surface states, large surface area, and possibility of nanosurface functionalization, high electron mobility, and facile synthesis (Cai et al., 2018). Photocatalytic degradation of chloramphenicol antibiotic was studied using nano-TiO_2 (TiO_2 P25) and nano-ZnO in aqueous heterogeneous solutions. Similar photocatalytic activity was observed for both materials which gave complete mineralization in 90 min (Chatzitakis et al., 2008). Photocatalysis using nanophotocatalysts exhibited great potential for degradation of different PhAc and personal care products in wastewater.

3.1.2 Adsorption

Adsorption process is one of the efficient and sustainable technologies for the treatment and recovery of HWW. The main advantage of adsorption technology is that it can be applied on both small and large scales. Along with this, adsorbents can be prepared from waste material and can be regenerated after use. Because of these reasons, adsorption-based treatment is considered as one of the promising treatment methods for HWW. Adsorption is generally applied after biological treatment of wastewater. Conventionally, different carbon materials along with some clays kaolin, attapulgite, silica, and zeolites are used for adsorptive removal of a class of pharmaceuticals from HWW. But the conventional adsorbents limited by their low adsorption rates and capacities (Bui and Choi, 2009). These limitations of convention adsorbents were overcome by development of different nanoadsorbents such as multiwall carbon nanotubes, carbon nanofibers, and nanometal oxides. These nanoadsorbents exhibit unique advantages such as presence of plentiful adsorption sites due to large surface area, tunable pore size, easy regeneration capability, and good selectivity (Qu et al., 2013). Adsorption potential of these nanoadsorbents for various pharmaceuticals and organic pollutants present in HWW have been explored by different research groups (Álvarez-Torrellas et al., 2017; Macías-García et al., 2019; Rakić et al., 2015). Along with these adsorbents, researchers are working for the development of adsorbents from waste materials to decrease the cost of wastewater treatment in a sustainable way; for this waste from different industries (bottom ash from thermal power plants, deoiled soya), agriculture (rice, wheat, coconut shell, tea, fruit peels), and municipal (sewage sludge) is collected and employed for adsorptive removal of different contaminants from wastewater (Yadav et al., 2019). Adsorption of contaminants on the surface of the adsorbents can be attributed to different interactions such as hydrogen bonding, covalent or ionic interactions, hydrophobic interactions, and $\pi-\pi$ stacking depending upon nature of adsorbent and adsorbates (Wang et al., 2017; Jung et al., 2013; Smith and Rodrigues, 2015). Adsorption of two antibiotics namely ciprofloxacin and sulfamethoxazole on the surface of graphene oxide is controlled by different mechanisms. Electrostatic attractions are observed between graphene oxide and ciprofloxacin, while $\pi-\pi$ stacking was responsible for adsorption of sulfamethoxazole on GO surface (Chen et al., 2015).

3.1.3 Membrane filtration

Membranes act as physical barrier that selectively resists the passage of contaminants depending on their size, shape, and other characteristics. Reverse osmosis, microfiltration, ultrafiltration, and nanofiltration membranes are commonly used as these are easy to operate and show high efficacy and good removal capacity for organic pollutants (Taheran et al., 2016). Owing to these properties, membranes can be employed for possible solution to the treatment of HWW. It is a pressure-driven technology in which pressure is applied to the solution to separate the contaminants. The choice of membrane greatly depends upon the type and size of pollutants (Mohammad et al., 2015). Membrane filtration is used in combination with other methods for the complete removal of the pollutants. Reverse osmosis–based filtration system was employed in a pilot-scale membrane bioreactor (MBR) and a good removal efficiency was observed. But the need of high pumping pressures and fouling problems limits the application of these membrane filters (Yadav et al., 2019). Nanotechnology offers a possible solution for these limitations of the membrane filtration method. By incorporating nanomaterials into membranes, properties such as stability and permeability can be effectively enhanced with lower fouling rates. Self-cleaning properties can be introduced into the membranes. For this, different methodologies such as use of composite materials, surface coating, and chemical treatment are employed (Qu et al., 2013). Modifications of surface of membranes with the help of nanomaterials help to enhance the removal efficiency of membranes for different pharmaceuticals and organic contaminants present in HWW. Different nanomaterials incorporate different properties in the membrane. Depending on the need of pollutant load and nature, coating material can be chosen. For example, to improve selectivity and permeability of the membranes CNTs are better option and on other hand self-cleaning properties can be incorporated using antibacterial materials such as nano-TiO_2 (Abdelmelek et al., 2011).

3.2 Biological processes

Due to cost-effectiveness, green, and eco-friendly nature, biological processes are of huge interest for scientists and environment concerned personnel nowadays. In bioremediation technologies, microorganisms are employed to remove contaminants from wastewater. Microorganisms use organic matter of wastewater to grow and convert it into harmless products (Borkar et al., 2013) Some of the biological techniques used for the treatment of HWW are discussed below.

3.2.1 Conventional activated sludge

Conventional activated sludge is one of the oldest biological techniques used for the treatment of wastewater. In this technique, microorganisms are used to feed the organic pollutants present in wastewater. In CAS method, first of all, aeration of HWW containing organic matter is performed. The microorganisms metabolize the suspended organic matter and grow in form of a floc also called activated sludge. This activated sludge is used to get settled in settling tank, leaving a clear and organic matter—free liquid. The clear water is then separated from activated sludge and the settled sludge is recycled to aeration tank to reuse again. This technique basically lowers down the amount of SSs and COD but is unable to remove micropollutants such as pharmaceuticals, pesticides, and heavy metals. Therefore, use of only CAS methods for treatment of HWW is not sufficient (Verlicchi et al., 2012; Pauwels et al., 2006; Lindberg et al., 2006; Göbel et al., 2005). Various researchers studied the removal of pharmaceuticals from wastewater using CAS method. Drugs such as ibuprofen, onalide, naproxen, galaxolide, and 17β-estradiol showed good removal rate through conventional method but no significant removal rate was observed for some drugs such as iopromide and diclofenac (Carballa et al., 2004; Clara et al., 2005). The schematic of activated sludge is shown in Fig. 9.3. To overcome the limitations of CAS method, new biological methods with some modifications in CAS such as MBRs, aerobic granular sludge (AGS) systems are explored (Hait and Mazumder, 2008; Karami et al., 2018).

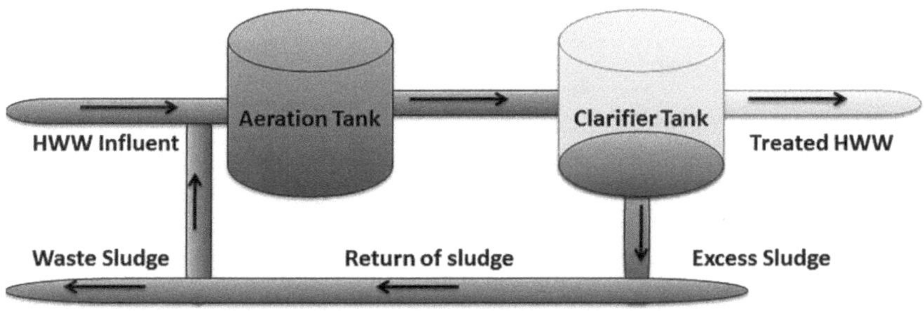

FIG. 9.3 Schematic of activated sludge method.

3.2.2 Sequencing batch reactor

Sequencing batch reactor (SBR) is a fill and draw type of activated sludge wastewater treatment method. SBR is similar to CAS in mechanism except all the treatment steps are performed in same tank in SBR method. Wastewater is treated in batches and for each batch, a series of treatment stages are sequenced. For this, wastewater is pumped into a tank already containing biomass and kept here for sufficient time to consume organic matter using biomass and then get settled and decanted off. After a rest step, the next sequence of batch reactor continued. Filling conditions of the reactor can be varied depending upon the type of sample. Mixing and aeration step can be skipped if there is no need for nitrification or denitrification and this condition is called static fill. If there is requirement of denitrification step, then mixed fill conditions are applied in which anoxic environment is provided without aeration step. In case of requirement of nitrification step both aeration and mixing are carried out (aerated fill). SBRs give excellent performance in treatment of wastewater from different sectors such as hospitals, pharmaceutical industries, chemical industries, and tannery and textile industries. But more sophisticated control systems in SBR need additional maintenance compared to conventional sludge treatment (Patil et al., 2013; Elmolla et al., 2012). The application of SBR method for treatment of pharmaceutical wastewater was explored in a study. The removal efficiency of 94 $\pm$ 1.3% for BOD_5 and 83.9 $\pm$ 1.7% for COD under optimized conditions was reported (Elmolla et al., 2012). Fig. 9.4 shows the fill and draw type SBR method.

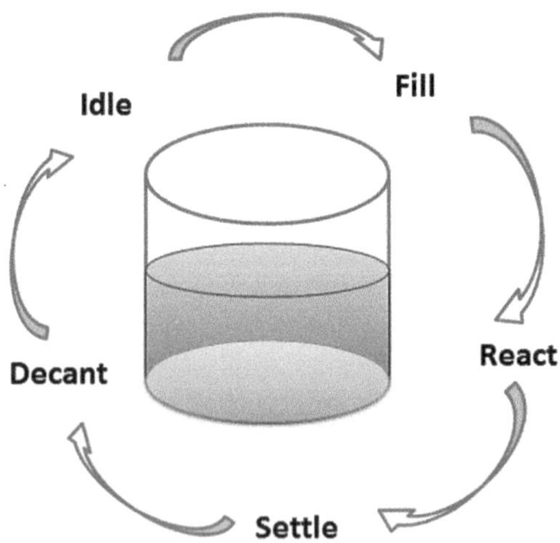

FIG. 9.4 Cycle of sequencing batch reactor.

3.2.3 Moving bed bioreactor

Moving bed bioreactor (MBBR) wastewater treatment technique is a combination of conventional activated sludge method with biofilm media. In MBBR method, thousands of biofilms carriers made up of polyethylene are operated in aerated activated sludge wastewater chamber with mixed motion. High surface area of biocarriers provides site to grow bacteria and thus increases productivity of the method. MBBR treatment is utilized at high volume loads in both aerobic and anaerobic conditions (Jasem et al., 2018; Ooi et al., 2018). A three-stage MBBR treatment system was installed at Aarhus University Hospital to treat wastewater generated at oncology department. The removal efficiency of the MBBR treatment method was compared with other techniques (activated sludge and MBR). Removal efficiency for different micropollutants (beta-blockers, X-ray contrast media, sulfonamides and macrolide antibiotics, analgesics, and antidepressants, etc.) was monitored. The MBBR treatment showed better removal efficiency than MBR and activated sludge methods (Casas et al., 2015). Schematic of MBBR unit is shown in Fig. 9.5.

3.2.4 Membrane bioreactor

MBR method is also a combination of activated sludge process and membrane filtration. Use of membrane filter replaces the sedimentation step of conventional biological wastewater treatment method. Membranes can be used within the bioreactors and externally also. MBRs can be of different types, i.e., submerged MBR, sidestream MBR, and extractive MBR (Bhattacharya et al., 2018). The efficiency of MBR technology depends on many factors such as concentration and age of sludge, nature of wastewater, temperature, conductivity, pH etc. Compared with simple sludge methods, in MBR method higher sludge concentrations and lower reactor volumes are processed. Therefore, it is more efficient and advanced than conventional sludge treatment method. Membranes being selective in nature help to remove microcontaminants and pathogens along with SSs and thus results in significantly better quality of effluent (Kovalova et al., 2012; Beier et al., 2011). The efficiency of pilot-scale MBR was checked in a swiss hospital for one year for removal of 68 micropollutants including 56 pharmaceuticals, 10 metabolites, and 2 corrosion inhibitors. Different removal efficiencies were observed for different class of

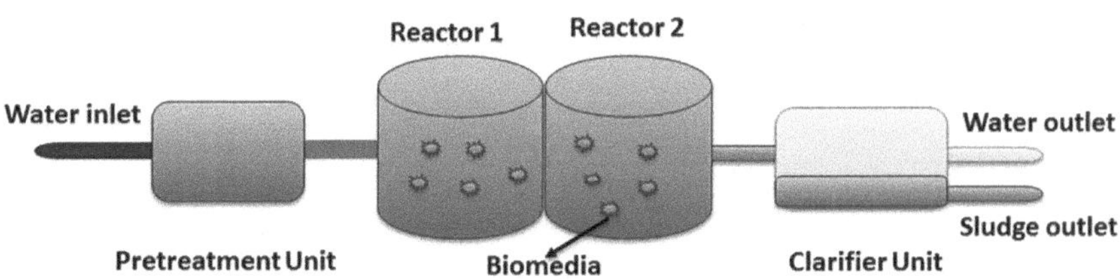

FIG. 9.5 Schematic of moving bed bioreactor unit (MBBR).

chemicals, for example, 31% for iopromide, 60% for antibiotics, and 96% for trimethoprim. Overall elimination of 22% was observed for all the micropollutants (Kovalova et al., 2012). But installation cost of MBR technology is high compared with CAS method. Secondly, higher power consumption is observed in MBR method. Fig. 9.6 shows the schematic of different types of MBR methods. Efficiency and stability of these bioreactors can be increased in terms of mineralization or separation of organic pollutants with the aid of nanofiltration and RO membranes modified with different nanoparticles (Kootenaei and Rad, 2013).

3.2.5 Constructed wetlands

Constructed wetlands (CWs) are engineered wetlands, where different physical, chemical, and biological processes are employed for treatment of wastewater. These are consisted of shallow ponds with aquatic plants (aquatic macrophytes). CWs are type of attached growth bioreactors where leaves, roots, and stems of aquatic vegetation provide the surface for attachment of microbes. There are three major components of CWs, first one is an impermeable layer of clay, second is a substrate layer of gravel, and third one is the vegetation zone. CWs are decentralized systems for treatment of wastewater. These are designed to mimic the natural wetland ecosystems in terms of water pollution remediation. In CWs, some of the pollutants are filtered and others are degraded depending upon type of pollutant. Different mechanisms such as diffusion, volatilization, adsorption, and biological degradation are employed in CWs. The performance of CWs depends on different factors such as wastewater quality, pollutant load, and climate. These are easy to operate and provide low cost and eco-friendly solution for treatment of HWW. Due to their low cost and maintenance, these can be used as a viable solution of HWW treatment in developing countries (Rørå, 2017; Sayadi et al., 2012).

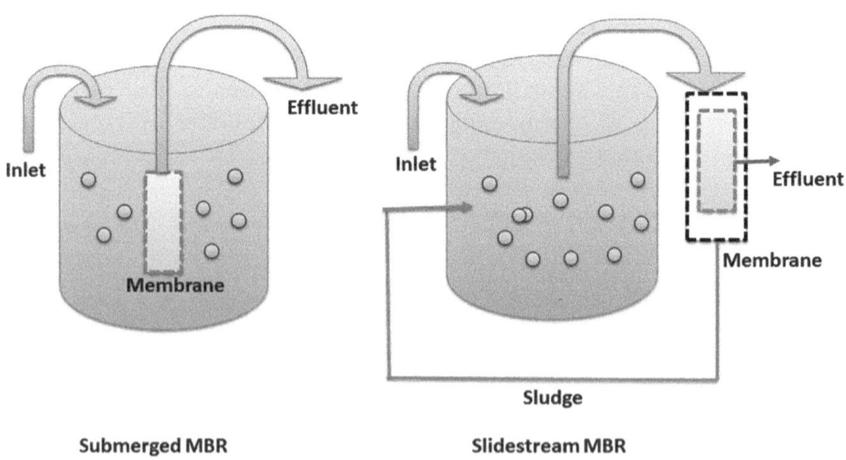

FIG. 9.6 Schematic of submerged MBR and slidestream MBR.

3.2.6 Submerged aerated fix film reactor (SAFF)

Submerged aerated fix film reactor (SAFF) is a fixed film-based technology which is combination of activated sludge process with submerged attached growth. For microbial growth, a rigid block type media with high surface area is used. In SAFF method, raw effluent is filled in SAFF reactor, where polymer-based biomedia are kept for attached growth of microbes. Aeration equipment helps to keep aerobic conditions in the reactor. Continuous supply of oxygen in the reactor helps to grow microorganisms on submerged rigid media surface. This technique is carried out in two compartments, in the first compartment majority of BOD removal observed and polishing of BOD occurs in the second compartment. In this technology, different size plants can be installed depending upon flow and loads. SAFF systems generate low sludge and can take higher loads. Chances of septic conditions are lower in SAFF method and there is no need of sludge recycling. But there is need for continuous monitoring in SAFF method as there is chances of clogging of biomedia due to excess sludge (Ahsan, 2012; Hamoda, 1995).

4. Conclusions and future scope

Hospitals are great source of wastewater and the water generated by different practices in hospitals is highly toxic due to presence of various hazardous chemicals. Treatment of this toxic water along with UWW in WWTPs is not praised by different researchers because of complex nature of HWW. Most of the micropollutants persist in WWTP effluent as these plants are not designed to remove these chemicals. In this scenario, it is urgent to explore new treatment methods for efficient removal of the hazardous micropollutants from HWW in most sustainable and cost-effective manners.

Over the time, green chemistry has offered a range of solutions to address the problems of wastewater treatment. Green chemistry principles administer the usage of cost-effective, eco-friendly, and sustainable practices which incorporate enzymes, microorganisms, sunlight, and industrial and agricultural wastes to capture or destroy different pollutants present in water in the most effective manner. Tertiary and advanced treatments such as AOPs, membrane filtration, adsorption, and different biological treatment methods have been exposed as promising green and sustainable methods for HWW reclamation. But a single practice for wastewater treatment with universal applicability is quite unrealistic, consequently a combination of different techniques is preferable to treat HWW efficiently and cost-effectively. Therefore, integrated techniques such as a combination of AOPs with biological methods should be used to increase the removal efficiency for wide range of micropollutants. Besides this, the mineralization efficiency of different conventional methods can be improved by incorporating nanotechnology. Different forms of nanotechnology can be employed depending upon the requirement. Catalytic ozonation can help to achieve mineralization of organic pollutants compared with ozonation alone. Similarly, use of different nanophotocatalysts can help to enhance efficiency of photocatalysis. Nanofiltration and RO membranes with nanomaterial coatings provide better filtration rates and can be employed in MBR technology. Usage of nanoadsorbents with lager surface area, tunable pore size, and functionality can help to achieve better filtration rates with good selectivity for the pollutants. In this way, nanotechnology in collaboration with green chemistry would be a potential solution to worldwide problem of HWW. But till now, most of the nanotechniques for reclamation and treatment of wastewater are under laboratory development. Therefore, more studies should be promoted to be conducted and evaluated in real conditions. An efficient approach should be developed to assess the feasibility of these techniques.

Along with this, guidelines should be established at national and international levels to set limit of different parameters of wastewater discharge into sewer systems. Hospitals should follow the guidelines set by various national and international agencies for discharge of HWW into environment to ensure minimum harm to both human and plant health. Besides this, disinfection of wastewater generated from infectious diseases wards should be performed before releasing it into sewage. HWW should be collected separately than UWW at WWTPs so that specific treatment can be given to HWW to remove micropollutants. The conventional approach of HWW treatment should be substituted with a decentralized method.

Acknowledgments

The authors are thankful to the Director, CSIR CSIO, Chandigarh, for his constant support and encouragement. Financial support from Council of Scientific and Industrial Research (CSIR-HRDG), New Delhi, for granting the Senior Research Fellowship (SRF) to Reetu Rani is greatly acknowledged.

References

Abdelmelek, S.B., et al., 2011. Removal of pharmaceutical and personal care products from reverse osmosis retentate using advanced oxidation processes. Environ. Sci. Technol. 45 (8), 3665–3671.

Achilleos, A., et al., 2010. UV-A and solar photodegradation of ibuprofen and carbamazepine catalyzed by TiO2. Separ. Sci. Technol. 45 (11), 1564–1570.

Affam, A.C., Chaudhuri, M., 2014. Optimization of Fenton treatment of amoxicillin and cloxacillin antibiotic aqueous solution. Desalin. Water Treat. 52 (10–12), 1878–1884.

Ahsan, N., 2012. Study of widely used treatment technologies for hospital wastewater and their comparative analysis. Int. J. Adv. Eng. Technol. 5 (1), 227.

Álvarez-Torrellas, S., et al., 2017. Effective adsorption of non-biodegradable pharmaceuticals from hospital wastewater with different carbon materials. Chem. Eng. J. 320, 319–329.

Amouei, A., et al., 2010. Characteristic of wastewater in the hospitals of babol medical university (Iran). J. Univ. Med. Sci. 12 (2), 77–82.

Anastas, P., Eghbali, N., 2010. Green chemistry: principles and practice. Chem. Soc. Rev. 39 (1), 301–312.

Avisar, D., Lester, Y., Mamane, H., 2010. pH induced polychromatic UV treatment for the removal of a mixture of SMX, OTC and CIP from water. J. Hazard Mater. 175 (1–3), 1068–1074.

Aydin, S., et al., 2019. Antibiotics in hospital effluents: occurrence, contribution to urban wastewater, removal in a wastewater treatment plant, and environmental risk assessment. Environ. Sci. Pollut. Control Ser. 26 (1), 544–558.

Balakrishna, K., et al., 2017. A review of the occurrence of pharmaceuticals and personal care products in Indian water bodies. Ecotoxicol. Environ. Saf. 137, 113–120.

Beier, S., et al., 2011. Full scale membrane bioreactor treatment of hospital wastewater as forerunner for hot-spot wastewater treatment solutions in high density urban areas. Water Sci. Technol. 63 (1), 66–71.

Ben, W., et al., 2017. Distribution of antibiotic resistance in the effluents of ten municipal wastewater treatment plants in China and the effect of treatment processes. Chemosphere 172, 392–398.

Benitez, F.J., et al., 2011. Comparison of different chemical oxidation treatments for the removal of selected pharmaceuticals in water matrices. Chem. Eng. J. 168 (3), 1149–1156.

Benotti, M.J., et al., 2008. Pharmaceuticals and endocrine disrupting compounds in US drinking water. Environ. Sci. Technol. 43 (3), 597–603.

Bhattacharya, J., Dev, S., Das, B., 2018. Chapter 1 - microbial treatment of industrial wastewater. In: Bhattacharya, J., Dev, S., Das, B. (Eds.), Low Cost Wastewater Bioremediation Technology. Butterworth-Heinemann, pp. 1–42.

Boillot, C., et al., 2008. Daily physicochemical, microbiological and ecotoxicological fluctuations of a hospital effluent according to technical and care activities. Sci. Total Environ. 403 (1–3), 113–129.

Bora, T., Dutta, J., 2014. Applications of nanotechnology in wastewater treatment—a review. J. Nanosci. Nanotechnol. 14 (1), 613–626.

Borkar, R., Gulhane, M., Kotangale, A., 2013. Moving bed biofilm reactor–a new perspective in wastewater treatment. J. Environ. Sci. Toxicol. Food Technol. 6 (6), 15–21.

Borowska, E., Felis, E., Żabczyński, S., 2015. Degradation of iodinated contrast media in aquatic environment by means of UV, UV/TiO_2 process, and by activated sludge. Water Air Soil Pollut. 226 (5), 151.

Bui, T.X., Choi, H., 2009. Adsorptive removal of selected pharmaceuticals by mesoporous silica SBA-15. J. Hazard Mater. 168 (2), 602–608.

Cai, Z., et al., 2018. Application of nanotechnologies for removing pharmaceutically active compounds from water: development and future trends. Environ. Sci. J. Integr. Environ. Res.: Nano 5 (1), 27–47.

Carballa, M., et al., 2004. Behavior of pharmaceuticals, cosmetics and hormones in a sewage treatment plant. Water Res. 38 (12), 2918–2926.

Carraro, E., et al., 2016. Hospital effluents management: chemical, physical, microbiological risks and legislation in different countries. J. Environ. Manag. 168, 185–199.

Casas, M.E., et al., 2015. Biodegradation of pharmaceuticals in hospital wastewater by staged moving bed biofilm reactors (MBBR). Water Res. 83, 293–302.

Chatzitakis, A., et al., 2008. Photocatalytic degradation and drug activity reduction of Chloramphenicol. Water Res. 42 (1), 386–394.

Chen, H., Gao, B., Li, H., 2015. Removal of sulfamethoxazole and ciprofloxacin from aqueous solutions by graphene oxide. J. Hazard Mater. 282, 201–207.

Clara, M., et al., 2005. Removal of selected pharmaceuticals, fragrances and endocrine disrupting compounds in a membrane bioreactor and conventional wastewater treatment plants. Water Res. 39 (19), 4797–4807.

Danchaivijitr, S., et al., 2005. Microbial and heavy metal contamination of treated hospital wastewater in Thailand. J. Med. Assoc. Thai. 88 (10), S59–S64.

Diao, Z.-H., et al., 2017. Enhanced catalytic degradation of ciprofloxacin with FeS_2/SiO_2 microspheres as heterogeneous Fenton catalyst: kinetics, reaction pathways and mechanism. J. Hazard Mater. 327, 108–115.

Diwan, V., et al., 2009. Detection of antibiotics in hospital effluents in India. Curr. Sci. 1752–1755.

Dudziak, M., Burdzik, E., 2016. Oxidation of bisphenol A from simulated and real urban wastewater effluents by UV, O_3 and UV/O_3. Desalin. Water Treat. 57 (3), 1075–1083.

El-Ogri, F., et al., 2016. A survey of wastewaters generated by a hospital in Marrakech city and their characterization. Desalin. Water Treat. 57 (36), 17061–17074.

Elmolla, E.S., Chaudhuri, M., 2009. Degradation of the antibiotics amoxicillin, ampicillin and cloxacillin in aqueous solution by the photo-Fenton process. J. Hazard Mater. 172 (2–3), 1476–1481.

Elmolla, E.S., Ramdass, N., Chaudhun, M., 2012. Optimization of sequencing batch reactor operating conditions for treatment of high-strength pharmaceutical wastewater. J. Environ. Sci. Technol. 5 (6), 452–459.

Emmanuel, E., et al., 2005. Ecotoxicological risk assessment of hospital wastewater: a proposed framework for raw effluents discharging into urban sewer network. J. Hazard Mater. 117 (1), 1–11.

Fraiese, A., et al., 2019. Removal of emerging contaminants in wastewater by sonolysis, photocatalysis and ozonation. Global NEST J. 21, 98–105.

Frédéric, O., Yves, P., 2014. Pharmaceuticals in hospital wastewater: their ecotoxicity and contribution to the environmental hazard of the effluent. Chemosphere 115, 31–39.

Ganzenko, O., et al., 2014. Electrochemical advanced oxidation and biological processes for wastewater treatment: a review of the combined approaches. Environ. Sci. Pollut. Control Ser. 21 (14), 8493–8524.

Göbel, A., et al., 2005. Occurrence and sorption behavior of sulfonamides, macrolides, and trimethoprim in activated sludge treatment. Environ. Sci. Technol. 39 (11), 3981–3989.

Gong, J., Liu, Y., Sun, X., 2008. O_3 and UV/O_3 oxidation of organic constituents of biotreated municipal wastewater. Water Res. 42 (4–5), 1238–1244.

Hait, S., Mazumder, D., 2008. Scope of improvement of treatment capacity of activated sludge process by hybrid modification. J. Environ. Eng. Sci. 7 (2), 147–158.

Hamoda, M., 1995. Biotreatment of wastewaters using aerated submerged fixed-film reactors. In: Environmental Biotechnology. Springer, pp. 431–449.

Hansen, K.M., et al., 2016. Ozonation for source treatment of pharmaceuticals in hospital wastewater—ozone lifetime and required ozone dose. Chem. Eng. J. 290, 507–514.

Hernando, M.D., et al., 2004. Analysis by liquid chromatography—electrospray ionization tandem mass spectrometry and acute toxicity evaluation for β-blockers and lipid-regulating agents in wastewater samples. J. Chromatogr. A 1046 (1–2), 133–140.

Horváth, I.T., Anastas, P.T., 2007. Introduction: Green Chemistry. ACS Publications.

Huber, M.M., et al., 2003. Oxidation of pharmaceuticals during ozonation and advanced oxidation processes. Environ. Sci. Technol. 37 (5), 1016–1024.

Huerta-Fontela, M., Galceran, M.T., Ventura, F., 2011. Occurrence and removal of pharmaceuticals and hormones through drinking water treatment. Water Res. 45 (3), 1432–1442.

Jasem, Y.I., Jumaha, G.F., Ghawi, A.H., 2018. Treatment of medical wastewater by moving bed bioreactor system. J. Ecol. Eng. 19 (3).

Jjemba, P.K., 2006. Excretion and ecotoxicity of pharmaceutical and personal care products in the environment. Ecotoxicol. Environ. Saf. 63 (1), 113–130.

Johnson, A., et al., 2005. Comparing steroid estrogen, and nonylphenol content across a range of European sewage plants with different treatment and management practices. Water Res. 39 (1), 47–58.

Joss, A., et al., 2004. Removal of estrogens in municipal wastewater treatment under aerobic and anaerobic conditions: consequences for plant optimization. Environ. Sci. Technol. 38 (11), 3047–3055.

Jung, C., et al., 2013. Adsorption of selected endocrine disrupting compounds and pharmaceuticals on activated biochars. J. Hazard Mater. 263, 702–710.

Kanakaraju, D., Glass, B.D., Oelgemöller, M., 2014. Titanium dioxide photocatalysis for pharmaceutical wastewater treatment. Environ. Chem. Lett. 12 (1), 27–47.

Karami, N., et al., 2018. High rate treatment of hospital wastewater using activated sludge process induced by high-frequency ultrasound. Ultrason. Sonochem. 46, 89–98.

Kim, Y., et al., 2007. Aquatic toxicity of acetaminophen, carbamazepine, cimetidine, diltiazem and six major sulfonamides, and their potential ecological risks in Korea. Environ. Int. 33 (3), 370–375.

Kim, I., Yamashita, N., Tanaka, H., 2009. Performance of UV and UV/H_2O_2 processes for the removal of pharmaceuticals detected in secondary effluent of a sewage treatment plant in Japan. J. Hazard Mater. 166 (2–3), 1134–1140.

Kootenaei, F.G., Rad, H.A., 2013. Treatment of hospital wastewater by novel nano-filtration membrane bioreactor (NF-MBR). Iran. J. Energy Environ. 4 (1), 60.

Kosma, C.I., Lambropoulou, D.A., Albanis, T.A., 2010. Occurrence and removal of PPCPs in municipal and hospital wastewaters in Greece. J. Hazard Mater. 179 (1–3), 804–817.

Kostich, M.S., Batt, A.L., Lazorchak, J.M., 2014. Concentrations of prioritized pharmaceuticals in effluents from 50 large wastewater treatment plants in the US and implications for risk estimation. Environ. Pollut. 184, 354–359.

Kovalova, L., et al., 2012. Hospital wastewater treatment by membrane bioreactor: performance and efficiency for organic micropollutant elimination. Environ. Sci. Technol. 46 (3), 1536–1545.

Kovalova, L., et al., 2013. Elimination of micropollutants during post-treatment of hospital wastewater with powdered activated carbon, ozone, and UV. Environ. Sci. Technol. 47 (14), 7899–7908.

Kumarathilaka, P., et al., 2015. General characteristics of hospital wastewater from three different hospitals in Sri Lanka. In: 6th International Conference on Structural Engineering and Construction Management, 2015.

Lee, Y., et al., 2014. Prediction of micropollutant elimination during ozonation of a hospital wastewater effluent. Water Res. 64, 134–148.

Lindberg, R.H., et al., 2006. Behavior of fluoroquinolones and trimethoprim during mechanical, chemical, and active sludge treatment of sewage water and digestion of sludge. Environ. Sci. Technol. 40 (3), 1042–1048.

Liu, Q., et al., 2010. Application of MBR for hospital wastewater treatment in China. Desalination 250 (2), 605–608.

Lusic, H., Grinstaff, M.W., 2012. X-ray-computed tomography contrast agents. Chem. Rev. 113 (3), 1641–1666.

Machado, E., et al., 2007. Secondary hospital wastewater detoxification and disinfection by advanced oxidation processes. Environ. Technol. 28 (10), 1135–1143.

Macías-García, A., et al., 2019. Adsorption of paracetamol in hospital wastewater through activated carbon filters. Sustainability 11 (9), 2672.

Malik, S.N., et al., 2019. Treatment of pharmaceutical industrial wastewater by nano-catalyzed ozonation in a semi-batch reactor for improved biodegradability. Sci. Total Environ. 678, 114–122.

Mendoza, A., et al., 2015. Pharmaceuticals and iodinated contrast media in a hospital wastewater: a case study to analyse their presence and characterise their environmental risk and hazard. Environ. Res. 140, 225–241.

Mesdaghinia, A., et al., 2009. Wastewater characteristics and appropriate method for wastewater management in the hospitals. Iran. J. Public Health 34–40.

Mhemdi, A., et al., 2013. Electrochemical advanced oxidation of 2-chlorobenzoic acid using BDD or Pt anode and carbon felt cathode. J. Electroanal. Chem. 709, 111–117.

Mishra, N.S., et al., 2017. A review on advanced oxidation processes for effective water treatment. Curr. World Environ. 12 (3), 470.

Mohammad, A.W., et al., 2015. Nanofiltration membranes review: recent advances and future prospects. Desalination 356, 226–254.

Motoc, S., et al., 2012. Electrochemical degradation of pharmaceutical effluents on carbon-based electrodes. Environ. Eng. Manag. J. 11 (3), 627–634.

Naddeo, V., et al., 2012. Degradation of antibiotics in wastewater during sonolysis, ozonation, and their simultaneous application: operating conditions effects and processes evaluation. Int. J. Photoenergy 2012.

Nyamukamba, P., et al., 2019. Evaluating physicochemical parameters, heavy metals, and antibiotics in the influents and final effluents of South African wastewater treatment plants. Pol. J. Environ. Stud. 28 (3), 1305–1312.

Ooi, G.T., et al., 2018. Biological removal of pharmaceuticals from hospital wastewater in a pilot-scale staged moving bed biofilm reactor (MBBR) utilising nitrifying and denitrifying processes. Bioresour. Technol. 267, 677–687.

Ouarda, Y., et al., 2019. Electrochemical treatment of real hospital wastewaters and monitoring of pharmaceutical residues by using surrogate models. J. Environ. Chem. Eng. 7 (5), 103332.

Pasternak, J.J., Williamson, E.E., 2012. Clinical pharmacology, uses, and adverse reactions of iodinated contrast agents: a primer for the non-radiologist. In: Mayo Clinic Proceedings. Elsevier.

Patil, P., et al., 2013. Aerobic sequencing batch reactor for wastewater treatment: a review. Int. J. Eng. Res. Technol. 2 (10), 534–550.

Pauwels, B., et al., 2006. Effluent quality of a conventional activated sludge and a membrane bioreactor system treating hospital wastewater. Environ. Technol. 27 (4), 395–402.

Pauwels, B., Verstraete, W., 2006. The treatment of hospital wastewater: an appraisal. J. Water Health 4 (4), 405–416.

Pogacean, F., et al., 2015. Electrochemical degradation of carbamazepine using modified electrode with graphene-AuAg composite. AIP Conf. Proc. 1700 (1), 060003.

Prayitno, Z.K., Yanuwiadi, B., Laksmono, R., 2013. Study of hospital wastewater characteristic in Malang City. Int. J. Eng. Sci. 2 (2), 13–16.

Qu, X., et al., 2013. Nanotechnology for a safe and sustainable water supply: enabling integrated water treatment and reuse. Acc. Chem. Res. 46 (3), 834–843.

Qu, X., Alvarez, P.J.J., Li, Q., 2013. Applications of nanotechnology in water and wastewater treatment. Water Res. 47 (12), 3931–3946.

Rakić, V., et al., 2015. The adsorption of pharmaceutically active compounds from aqueous solutions onto activated carbons. J. Hazard Mater. 282, 141–149.

Rame, R., et al., 2018. Catalytic ozonation based advanced oxidation process for effective treating wastewater from hospital and community health centre facility by FLASH WWT catalyst system in Indonesia. In: Journal of Physics: Conference Series. IOP Publishing.

Rodriguez-Mozaz, S., et al., 2015. Occurrence of antibiotics and antibiotic resistance genes in hospital and urban wastewaters and their impact on the receiving river. Water Res. 69, 234–242.

Rørå, A., 2017. Wetland Treatment of Hospital Wastewater in Far Western Nepal Including a Knowledge Attitude and Practice (KAP) Study on Sanitation. Norwegian University of Life Sciences. Ås.

Sanches, S., Crespo, M.T.B., Pereira, V.J., 2010. Drinking water treatment of priority pesticides using low pressure UV photolysis and advanced oxidation processes. Water Res. 44 (6), 1809–1818.

Santos, L.H., et al., 2013. Contribution of hospital effluents to the load of pharmaceuticals in urban wastewaters: identification of ecologically relevant pharmaceuticals. Sci. Total Environ. 461, 302–316.

Sayadi, M., et al., 2012. Hybrid constructed wetlands for wastewater treatment: a worldwide review. Proc. Int. Acad. Ecol. Environ. Sci. 2 (4), 204.

Schrap, S.M., et al., 2003. Humane en veterinaire geneesmiddelen in Nederlands oppervlaktewater en afvalwater. RIZA report, p. 86.

Smith, S.C., Rodrigues, D.F., 2015. Carbon-based nanomaterials for removal of chemical and biological contaminants from water: a review of mechanisms and applications. Carbon 91, 122–143.

Sopaj, F., et al., 2015. Influence of the anode materials on the electrochemical oxidation efficiency. Application to oxidative degradation of the pharmaceutical amoxicillin. Chem. Eng. J. 262, 286–294.

Taheran, M., et al., 2016. Membrane processes for removal of pharmaceutically active compounds (PhACs) from water and wastewaters. Sci. Total Environ. 547, 60–77.

Ternes, T.A., et al., 2003. Ozonation: a tool for removal of pharmaceuticals, contrast media and musk fragrances from wastewater? Water Res. 37 (8), 1976–1982.

Trovó, A.G., et al., 2011. Degradation of the antibiotic amoxicillin by photo-Fenton process—chemical and toxicological assessment. Water Res. 45 (3), 1394–1402.

Umadevi, V., 2015. Fenton process-A pre-treatment option for hospital wastewater. Int. J. Innov. Eng. Technol. 5 (2), 306–312.

Velichkova, F., et al., 2017. Heterogeneous fenton and photo-fenton oxidation for paracetamol removal using iron containing ZSM-5 zeolite as catalyst. AIChE J. 63 (2), 669–679.

Verlicchi, P., et al., 2010. Hospital effluents as a source of emerging pollutants: an overview of micropollutants and sustainable treatment options. J. Hydrol. 389 (3–4), 416–428.

Verlicchi, P., et al., 2012. Hospital effluent: investigation of the concentrations and distribution of pharmaceuticals and environmental risk assessment. Sci. Total Environ. 430, 109–118.

Verlicchi, P., Al Aukidy, M., Zambello, E., 2015. What have we learned from worldwide experiences on the management and treatment of hospital effluent? — an overview and a discussion on perspectives. Sci. Total Environ. 514, 467–491.

Vieno, N., Tuhkanen, T., Kronberg, L., 2007. Elimination of pharmaceuticals in sewage treatment plants in Finland. Water Res. 41 (5), 1001–1012.

Wang, F., et al., 2017. Interaction mechanisms of antibiotic sulfamethoxazole with various graphene-based materials and multiwall carbon nanotubes and the effect of humic acid in water. Carbon 114, 671–678.

Warner John, C., Cannon Amy, S., Dye Kevin, M., 2004. Environment and health: new answers, new questions. Environ. Impact Assess. Rev. 24, 775–799.

Weinstein, R.A., Hota, B., 2004. Contamination, disinfection, and cross-colonization: are hospital surfaces reservoirs for nosocomial infection? Clin. Infect. Dis. 39 (8), 1182–1189.

Yadav, M., Gupta, R., Sharma, R.K., 2019. Green and sustainable pathways for wastewater purification. In: Advances in Water Purification Techniques. Elsevier, pp. 355–383.

Chapter 10

Recent developments in sonochemical treatments of contaminated wastewaters

Efraím A. Serna-Galvis, Ricardo A. Torres-Palma

Grupo de Investigación en Remediación Ambiental y Biocatálisis (GIRAB), Instituto de Química, Facultad de Ciencias Exactas y Naturales, Universidad de Antioquia UdeA, Medellín, Colombia

1. Some fundamental aspects of sonochemistry

Ultrasound (US, also represented as ")))") is a process that uses mechanical waves ranging from 20 to 10,000 kHz (10 MHz) in frequency. Such a frequency spectrum is usually segmented into three regions: low frequency (20−100 kHz), high frequency (100 kHz−1 MHz), and very high frequency (1−10 MHz) (Gonçalves et al., 2015). For the degradation of organic pollutants, frequencies between 20 kHz and 2 MHz are commonly used.

Pollutant degradation by US is mainly associated with its chemical effects (sonochemistry); i.e., pyrolysis of pollutants or water (and oxygen), which leads to the generation of radical species such as hydroxyl radical (HO•). The pyrolysis of chemicals occurs through the acoustic cavitation phenomenon, which comprises the creation of bubbles (which contain oxygen gas and/or water vapor) by passing US waves along the aqueous medium. The bubbles grow achieving an unstable size and then they violently implode (Adewuyi, 2001; Riesz et al., 1985). This bubble collapse leads to homolysis of water vapor and oxygen molecules generating radical species (Eqs. 10.1−10.4). In addition, volatile organic compounds can be pyrolyzed. Besides radicals formation, cavitation can also produce sonoluminescence (i.e., light emission) (McNamara et al., 1999; Beckett and Hua, 2001).

$$H_2O +))) \rightarrow H\bullet + HO\bullet \tag{10.1}$$

$$O_2 +))) \rightarrow 2O\bullet \tag{10.2}$$

Green Chemistry and Water Remediation: Research and Applications.
https://doi.org/10.1016/B978-0-12-817742-6.00010-4

$$H_2O + O\bullet \rightarrow 2HO\bullet \quad (10.3)$$

$$O_2 + H\bullet \rightarrow O\bullet + HO\bullet \quad (10.4)$$

HO• is a strong oxidant species (E°: 2.80 V) that potentially reacts with almost any organic compound. As US is able to produce HO• and this can be used for degrading organic pollutants in aqueous matrices, it is considered an advanced oxidation process (Mahamuni and Adewuyi, 2010). In addition to attacks on organic substances, radicals can combine themselves producing hydrogen peroxide (Eq. 10.5). Indeed, the measurement of accumulated H_2O_2 is an indirect way to evidence radical generation from sonication (Tran et al., 2015; Montoya-Rodríguez et al., 2020).

$$2HO\bullet \rightarrow H_2O_2 \quad (10.5)$$

The sonochemical system is subdivided in three reaction zones: (1) bulk of the liquid, (2) interior of the bubble, and (3) interface of the bubble/liquid. Volatile substances are pyrolyzed inside the bubble, while hydrophobic nonvolatile compounds and hydrophilic pollutants are mainly degraded at the interface and bulk of the liquid, respectively (Adewuyi, 2001).

The primary operational parameters of US influencing the efficiency of pollutant degradation are frequency, power (or power density), and reactor configuration (Wood et al., 2017). For volatile compounds, which are eliminated by pyrolysis-based reactions, the degradation efficiency is higher with the increasing of frequency. Meanwhile, for nonvolatile substances there is an optimum frequency of degradation by free radical attacks (Wang et al., 2019). For example, treatment of phenol (a nonvolatile compound) and carbon tetrachloride (a volatile substance) at frequencies between 20 and 800 kHz showed that carbon tetrachloride is more efficiently eliminated at 800 kHz. In turn, phenol is degraded faster at 200 kHz (Pétrier and Francony, 1997) (Fig. 10.1).

As the frequency is increased an increase in volatile substance flux toward the bubble causes degradation of compounds such as carbon tetrachloride (which mainly occurs inside the cavitation bubble via pyrolysis). In contrast, for nonvolatile compounds such as phenol that are degraded outside of the cavitation bubbles by hydroxyl radicals, their elimination is faster at the frequency of highest HO• generation (Pétrier and Francony, 1997). In line with these observations, frequencies between 200 and 600 kHz are more efficient for the production of radicals (Torres-Palma and Serna-Galvis, 2018).

Regarding power (or power density), it can be shown that the degradation of organic pollutants in aqueous matrices is higher as this parameter increases (Table 10.1). US power is directly related to the number of cavitation bubbles, the temperature of collapse, and the sonochemical yield (Wood et al., 2017; Wang et al., 2019). However, this parameter has upper and lower limits. Below the lower limit the US field is very small to induce nucleation or bubble growth resulting in little active cavitation. Above the upper limit, coalescence and degassing occur, decreasing the active cavitation (Wood et al., 2017).

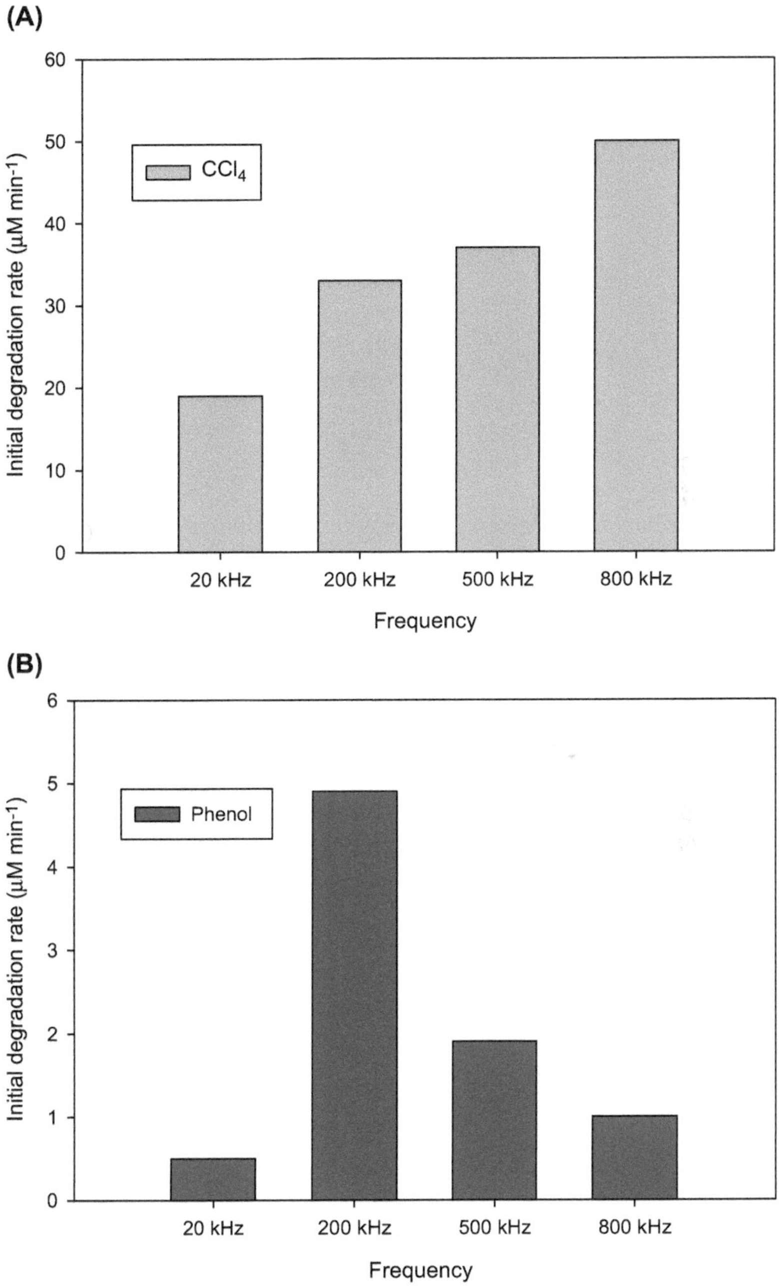

FIG. 10.1 Effect of frequency on degradation of carbon tetrachloride (A) and phenol (B). *Figure made by the authors based on data from Pétrier, C., Francony, A., 1997. Ultrasonic wastewater treatment: incidence of ultrasonic frequency on the rate of phenol and carbon tetrachloride degradation, Ultrason. Sonochem. 4, 295–300. doi:10.1016/S1350-4177(97)00036-9.*

TABLE 10.1 Some examples of the ultrasound power effect.

Pollutant	Tested powers	Best degradation	References
Acetaminophen (pharmaceutical)	20, 30, 40, 50, and 60 W	At 60 W, ~45% of acetaminophen elimination in 40 min	Villaroel et al. (2014)
Crystal violet (dye)	20, 40, 60, and 80 W	At 80 W, ~90% of dye degradation after 120 min of treatment	Guzman-Duque et al. (2011)
4-Cumylphenol (endocrine disruptor)	20, 40, 60, 80, and 100 W	At 100 W, ~90% of pollutant degradation after 30 min of sonication	Chiha et al. (2011)

US waves are introduced to the aqueous medium through a transducer powered by an amplifier. The transducer can be attached to a vibrating plate of a bath (Fig. 10.2A), to a tip in ultrasonic horns (Fig. 10.2B), or directly glued to the bottom of the reactor (Fig. 10.2C) (Ashokkumar et al., 2009). Ultrasonic baths and horns work at low frequencies (20–100 kHz), while a specialized reactor arrangement coupled to a high-power amplifier is needed to operate at frequencies between 100 and 2000 kHz.

Lab-scale reactors generally use one ultrasonic horn or a single plate transducer to treat volumes lower than 1.0 L in batch mode. Meanwhile, for large-scale applications, multiple horns or plate transducers are utilized. Until now, there have been few studies about large-scale reactors for pollutant elimination. The highest ultrasonic reactor (91 L) was reported for eliminating per-fluorinated alkyl sulfonates (Gole et al., 2018). Although, another US large-scale reactor (140 L) able to generate free radicals at a rate of 5.2 μmol L^{-1} min^{-1} has been reported, its application for degrading specific organic pollutants was not specified (Caretti et al., 2011).

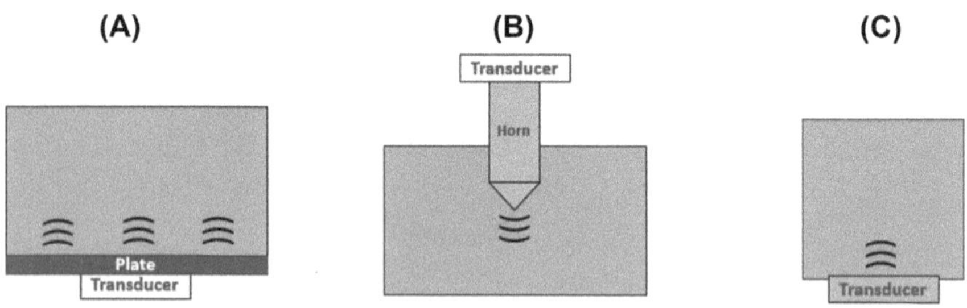

FIG. 10.2 Simplified schemes of typical setups for ultrasound reactors at lab-scale. (A) Vibrating plate bath reactor, (B) horn-tip reactor, and (C) bottom plate reactor. *Figure made by the authors.*

It is recognized that the field of US reactors is dependent on liquid volume, liquid height, and vessel geometry. Consequently, these parameters of reactor design influence cavitation activity, and this in turn modifies sonochemical efficiency (Lee, 2016). Hence, the upscaling of reactors must consider optimization of engineering factors (which is still scarce in the current literature) in addition to chemical aspects. Studies on upscaling of US reactors, mainly at high frequencies (100–1000 kHz), should be developed in the future.

On the other hand, recent advances and research on sonochemical degradation of organic compounds have been focused on aspects such as improving pollutant removal by the addition of reagents (such as persulfate[PS]), new catalysts (such as carbon-based materials), and application of sonochemistry (alone or in hybrid systems) to remove pollution from complex matrices. These topics are presented in the following sections.

2. Persulfate-enhanced ultrasonic action

Ultrasonication has been applied to eliminate diverse organic pollutants from water. This process successfully degrades dyes, pharmaceuticals, phenolic compounds, pesticides, etc. (Tran et al., 2015; Wang et al., 2019; Chowdhury and Viraraghavan, 2009; Rayaroth et al., 2016; Camargo-Perea et al., 2020). However, elimination of hydrophilic compounds is slow. Sometimes high energy/treatment times for complete mineralization (which is economically unfavorable) are required. Hence, a facile strategy to enhance/accelerate the degrading/mineralizing ability of the US systems is the addition of substances to the reaction vessel. Such is the case of PS anion.

PS has gained attention in recent years as an alternative to hydrogen peroxide for advanced oxidation processes. It a precursor of sulfate anion radical that also has a high redox potential ($SO_4^{\bullet-}$, 2.5–3.1 V) and a longer lifetime than hydroxyl radical. US waves can activate PS to generate $SO_4^{\bullet-}$ (Eq. 10.6). It can be shown that PS has been added to both low ($<$100 kHz) and high (100–2000 kHz) US frequencies (Yang et al., 2019).

$$S_2O_8^{2-} +))) \rightarrow SO_4^{\bullet-} \quad (10.6)$$

Apart from activation, US arouses violent turbulence in aqueous media, which enhances the mass transfer of chemical reactions in the solution, accelerating the reaction processes in the solution. The combined action of US and PS has shown to be a promising alternative for organic wastewater treatment. Although PS itself has a degrading ability because it is an oxidant substance (E°: 2.1. V), in many cases, the addition of a PS to US (US + PS) system has synergistic effects and reduces the time required to achieve complete degradation of pollutants.

Fig. 10.3 illustrates the degradation of 1,1,1-trichloroethane (TCA) in field groundwater by different treatments: PS oxidation, ultrasonication alone, and their combination (US + PS). The concentrations of TCA, PS, and US frequency were 50 mg L^{-1}, 0.94 mmol L^{-1}, and 400 kHz, respectively. From this figure the synergistic effect of joining US and PS, which achieved complete TCA removal at 60 min of treatment is evident, whereas PS and US individually degraded <5% and ~15%, respectively. The US + PS system can generate more sulfate and hydroxyl radicals than the individual components, leading to an enhanced and faster degradation (Li et al., 2013).

In the US + PS process, a very relevant variable is the initial concentration of PS, which strongly influences the degradation of pollutants. Typically, the increasing of PS concentration leads to higher degradation of the contaminants due to generation of more sulfate radicals. However, an excess of PS may produce scavenging/recombination of radicals leading to species less active than hydroxyl or sulfate radicals (Eqs. 10.7–10.9). Therefore a detrimental effect on pollutant elimination can be observed (Yang et al., 2019; Li et al., 2013; Hao et al., 2014). Other experimental factors such as pH and

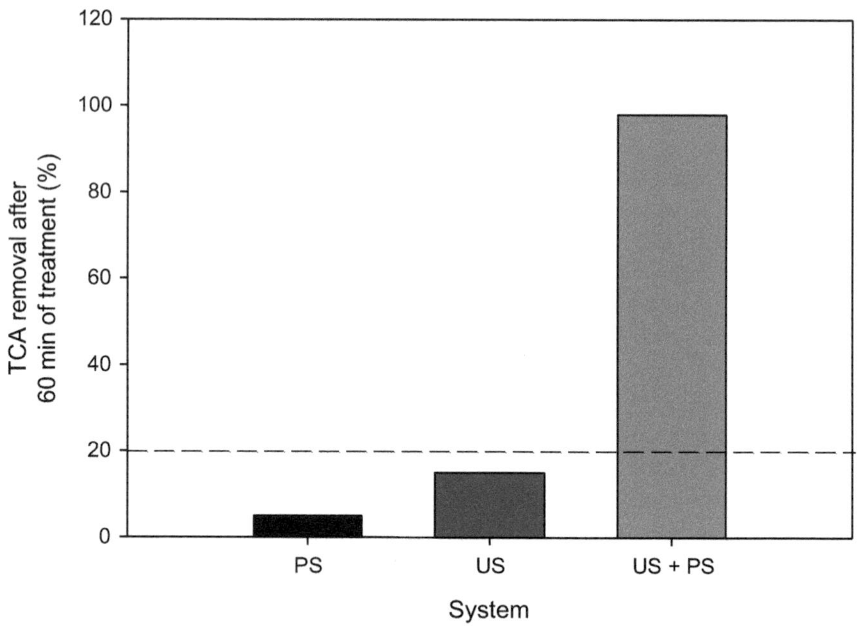

FIG. 10.3 1,1,1-Trichloroethane (TCA) degradation by individual systems (persulfate or ultrasound [PS or US]) and the hybrid process (US + PS). *Note*: dashed line represents the arithmetic addition of the individual systems. *Figure made by the authors based on data from Li, B., Li, L., Lin, K., Zhang, W., Lu, S., Luo, Q., 2013. Removal of 1,1,1-trichloroethane from aqueous solution by a sono-activated persulfate process, Ultrason. Sonochem. 20, 855–863. doi:10.1016/j.ultsonch.2012.11.014.*

temperature also modify the performance of the US + PS process. Information on these parameter can be found in the references (Yang et al., 2019; Wang and Zhou, 2016; Fagan et al., 2020).

$$SO_4^{\bullet -} + SO_4^{\bullet -} \rightarrow S_2O_8^{2-} \quad (10.7)$$

$$SO_4^{\bullet -} + S_2O_8^{2-} \rightarrow SO_4^{2-} + S_2O_8^{2-} \quad (10.8)$$

$$S_2O_8^{2-} + HO\bullet \rightarrow S_2O_8\bullet^{-} + HO^{-} \quad (10.9)$$

The degradation of organic pollutants by US + PS has also been applied to aqueous matrices containing inorganic ions (which are commonly present in natural water bodies or some industrial wastewater). Generally, anions such as chloride, nitrate, sulfate, and phosphate inhibit decomposition of target contaminants because these anions trap sulfate and hydroxyl radicals forming less powerful oxidizing species. Also, organic components such as humic acids or surfactants at elevated concentrations in the water medium affect pollutant elimination rate owing to a competence by the radicals (Yang et al., 2019; Li et al., 2013).

On the other hand, the treatment of complex wastewater by US + PS has also been reported. For example, wastewater from the dinitrotoluene industry showed almost complete elimination of the nitrotoluene contaminants by virtue of sono-activated PS oxidation. Additionally, the process was able to induce a significant mineralization of this industrial wastewater (Chen and Su, 2012).

3. Ultrasound combined with novel (photo)catalysts

Homogeneous or heterogeneous catalysts can be added to the sono-reactor to enhance/accelerate pollutant removal. One of the first reported methods of sonocatalysis was the sono-Fenton process (i.e., US plus Fe^{2+} ions) (Joseph et al., 2000; Torres et al., 2007). The sono-Fenton process takes advantage of sono-generated hydrogen peroxide (mainly at frequencies between 100 and 2000 kHz) for producing extra radicals through Fenton reaction (Eqs. 10.10–10.11). When the system is complemented with UV light (sono-photo-Fenton process), extra hydroxyl radicals can be formed by photoreduction of ferric ions (Eq. 10.12) (Torres et al., 2007).

$$Fe^{2+} + H_2O_2 \rightarrow Fe^{3+} + HO\bullet + HO^{-} \quad (10.10)$$

$$Fe^{3+} + H_2O_2 \rightarrow Fe^{2+} + HOO\bullet + H^{+} \quad (10.11)$$

$$Fe^{3+} + H_2O + hv \rightarrow Fe^{2+} + HO\bullet + H^{+} \quad (10.12)$$

The sono-Fenton process is also generated by the addition of both ferrous ions and hydrogen peroxide at sono-reactors of low frequencies (<100 kHz).

Alternatively, the sono-Fenton process can be done by utilizing iron oxides, zero-valent iron, or iron composites with other metals. These solid materials of iron make the process heterogeneous (but if iron ions are released from the solids, homogeneous Fenton reactions also contribute to the process). It is reported that these kinds of treatment effectively degrade water pollutants like dyes, pharmaceuticals, and industrial reagents (Zhang et al., 2018; Hou et al., 2016).

The presence of the solid catalyst in the US system (as in heterogenous sono-Fenton) induces weaker points to the liquid, facilitating the generation of cavitation bubbles. Also, the mechanical effects of US shake the liquid and hit the solid surface, enhancing mass transfer and reaction rate. Consequently, pollutant degradation is increased (Gholami et al., 2019a).

Besides sono-Fenton processes, the addition of TiO_2 plus UV light is another well-known approach for improving sonochemical effects on organic pollutants. Sono-photocatalysis using TiO_2 has the advantages of sonocatalysis (i.e., more cavitation events and an increase in mass transfer), in addition to the formation of extra oxidizing agents by light action (Eqs. 10.13–10.17), which are useful for extra pollutant elimination (Torres et al., 2008).

$$TiO_2 + h\nu_{(\lambda<387nm)} \rightarrow TiO_2(e^- - h^+) \tag{10.13}$$

$$h^+ + H_2O \rightarrow H^+ + HO\bullet \tag{10.14}$$

$$h^+ + HO^- \rightarrow HO\bullet \tag{10.15}$$

$$e^- + O_2 \rightarrow \bullet O_{2^-} \rightarrow \rightarrow \rightarrow HO\bullet, HOO\bullet, H_2O_2 \tag{10.16}$$

$$e^- + H_2O_2 \rightarrow HO\bullet + HO^- \tag{10.17}$$

More recently, novel materials such as carbon-based catalysts have been considered for sonocatalysis. Sonocatalytic treatment of organic pollutants with carbon-based catalysts includes carbon nanotubes (CNTs), graphene, graphene oxide (GO), reduced graphene oxide, activated carbon (AC), biochar (BC), graphitic carbon nitride, carbon-doped materials, buckminsterfullerene, and mesoporous carbon (Gholami et al., 2019a). Carbon-based catalysts have unique characteristics such as high chemical stability, mechanical strength, and porous structures that make them useful materials for sonocatalytic degradation of contaminants. The carbonaceous materials augment the formation of radicals and adsorb the pollutants on the particles, surface (Gholami et al., 2019a; Im et al., 2013, 2014).

For example, degradation of pharmaceuticals (acetaminophen, naproxen, ibuprofen, and sulfamethoxazole) by the sonocatalytical process with CNTs demonstrated that pollutant removal is associated with the combined effects of oxidation and adsorption pathways. In fact, synergies observed due to adsorption on CNTs are enhanced by US irradiation, and because cavitation capability is improved by the dispersed CNTs, which offer more nucleation sites during the sonocatalytic process (Im et al., 2013; Al-Hamadani et al., 2017).

The enhancement of pharmaceutical degradation by sonocatalysis using GO has also been studied. In this case, it was shown that at low frequency (28 kHz) GO was more stable than at higher frequencies (580 and 1000 kHz). The surface charges on GO were more negative at 28 kHz increasing the repulsion between GO particles, thereby the adsorption capacity of pollutants was higher. Again, the overall enhancement of pharmaceutical removal was due to an increase in cavitation activity, which induced a higher production of radicals, and an adsorption increment due to catalyst dispersion by US irradiation (Al-Hamadani et al., 2018).

Reports about sonocatalysis with AC show that contaminant removal also occurs via adsorption plus radical attacks (Im et al., 2014). As the application of sonocatalysis using novel carbon-based materials must consider both removal efficiency and economic costs, some BCs have been synthesized and utilized as an inexpensive alternative to conventional AC or other carbonaceous catalysts (Gholami et al., 2019a). For instance, acetaminophen and naproxen were submitted to sonocatalysis with BC and AC, and it was found that BC was more efficient for treating these pharmaceuticals than AC under sonication (Im et al., 2014). Likewise, rhodamine-B has been treated by using US (40 or 300 kHz) plus BC derived from peat moss. This sonocatalytic system presented a better performance at the lower frequency because BC dispersion in solution is higher thanks to the stronger sonophysical effects at 40 kHz (Kim et al., 2018). An illustrative example of improved removal of an organic pollutant by BC plus US is depicted in Fig. 10.4 (Gholami et al., 2019b).

Many carbonaceous materials have been composed with semiconductors (e.g., TiO_2, ZnO, and Fe_3O_4/TiO_2) and used for degrading pollutants by sonocatalysis. Such composites give the catalyst better separation of hole–electron pairs, according to the high electroconductivity of carbonaceous materials. Additionally, they are able to narrow the band gap of semiconductors, enhancing the photon absorption efficiency. It must be indicated that sonocatalysis with these composites also has a photocatalytic factor, which is initiated by the sonoluminescence phenomenon from US (Fig. 10.5). Sonoluminescence is able to activate the semiconductor for the production of radicals in an analogous way as presented in Eqs. (10.13)–(10.17); consequently, sonocatalytic efficacy is enhanced (Gholami et al., 2019a; Geng et al., 2019; Samanta et al., 2018). Fig. 10.6 exemplifies the removal of a pollutant by sonocatalysis using a semiconductor/biochar composite (Gholami et al., 2019b).

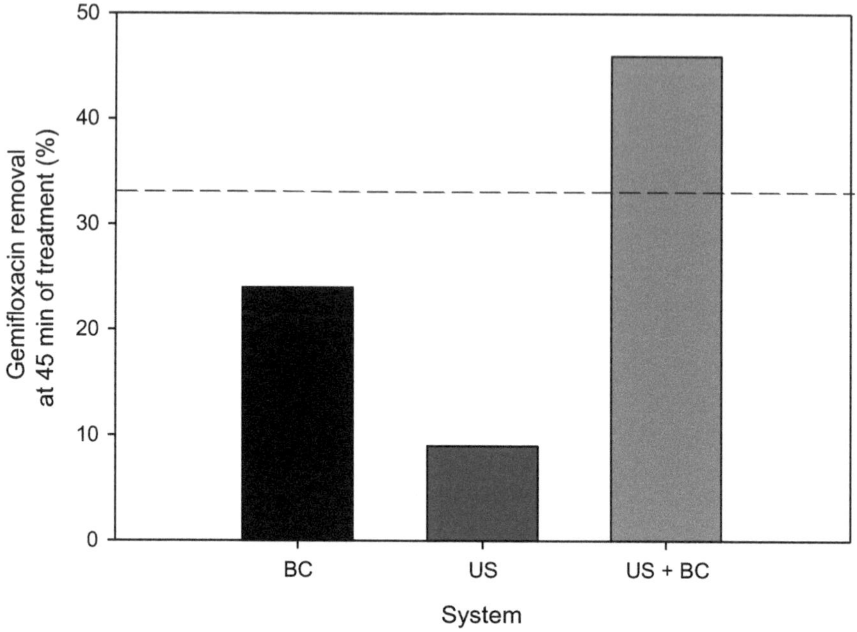

FIG. 10.4 Gemifloxacin degradation by individual systems (biochar or ultrasound [BC + US]) and the hybrid process (US + BC). *Note*: dashed line represents the arithmetic addition of the individual systems. *Figure made by the authors based on data from Gholami, P., Dinpazhoh, L., Khataee, A., Orooji, Y., 2019b. Sonocatalytic activity of biochar-supported ZnO nanorods in degradation of gemifloxacin: synergy study, effect of parameters and phytotoxicity evaluation, Ultrason. Sonochem. 55, 44–56. doi:10.1016/j.ultsonch.2019.03.001.*

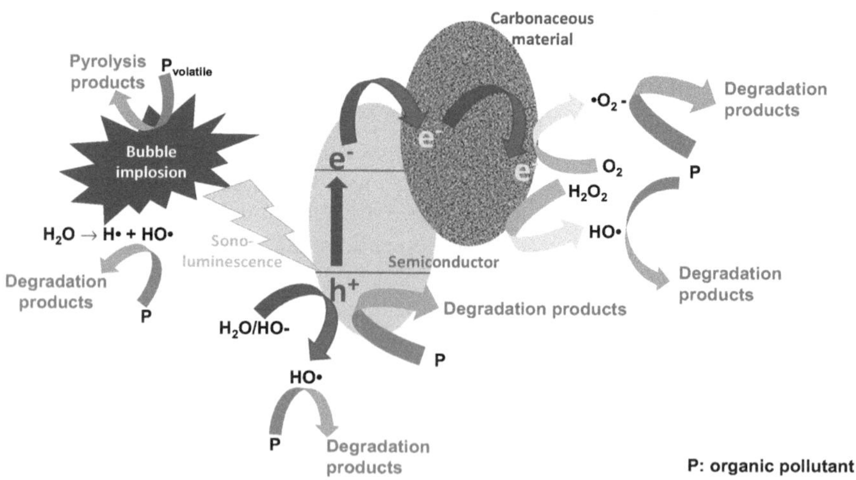

FIG. 10.5 Possible degradation routes for organic pollutants during sonocatalysis using a semiconductor/carbonaceous material composite. *Figure made by the authors.*

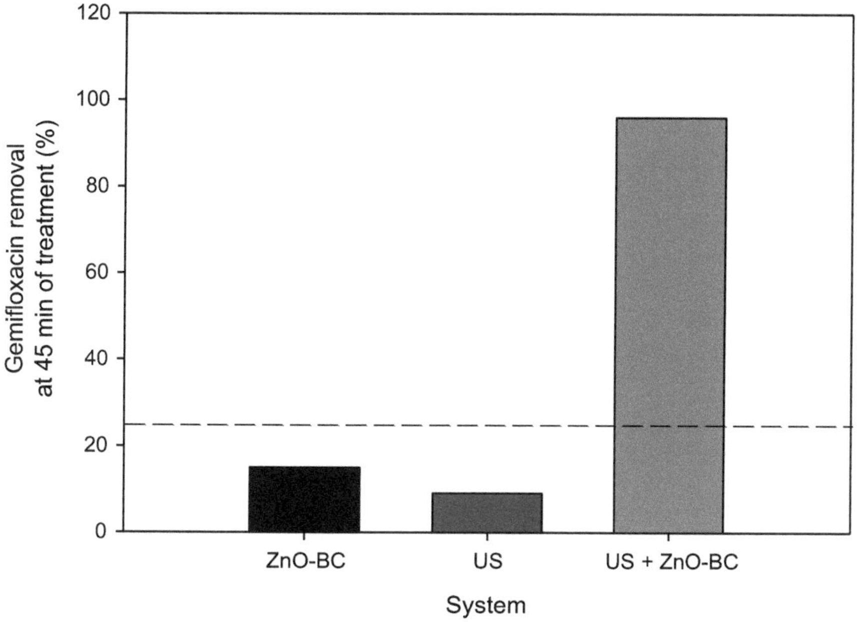

FIG. 10.6 Gemifloxacin degradation by individual systems (zinc oxide-biochar or ultrasound [ZnO-BC or US]) and the hybrid process (US + ZnO-BC). *Note*: dashed line represents the arithmetic addition of the individual systems. *Figure made by the authors based on data from Gholami, P., Dinpazhoh, L., Khataee, A., Orooji, Y., 2019b. Sonocatalytic activity of biochar-supported ZnO nanorods in degradation of gemifloxacin: synergy study, effect of parameters and phytotoxicity evaluation, Ultrason. Sonochem. 55, 44–56. doi:10.1016/j.ultsonch.2019.03.001.*

4. Treatment of complex matrices by ultrasound alone or by hybrid/combined systems

Another line of recent developments in the US area is its application to complex matrices. The sonolytic system has been utilized alone and in hybrid or combined combat water pollution (Anandan et al., 2020).

Pretreatment using USbased systems can be an alternative to improve the biological oxidation performance of wastewater. Gogate et al. researched the processing of real municipal wastewater by US alone and under hydrogen peroxide addition (US + H_2O_2), reporting that US alone reduced 11.3% of wastewater chemical oxygen demand (COD), whereas under optimized conditions the US + H_2O_2 system resulted in 24.9% of COD removal (because extra production of radicals from H_2O_2 were broken down by US, Eqs. 10.18–10.19). Additionally, the biodegradability index of the wastewater was enhanced to 0.6 and this pretreatment improved the performance of biological oxidation, which meant lower treatment times to achieve the desired COD reduction (Gogate et al., 2020).

$$H_2O_2 +))) \rightarrow 2HO\bullet \quad (10.18)$$

$$H\bullet + H_2O_2 \rightarrow HO\bullet + H_2O \tag{10.19}$$

The effect of US on effluents of municipal wastewater and distilled water spiked with pharmaceuticals (such as losartan and valsartan) has also been published, showing that degradations of these pollutants in both matrices are very similar. This because of the high hydrophobicity of such pharmaceuticals, which limited competence by matrix components (Serna-Galvis et al., 2019a). Besides, treatment of wastewater effluents containing emerging contaminants at realistic levels ($\mu g\ L^{-1}$) has shown that due to the physical action of US (i.e., turbulence and high shearing in the aqueous medium) some sorbed pharmaceuticals were released from the solid present in the effluents. At the same time, it was shown that the action of hydroxyl radicals (chemical effects of the process) led to high degradations of most contaminants in the effluent (Serna-Galvis et al., 2019b).

In the same way, the sonication of a real hospital effluent pretreated by biological oxidation led to a significant decrease in the concentrations of most pollutants (~59% removed after 90 min of treatment). The addition of ferrous ions and UVC light to the ultrasound system (forming a sono-photo-Fenton process) increased the degradation of contaminants from ~59% to ~83%. The sono-photo-Fenton system produced additional radical species through interaction between iron, hydrogen peroxide, and light (Eqs. 10.10–10.12). Moreover, the UVC radiation induced photo degradation of the pollutant because of its high energy. Such light also promoted the homolytic rupture of sonogenerated hydrogen peroxide (Eq. 10.20) (Serna-Galvis et al., 2019c).

$$H_2O_2 + UVC \rightarrow 2HO\bullet \tag{10.20}$$

On the other hand, hybrid or combined processes involving US and electrochemistry (i.e., sonoelectrochemistry) are gaining interest for treating contaminated wastewater because they offer advantages such as fast reaction rates, enhanced electrode surface activation, and mass transport (Theerthagiri et al., 2020). Sonoelectrochemistry can utilize electrodes that promote the formation of radicals or other oxidizing species (e.g., active chlorine or hydrogen peroxide, Eqs. 10.21–10.23), which may accelerate both pollutant elimination and mineralization (Theerthagiri et al., 2020; Guzman-Duque et al., 2016).

$$\text{Anode} + H_2O \rightarrow HO\bullet + H^+ + e^- \tag{10.21}$$

$$\text{Anode} + 2Cl^- + H_2O \rightarrow HOCl + HCl + 2e^- \tag{10.22}$$

$$\text{Cathode} + O_2 + 2H^+ + 2e^- \rightarrow H_2O_2 \tag{10.23}$$

Methyl paraben synthetic wastewater has been treated by a coupling between US and electrochemistry, where the influence of supporting electrolyte was evaluated. This study showed that removal of target pollutant fits well with pseudo-first-order kinetics and it is faster in the presence of chloride; its mineralization is more efficient in sulfate medium, due to the nature of the oxidant species generated in each condition. Furthermore, the US/electrochemistry combination has a synergistic effect on the abatement of organic matter, due to the improvement in mediated oxidation by electrogenerated species (Dionisio et al., 2019).

Sonoelectrochemical processes have also been tested for the treatment of phenolic wastewater. The effects of ultrasonic frequency and current were investigated, and it was found that the process removes ~50% of COD after 60 min of treatment at 40 kHz and 1.4 A of frequency and current, respectively (Zhang et al., 2020). In the same way, sonoelectrochemistry showed significant synergistic effects for the degradation of carbamazepine (a highly consumed pharmaceutical) in tap water. It was shown that the increasing of US power from 20 to 40 W and current from 1 to 3 A significantly augmented the degradation rates due to hydroxyl radical production, which is higher as these parameters increase (Tran et al., 2017).

It should be indicated that the foregoing works on pollutant degradation in complex matrices present a lot of information about parameter effects and kinetic aspects. However, toxicity or biodegradability analyses are needed to demonstrate the actual environmental impact of the processes. Moreover, details on the transformation products from target pollutant are scarce. Thus future work should consider such topics.

5. Concluding remarks

Based on the information given in this chapter the following aspects can be presented.

- US has shown a high ability for degrading diverse pollutants and treating several kinds of wastewater at laboratory scale (small volumes). However, until now, there are few studies about large-scale reactors for pollutant elimination. Thus studies on the upscaling of US reactors, mainly at high frequencies (100–1000 kHz) to eliminate pollution from wastewater, should be developed in the future.
- Recent hybrid processes based on US plus PS have great potential to degrade organic pollutants with synergistic effects. Nevertheless, more research to address the treatment of wastewater samples is needed. Furthermore, comprehensive evaluation of the effects on salinity of treated samples by PS addition are indispensable in further work.
- The addition of carbon-based materials to sonochemical systems is a good alternative. Indeed, inexpensive catalysts such as biochar can be used to enhance the contaminants, sonodegradation. It is expected that further publications should include information on the application to aqueous complex matrices, treatment extent (assessments of toxicity and biodegradability of treated water), reuse cycles of carbon-based catalysts, and in-depth discussion on limitations and alternatives for the separation of solid carbonaceous material from the liquid phase at the end of the process.

- US in hybrid/combined systems such as sono-photo-Fenton or sonoelectrochemistry allows the limitation of individual techniques to be overcome, making this is an interesting option to deal with the pollution of complex matrices. However, upcoming work should demonstrate the challenges of the environmental impact of these processes and study transformation products of pollutants in wastewater.

Acknowledgments

The authors thank the financial support provided by the Royal Society (UK) through the project: "Sound" methods of remediating emerging contaminants in hospital wastewater (RA4056, ICA/R1/191053); the Universidad de Antioquia UdeA for the support provided to their research group through "PROGRAMA DE SOSTENIBILIDAD"; and financing from COLCIENCIAS through project No. 111577757323.

References

Adewuyi, Y.G., 2001. Sonochemistry: environmental science and engineering applications. Ind. Eng. Chem. Res. 40, 4681–4715. https://doi.org/10.1021/ie010096l.

Al-Hamadani, Y.A.J., Jung, C., Im, J.-K., Boateng, L.K., Flora, J.R.V., Jang, M., Heo, J., Park, C.M., Yoon, Y., 2017. Sonocatalytic degradation coupled with single-walled carbon nanotubes for removal of ibuprofen and sulfamethoxazole. Chem. Eng. Sci. 162, 300–308. https://doi.org/10.1016/j.ces.2017.01.011.

Al-Hamadani, Y.A.J., Lee, G., Kim, S., Park, C.M., Jang, M., Her, N., Han, J., Kim, D.-H., Yoon, Y., 2018. Sonocatalytic degradation of carbamazepine and diclofenac in the presence of graphene oxides in aqueous solution. Chemosphere 205, 719–727. https://doi.org/10.1016/j.chemosphere.2018.04.129.

Anandan, S., Kumar Ponnusamy, V., Ashokkumar, M., 2020. A review on hybrid techniques for the degradation of organic pollutants in aqueous environment. Ultrason. Sonochem. 67, 105130. https://doi.org/10.1016/j.ultsonch.2020.105130.

Ashokkumar, M., Lee, J., Iida, Y., Yasui, K., Kozuka, T., Tuziuti, T., Towata, A., 2009. The detection and control of stable and transient acoustic cavitation bubbles. Phys. Chem. Chem. Phys. 11, 10118–10121. https://doi.org/10.1039/B915715H.

Beckett, M.A., Hua, I., 2001. Impact of ultrasonic frequency on aqueous sonoluminescence and sonochemistry. J. Phys. Chem. 105, 3796–3802. https://doi.org/10.1021/jp003226x.

Camargo-Perea, A.L., Rubio-Clemente, A., Peñuela, G.A., 2020. Use of ultrasound as an advanced oxidation process for the degradation of emerging pollutants in water. Water 12, 1068. https://doi.org/10.3390/w12041068.

Caretti, C., Coppini, E., Fatarella, E., Lubello, C., 2011. Membrane filtration and sonication for industrial wastewater reuse. Water Sci. Technol. 64, 2500–2507. https://doi.org/10.2166/wst.2011.800.

Chen, W.-S., Su, Y.-C., 2012. Removal of dinitrotoluenes in wastewater by sono-activated persulfate, Ultrason. Sonochem 19, 921–927. https://doi.org/10.1016/j.ultsonch.2011.12.012.

Chiha, M., Hamdaoui, O., Baup, S., Gondrexon, N., 2011. Sonolytic degradation of endocrine disrupting chemical 4-cumylphenol in water. Ultrason. Sonochem. 18, 943–950. https://doi.org/10.1016/j.ultsonch.2010.12.014.

Chowdhury, P., Viraraghavan, T., 2009. Sonochemical degradation of chlorinated organic compounds, phenolic compounds and organic dyes - a review. Sci. Total Environ. 407, 2474–2492. https://doi.org/10.1016/j.scitotenv.2008.12.031.

Dionisio, D., Motheo, A.J., Sáez, C., Canizares, P., Rodrigo, M.A., 2019. Coupling ultrasound to the electro-oxidation of methyl paraben synthetic wastewater: effect of frequency and supporting electrolyte. ChemElectroChem 6, 1199–1205. https://doi.org/10.1002/celc.201801199.

Fagan, W.P., Zhao, J., Villamena, F.A., Zweier, J.L., Weavers, L.K., 2020. Synergistic, aqueous PAH degradation by ultrasonically-activated persulfate depends on bulk temperature and physicochemical parameters. Ultrason. Sonochem. 67, 105172. https://doi.org/10.1016/j.ultsonch.2020.105172.

Geng, N., Chen, W., Xu, H., Ding, M., Liu, Z., Shen, Z., 2019. A sono-photocatalyst for humic acid removal from water: operational parameters, kinetics and mechanism. Ultrason. Sonochem. 57, 242–252. https://doi.org/10.1016/j.ultsonch.2019.03.022.

Gholami, P., Khataee, A., Darvishi, R., Soltani, C., Bhatnagar, A., 2019. A review on carbon-based materials for heterogeneous sonocatalysis: fundamentals, properties and applications. Ultrason. Sonochem. 58, 104681. https://doi.org/10.1016/j.ultsonch.2019.104681.

Gholami, P., Dinpazhoh, L., Khataee, A., Orooji, Y., 2019. Sonocatalytic activity of biochar-supported ZnO nanorods in degradation of gemifloxacin: synergy study, effect of parameters and phytotoxicity evaluation. Ultrason. Sonochem. 55, 44–56. https://doi.org/10.1016/j.ultsonch.2019.03.001.

Gogate, P.R., Thanekar, P.D., Oke, A.P., 2020. Strategies to improve biological oxidation of real wastewater using cavitation based pre-treatment approaches. Ultrason. Sonochem. 64, 105016. https://doi.org/10.1016/j.ultsonch.2020.105016.

Gole, V.L., Fishgold, A., Sierra-Alvarez, R., Deymier, P., Keswani, M., 2018. Treatment of perfluorooctane sulfonic acid (PFOS) using a large-scale sonochemical reactor. Separ. Purif. Technol. 194, 104–110. https://doi.org/10.1016/j.seppur.2017.11.009.

Gonçalves, I., Silva, C., Cavaco-Paulo, A., 2015. Ultrasound enhanced laccase applications. Green Chem. 17, 1362–1374. https://doi.org/10.1039/c4gc02221a.

Guzman-Duque, F., Pétrier, C., Pulgarin, C., Peñuela, G., Torres-Palma, R., 2011. Effects of sonochemical parameters and inorganic ions during the sonochemical degradation of crystal violet in water. Ultrason. Sonochem. 18, 440–446.

Guzman-Duque, F., Pétrier, C., Pulgarin, C., Peñuela, G., Herrera-Calderón, E., Torres-Palma, R.A., 2016. Synergistic coupling between electrochemical and ultrasound treatments for organic pollutant degradation as a function of the electrode material (IrO2 and BDD) and the ultrasonic frequency (20 and 800 kHz). Int. J. Electrochem. Sci. 11, 7380–7394. https://doi.org/10.20964/2016.09.23.

Hao, F., Guo, W., Wang, A., Leng, Y., Li, H., 2014. Intensification of sonochemical degradation of ammonium perfluorooctanoate by persulfate oxidant, Ultrason. Sonochem. 21, 554–558. https://doi.org/10.1016/j.ultsonch.2013.09.016.

Hou, L., Wang, L., Royer, S., Zhang, H., 2016. Ultrasound-assisted heterogeneous Fenton-like degradation of tetracycline over a magnetite catalyst. J. Hazard Mater. 302, 458–467. https://doi.org/10.1016/j.jhazmat.2015.09.033.

Im, J.-K., Heo, J., Boateng, L.K., Her, N., Flora, J.R.V., Yoon, J., Zoh, K.-D., Yoon, Y., 2013. Ultrasonic degradation of acetaminophen and naproxen in the presence of single-walled carbon nanotubes. J. Hazard Mater. 254–255, 284–292. https://doi.org/10.1016/j.jhazmat.2013.04.001.

Im, J.-K., Boateng, L.K., Flora, J.R.V., Her, N., Zoh, K.-D., Son, A., Yoon, Y., 2014. Enhanced ultrasonic degradation of acetaminophen and naproxen in the presence of powdered activated carbon and biochar adsorbents. Separ. Purif. Technol. 123, 96–105. https://doi.org/10.1016/j.seppur.2013.12.021.

Joseph, J.M., Destaillats, H., Hung, H.-M., Hoffmann, M.R., 2000. The sonochemical degradation of azobenzene and related azo Dyes: rate enhancements via fenton's reactions. J. Phys. Chem. 104, 301–307. https://doi.org/10.1021/jp992354m.

Kim, J., Park, B., Son, Y., Khim, J., 2018. Peat moss-derived biochar for sonocatalytic applications, Ultrason. Sonochem. 42, 26–30. https://doi.org/10.1016/j.ultsonch.2017.11.005.

Lee, J., 2016. Importance of Sonication and Solution Conditions on the Acoustic Cavitation Activity BT - Handbook of Ultrasonics and Sonochemistry, in. Springer Singapore, Singapore, pp. 137–175. https://doi.org/10.1007/978-981-287-278-4_10.

Li, B., Li, L., Lin, K., Zhang, W., Lu, S., Luo, Q., 2013. Removal of 1,1,1-trichloroethane from aqueous solution by a sono-activated persulfate process. Ultrason. Sonochem. 20, 855–863. https://doi.org/10.1016/j.ultsonch.2012.11.014.

Mahamuni, N.N., Adewuyi, Y.G., 2010. Advanced oxidation processes (AOPs) involving ultrasound for waste water treatment: a review with emphasis on cost estimation. Ultrason. Sonochem. 17, 990–1003. https://doi.org/10.1016/j.ultsonch.2009.09.005.

McNamara, W.B., Didenko, Y.T., Suslick, K.S., 1999. Sonoluminescence temperatures during multi-bubble cavitation. Nature 401, 772–775. https://doi.org/10.1038/44536.

Montoya-Rodríguez, D.M., Ferraro, F., Serna-Galvis, E.A., Torres-Palma, R.A., 2020. Degradation of the emerging concern pollutant ampicillin in aqueous media by sonochemical advanced oxidation processes - parameters effect, removal of antimicrobial activity and pollutant treatment in hydrolyzed urine. J. Environ. Manag. https://doi.org/10.1016/j.jenvman.2020.110224 (in press).

Pétrier, C., Francony, A., 1997. Ultrasonic waste-water treatment: incidence of ultrasonic frequency on the rate of phenol and carbon tetrachloride degradation. Ultrason. Sonochem. 4, 295–300. https://doi.org/10.1016/S1350-4177(97)00036-9.

Rayaroth, M.P., Aravind, U.K., Aravindakumar, C.T., 2016. Degradation of pharmaceuticals by ultrasound-based advanced oxidation process. Environ. Chem. Lett. 14, 259–290. https://doi.org/10.1007/s10311-016-0568-0.

Riesz, P., Berdahl, D., Christman, C.L., 1985. Free radical generation by ultrasound in aqueous and nonaqueous solutions. Environ. Health Perspect. 64, 233–252. https://doi.org/10.2307/3430013.

Samanta, M., Mukherjee, M., Ghorai, U.K., Sarkar, S., Bose, C., Chattopadhyay, K.K., 2018. Ultrasound assisted catalytic degradation of textile dye under the presence of reduced Graphene Oxide enveloped Copper Phthalocyanine nanotube. Appl. Surf. Sci. 449, 113–121. https://doi.org/10.1016/j.apsusc.2018.01.118.

Serna-Galvis, E.A., Isaza-Pineda, L., Moncayo-Lasso, A., Hernández, F., Ibáñez, M., Torres-Palma, R.A., 2019. Comparative degradation of two highly consumed antihypertensives in water by sonochemical process. Determination of the reaction zone, primary degradation products and theoretical calculations on the oxidative process. Ultrason. Sonochem. 58, 104635. https://doi.org/10.1016/j.ultsonch.2019.104635.

Serna-Galvis, E.A., Botero-Coy, A.M., Martínez-Pachón, D., Moncayo-Lasso, A., Ibáñez, M., Hernández, F., Torres-Palma, R.A., 2019. Degradation of seventeen contaminants of emerging concern in municipal wastewater effluents by sonochemical advanced oxidation processes. Water Res. 154, 349–360. https://doi.org/10.1016/j.watres.2019.01.045.

Serna-Galvis, E.A., Silva-Agredo, J., Botero-Coy, A.M., Moncayo-Lasso, A., Hernández, F., Torres-Palma, R.A., 2019. Effective elimination of fifteen relevant pharmaceuticals in hospital wastewater from Colombia by combination of a biological system with a sonochemical process. Sci. Total Environ. 670, 623–632. https://doi.org/10.1016/j.scitotenv.2019.03.153.

Theerthagiri, J., Madhavan, J., Lee, S.J., Choi, M.Y., Ashokkumar, M., Pollet, B.G., 2020. Sonoelectrochemistry for energy and environmental applications, Ultrason. Sonochem 63, 104960. https://doi.org/10.1016/j.ultsonch.2020.104960.

Torres, R.A., Pétrier, C., Combet, E., Moulet, F., Pulgarin, C., 2007. Bisphenol A mineralization by integrated ultrasound-UV-iron (II) treatment. Environ. Sci. Technol. 41, 297–302. https://doi.org/10.1021/es061440e.

Torres, R.A., Nieto, J.I., Combet, E., Pétrier, C., Pulgarin, C., 2008. Influence of TiO2 concentration on the synergistic effect between photocatalysis and high-frequency ultrasound for organic pollutant mineralization in water. Appl. Catal. B Environ. 80, 168–175. https://doi.org/10.1016/j.apcatb.2007.11.013.

Torres-Palma, R.A., Serna-Galvis, E.A., 2018. Sonolysis. In: Ameta, S.C., Ameta, R. (Eds.), Adv. Oxid. Process. Wastewater Treat. - Emerg. Green Chem. Technol., A, London, pp. 177–213.

Tran, N., Drogui, P., Brar, S.K., 2015. Sonochemical techniques to degrade pharmaceutical organic pollutants. Environ. Chem. Lett. 13, 251–268. https://doi.org/10.1007/s10311-015-0512-8.

Tran, N., Drogui, P., Brar, S.K., De Coninck, A., 2017. Synergistic effects of ultrasounds in the sonoelectrochemical oxidation of pharmaceutical carbamazepine pollutant. Ultrason. Sonochem. 34, 380–388. https://doi.org/10.1016/j.ultsonch.2016.06.014.

Villaroel, E., Silva-Agredo, J., Petrier, C., Taborda, G., Torres-Palma, R., 2014. Ultrasonic degradation of acetaminophen in water: effect of sonochemical parameters and water matrix. Ultrason. Sonochem. 21, 1763–1769.

Wang, S., Zhou, N., 2016. Removal of carbamazepine from aqueous solution using sono-activated persulfate process. Ultrason. Sonochem. 29, 156–162. https://doi.org/10.1016/j.ultsonch.2015.09.008.

Wang, J., Wang, Z., Vieira, C.L.Z., Wolfson, J.M., Pingtian, G., Huang, S., 2019. Review on the treatment of organic pollutants in water by ultrasonic technology. Ultrason. Sonochem. 55, 273–278. https://doi.org/10.1016/j.ultsonch.2019.01.017.

Wood, R.J., Lee, J., Bussemaker, M.J., 2017. A parametric review of sonochemistry: control and augmentation of sonochemical activity in aqueous solutions. Ultrason. Sonochem. 38, 351–370. https://doi.org/10.1016/j.ultsonch.2017.03.030.

Yang, L., Xue, J., He, L., Wu, L., Ma, Y., Chen, H., Li, H., Peng, P., Zhang, Z., 2019. Review on ultrasound assisted persulfate degradation of organic contaminants in wastewater: influences, mechanisms and prospective. Chem. Eng. J. 378, 122146. https://doi.org/10.1016/j.cej.2019.122146.

Zhang, N., Xian, G., Li, X., Zhang, P., Zhang, G., Zhu, J., 2018. Iron based catalysts used in water treatment assisted by ultrasound: a mini review. Front. Chem. 6, 1–6. https://doi.org/10.3389/fchem.2018.00012.

Zhang, M., Zhang, Z., Liu, S., Peng, Y., Chen, J., Yoo Ki, S., 2020. Ultrasound-assisted electrochemical treatment for phenolic wastewater. Ultrason. Sonochem. 65, 105058. https://doi.org/10.1016/j.ultsonch.2020.105058.

Chapter 11

Modified titania, A green material for wastewater remediation

Sofia Sandhu, Rehmat Brar, Siddhant Gulati, Vasundhara Singh
Punjab Engineering College, Chandigarh, Punjab, India

1. Introduction

About three decades ago, titanium dioxide first came under the limelight when it was found that the oxidizing power of the photogenerated holes in TiO_2 is sufficient to decompose water and air pollutants. In a very short span of time, TiO_2 emerged as one of the most promising photocatalysts finding its usage from the conventional areas like pigments in foodstuffs, paint, and cosmetics to the advanced sectors such as catalysis, photovoltaic cell, sensors, and biomedical treatment due to its broad functionality, longer stability, and nontoxicity (Humayun et al., 2018). Titanium dioxide owing to its fast electron–hole pair recombination rate and large band gap of 3.2 eV has not yet evolved into its most functional self. Unmodified TiO_2 can only be excited by ultraviolet irradiations ($\lambda < 388$ nm) which constitute less than 7% of solar light. The energy of photon incident onto TiO_2 is directly proportional to the number of photoformed electrons and holes, and hence is one of the leading factors to improve the utilization of sunlight. Development of modified titania with lower band gap so as to make effective use of visible light spectrum of sunlight is the requirement of the hour. Several strategies have been investigated in the past few years to improve the visible light absorbance and increase the charge carrier's lifetime of TiO_2. To achieve a higher efficiency of a photocatalytic reaction instigated by TiO_2 upon visible light irradiation, two problems must be dealt with. The first being production of more number of photoformed electrons and holes; and the other is to improve the separation efficiency of these electrons and holes generated. Both of these issues can be resolved by tuning the band gap by synthesis of titania with exposed high reactive facet. Modification in band gap is possible by doping metals, nonmetal, metal/nonmetal codoping, coupling, and by making the use of green solvents like ionic liquids and deep eutectic

Green Chemistry and Water Remediation: Research and Applications.
https://doi.org/10.1016/B978-0-12-817742-6.00011-6

solvents (DESs) for synthesizing titania and a lot of research work has been done in this area (Kumaravel et al., 2019; Nagaraju et al., 2018).

In our lab, undoped, N-doped, N,Cu and N,Fe-codoped titania nanoparticles were synthesized by simple sol−gel method and nanoparticles with different crystallite size, phase, BET surface area, and band gap were obtained. The photocatalytic activities of the as-synthesized TiO_2 nanoparticles were analyzed by degrading four different Reactive dyes under visible light irradiation (Kaur et al., 2016a,b). Further, to encounter the issue of separation of N,Fe-codoped TiO_2 photocatalyst from the degraded dye solution, magnetically separable γ-Fe_2O_3/N,Fe-codoped TiO_2 nanoparticles were synthesized which exhibited superior photocatalytic activity (Kaur et al., 2015). In another work, first-transition metal series with nitrogen was codoped in TiO_2 using choline chloride, a biodegradable, nontoxic ionic salt. Among the entire synthesized codoped titania, N,Fe-codoped titania was found to be the most efficient one for the degradation of Reactive Black 5 dye (Kaur et al., 2016a,b). Kaur et al. reviewed the advancement that has occurred in the various synthesis methods of titania making use of all the generations of ionic liquids (ILs) which plays the role of templating agent, solvent media, and reactants leading to the controlled morphologies and improved physicochemical properties. The review emphasizes on the contribution of ILs/DES in synthesizing visible light active titania and stated that sometimes a single dopant source could be responsible for doping the substrate with different elements which is completely based on the synthetic method used (Kaur and Singh, 2017).

In the quest to synthesize titania via a greener method synthesis of modified nanocrystalline anatase TiO_2 was synthesized in an acidic room temperature ionic liquid (AIL) 1-butyl-3-methylimidazolium hydrogen sulfate [Bmim] HSO_4 and the nanoparticles showed good efficiency in photocatalytic degradation of methyl orange under UV light (Shahi et al., 2015). Further, another AIL 3-methyl-1-(3-sulfonylpropyl) imidazolium trifluoromethanesulfonate $[HO_3S(CH_2)_3MIM][CF_3SO_3]$ was used as an acidic hydrolyzing and templating agent for synthesis of TiO_2 via both sol−gel and hydrothermal route (Shahi et al., 2016). A comprehensive analysis of effective control of highly exposed {001} facet in the above synthesized titania nanoparticles was done (Sandhu et al., 2018). The photocatalytic activity of the catalysts was explained on the basis of percentage of {001} facet present in the nanocrystals. We also reported a low cost and greener solvent system, DES, and choline chloride/p-toluene sulfonic acid which play the dual role templating and hydrolyzing agent to synthesize TiO_2 nanoparticles via a simple sol−gel route (Shahi et al., 2017).

This chapter focuses on reaction mechanisms involved in band gap modification of TiO_2 with doping of metals and nonmetals and application of this modified titania in reactive dye degradation. Also, it details discussion about work done in our lab in the last decade for synthesis of modified titania in terms of visible light absorbance and reactivity via band gap and facet engineering. Further, the synthesis of titania with exposed {001} reactive facets via a biocompatible route using ILs and DES as acidic/basic reaction media has been explored.

2. Mechanism of photocatalytic degradation of dye using titanium dioxide

Based upon the modification in the electronic structure of titania, the mechanisms that are possible in the photocatalytic degradation of dyes can be summarized as explained below.

2.1 Based on band gap of the catalyst

The band gap of the catalyst decides which mechanism out of I and II (Table 11.1) will play a major role in the photocatalytic degradation of dyes under visible light. For the undoped TiO_2 having band gap greater than 2.9 eV the mechanism I is dominantly involved where upon visible light absorption, the adsorbed dye gets excited (D*) which is scavenged by O_2 and results in the formation of oxygen radicals. These oxygen radicals further form hydroxyl radicals which play a major role for the oxidation of the organic compound.

When photon with energy either equal or greater than the band gap of the photocatalyst is incident on the semiconductor, the degradation pathway follows the indirect mechanism II in which photoelectron is elevated from the filled valence band to the conduction band of catalyst which results in the formation of e^-/h^+ pair. The holes when react with water produce hydroxyl radical which being a strong oxidizing agent attacks on adsorbed organic molecules. Electrons that are promoted to conduction band are consumed and generate anionic superoxide radical which not only take part in oxidation process but also act as a barrier for the electron and hole recombination and thus maintain electron neutrality within TiO_2 matrix.

The photodegradation efficiency via direct mechanism I is dependent on the nature of the dye and its efficiency as a photosensitiser of the photocatalyst. According to the literature, indirect mechanism II is found to be more prevailing as compared with direct mechanism I and found to be the main contributor for the dyes in the visible light–initiated mechanism.

2.2 Based on the facet present in the TiO_2

Facets with higher energy have larger number of holes as compared with low-energy facets. This makes high-energy facets more active for photooxidation and low-energy facet for photoreduction. Presence of both high-energy facet (oxidation site) and low-energy facet (reduction site) (Table 11.1) permits the photogenerated electron and hole pairs to be separated for a longer time interval thus making titania more efficient in photocatalytic degradation of dyes.

TABLE 11.1 Different pathways involved in the photocatalytic degradation of dyes with TiO_2.

Direct mechanism I	Indirect mechanism II (visible light active modified TiO_2)	Mechanism III (TiO_2 with exposed high reactive facets)
	Photoexcitation:	
$D + h\nu(\text{visible}) \rightarrow D^*$	$^{D}TiO_2 + h\nu\ (\text{visible}) \rightarrow {}^{D}TiO_2\left(e^-_{CB} + h^+_{VB}\right)$	$^{D}TiO_2 + h\nu\ (\text{UV}) \rightarrow {}^{D}TiO_2\left(e^-_{CB} + h^+_{VB}\right)$
	Ionization of water:	
$D^* + TiO_2 \rightarrow D^{\cdot +} + TiO_2\ (e^-_{CB})$	$^{D}TiO_2\left(h^+_{VB}\right) + H_2O \rightarrow {}^{D}TiO_2 + H^+ + {}^{\cdot}OH$	$^{D}TiO_2\left(h^+_{VB}\right) + H_2O \rightarrow {}^{D}TiO_2 + H^+ + {}^{\cdot}OH$
$TiO_2(e^-_{CB}) + O_2 \rightarrow TiO_2 + O_2^{\cdot -}$	$^{D}TiO_2\left(h^+_{VB}\right) + OH^- \rightarrow {}^{D}TiO_2 + {}^{\cdot}OH$	$^{D}TiO_2\left(h^+_{VB}\right) + OH^- \rightarrow {}^{D}TiO_2 + {}^{\cdot}OH$
		$^{D}TiO_2(e^-_{CB}) + O_2 \rightarrow {}^{D}TiO_2 + O_2^{\cdot -}$
	Oxygen ionosorption:	
$O_2^{\cdot -} + 2H_2O \rightarrow {}^{\cdot}OH + OH^-$	$^{D}TiO_2(e^-_{CB}) + O_2 \rightarrow {}^{D}TiO_2 + O_2^{\cdot -}$	$O_2^{\cdot -} + H^+ \rightarrow HO_2^{\cdot}$
	Protonation of superoxide:	
$D^+ + O_2(\text{or } O_2^{\cdot -} \text{ or OH}) \rightarrow$ degraded products	$O_2^{\cdot -} + H^+ \rightarrow HO_2^{\cdot}$	Dye + OH $\left(\text{or } e^-_{CB} \text{ or } h^+_{VB}\right) \rightarrow$ degraded products

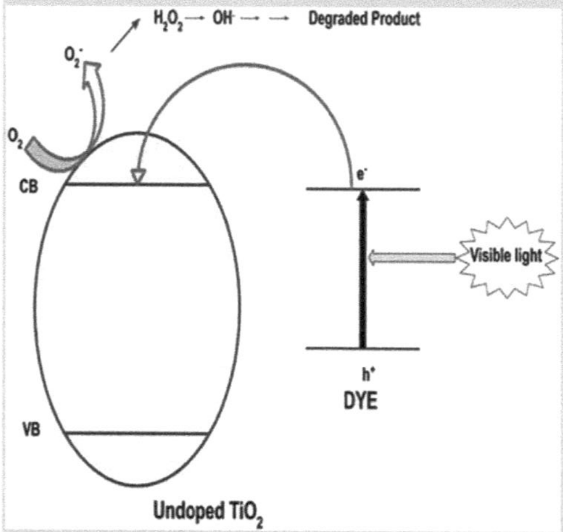

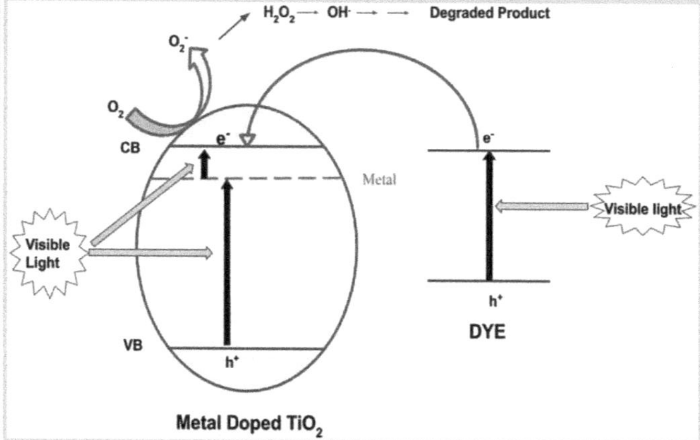

$$Dye + OH\left(\text{or } e^-_{CB} \text{ or } h^+_{VB}\right) \rightarrow \text{degraded products}$$

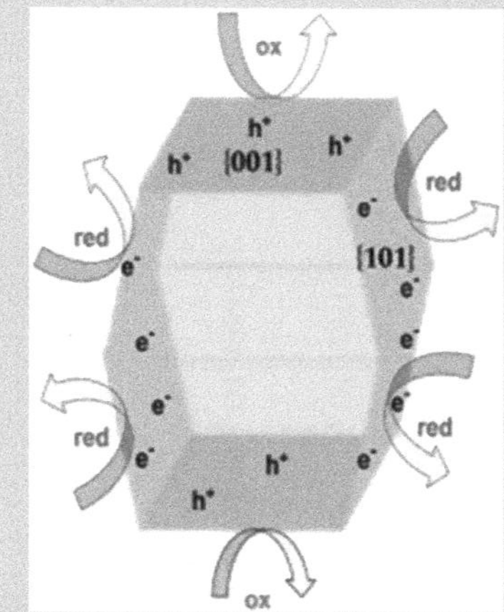

Thus, for the improved photocatalysis not only more reactive facet but also the presences of less reactive facets in the single crystal are important.

3. Different strategies for the synthesis of visible light–responsive titania

As explained in the above section, for the enhanced photocatalytic activity of a catalyst under visible light, the number of photoformed holes and electrons plays the major role by participating in the redox reaction. To make titanium dioxide more active and efficient in the visible light spectra doping with metal ions or nonmetal ions or codoped with metals and nonmetals, etc., could lead to the satisfactory activity.

3.1 Doping with metals

An optimum amount of transition metal, rare-earth metal, and other metal ions doped into TiO_2 can improve its photocatalytic efficiency due to following reasons:

i Doping with transition metals creates additional energy levels between the conduction and valence band of a semiconductor which leads to lowering in activation photon energy since the electron can transfer to one of the new lower energy bands resulting in a red shift in band gap absorption. Introducing transition metals into the TiO_2 matrix creates localized and delocalized impurity states within the band gap of metal oxide which effectively modulates the band gap energy, fermi levels, and d electron configuration of the catalyst. Due to the partially filled d orbitals, a spectrum of new energy levels is created below the conduction band which causes the red shift in the band gap of the catalyst making it visible light responsive.

ii Doping with metal reduces the recombination of charges since the metal scavengers the electrons ultimately resulting in hydroxyl radical formation which improvise the photocatalytic efficiency of TiO_2 (Table 11.2). Transition metal ions increase the degradation rates by changing the carrier equilibrium concentration as they play the role of traps for electrons and thus decreasing the recombination rate.

Among the various metal-doped TiO_2, Fe–TiO_2 has been found to show superior photocatalytic activity owing to its half-filled electronic configuration and facile trapping of e-/h + by reducing the trapping density and recombination of charges and thus enhancing adsorption. In the studies carried out by Reddy et al. for the degradation of Naphthol-Blue–Black (NBB) dye and real textile wastewater sonophotocatalytically by Fe, TiO_2, and Fe–TiO_2 exhibited higher color removal of 96% with Fe–TiO_2 compared with Fe and TiO_2 catalyst. In addition to the color removal, higher TOC reduction by 91% was

TABLE 11.2 Schematic representation of reactions involved in trapping electron–hole pair by metal ions.

Reactions involved in trapping of e^- and h^+	Schematic diagram
$M^{n+} + e^- \rightarrow M^{(n-1)+}$	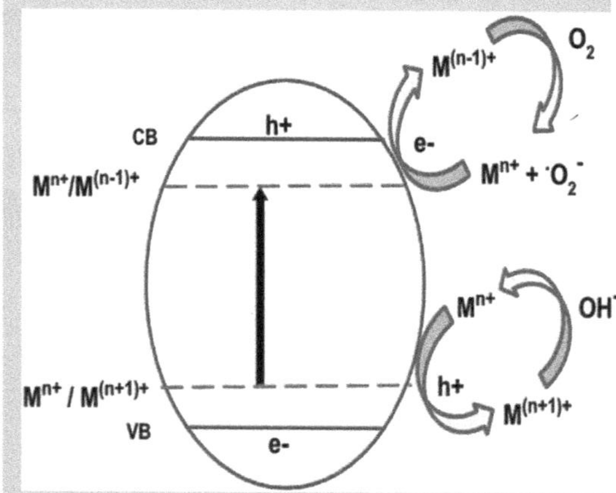
$M^{(n-1)+} + O_2 \rightarrow M^{n+} + O_2^{\cdot -}$	
$M^{n+} + h^+ \rightarrow M^{(n+1)+}$	
$M^{(n+1)+} + OH^- \rightarrow M^{n+} + {}^{\cdot}OH$	

obtained with lesser amount of Fe$-$TiO$_2$ catalyst. The catalyst and the methodology both proved to be very effective for the synthetic NBB dye and textile wastewater taken from industry (Reddy et al., 2016).

However, metal-doped TiO_2 also suffer from significant disadvantages. Decrease in photocatalytic activity of metal-doped TiO_2 has been reported even in UV irradiation owing to the reduction in thermal stability and increase in e^-/h^+ recombination rate. Especially for higher concentration of metal, metal centers attract holes after trapping electrons encouraging recombination of photogenerated electron/hole pairs. According to current research, nonmetal doping is more preferable than metal doped for band gap narrowing of titania (Ünlü et al., 2018).

3.2 Doping with nonmetal

Doping of TiO_2 with metal ions has both its pros and cons—although it has the desirable effect of inducing spectral shifts into the visible light region, it introduces additional recombination centers and leads to reduction in thermal stability of TiO_2, and the associated modification costs are also high. Research related to doping of nonmetals in TiO_2 has gained relevance which highlights enhanced photocatalytic activity and increased visible light absorption in TiO_2 as a result of nitrogen doping. Subsequently, many researchers have demonstrated the positive effects of nonmetal doping on visible light absorption ability and photocatalytic activity of TiO_2 in the last few years. Nonmetal doping is done with elements whose atomic radius is similar to the atomic radius of the O atom, and B, N, C, and F have been mainly used.

Among all nonmetals, doping with nitrogen has attracted the maximum attention, because of its small ionization energy, comparability of atomic size with oxygen, and stability, which make it easier to introduce N into the crystal lattice of TiO_2. In N-doped TiO_2, the N atom can either occupy an interstitial site or can substitute the oxygen atom.

Regarding the modification mechanism of N-doped TiO_2, three different explanations are available:

(a) Impurity energy levels: After doping with nitrogen, the O element in the TiO_2 lattice is substituted by N, which forms isolated impurity levels above the valence band. Whereas UV light irradiation leads to electron excitation in both the impurity energy levels and the VB, visible light illumination is able to excite electrons in the impurity energy level only.

(b) Band gap narrowing: Owing to the similar energies of N and O atoms, the 2p states of both the atoms hybrid, leading to the decrease of band gap, making N$-$TiO$_2$ visible light active.

(c) Oxygen vacancies: The grain boundaries of the metal oxide lattice have oxygen vacancies. Doping with nitrogen helps in maintaining charge neutrality in the TiO_2 lattice, as a consequence of which the visible light response of the TiO_2 increases (Nasirian et al., 2018).

Borolo et al. in their work studied the phenomenon of generation of surface charge carriers with the help of electron paramagnetic resonance (EPR) under various kind of irradiation (visible light and NIR) in N-doped TiO_2. The doped catalyst was found to be moderately active for the photoactive processes attributed to the formation of limited concentrations of NO^{x-} centers (Borolo et al., 2012).

Various theoretical and experimental studies indicate that the photo-catalytic activity of nonmetal-doped TiO_2 in the visible region is improved due the formation of a localized midgap state above the valence band. However, the oxidation power of the generated hole decreases as a result of this midgap localization. Another issue that is found to be responsible for the reduction in photocatalytic efficiency of N-doped TiO_2 is the creation of new recombination sites that emerge due to the formation of new oxygen vacancies as a consequence of N doping. These prominent drawbacks, which occur due to doping with nonmetallic anions in TiO_2, limit their use for visible light active photocatalysis (Joseph and Thiripuranthagan, 2017).

3.3 Co-doped and coupled TiO_2

Doping by metal and nonmetals in the TiO_2 lattice changes its electronic structure by creating new impurity levels within the band gap of TiO_2. The doping levels generated by metals are found to be below the CB of TiO_2 and in case of nonmetal ions, the doping levels are closer to the valence band of TiO_2. The absorption range of TiO_2 can be extended either by absorbing photons, having longer wavelengths or by accepting photogenerated e^- from the valence band by new doping levels. In the synergistic effect of metal and nonmetal dopants in TiO_2, the metal ions are responsible for the decrease in the e^-/h^+ recombination rate by introducing traps, while the nonmetal ions are responsible for the increase in absorption of metal oxide in the visible light spectrum. The photocatalytic degradation of reactive dyes under visible light was carried out by incorporating various metals with nitrogen in TiO_2 as reported by Kaur et al. which were synthesized by simple sol−gel method. The paper reported an anomalous trend of the reactivity of as synthesized photocatalysts in the degradation of various dyes and concluded that it is not just the degradation pathway or the physicochemical properties of the doped TiO_2 but also the substrate specificity should be taken into consideration (Elgendy et al., 2019; Jyothi et al., 2019).

To further improve the photocatalytic performance, various attempts have been made to study the combined effects of coupling and doping. Coupling with a magnetic component like γ-Fe_2O_3 not only overcomes the limitation of narrow light response but also helps in easy separation of catalyst (Ren et al., 2019; Wang et al., 2018). γ-Fe_2O_3/N,Fe codoped TiO_2 synthesized by Kaur et al. exhibited enhanced photocatalytic activity for the degradation of RB 4 dye in comparison with nonmagnetic N,Fe codoped TiO_2. In γ-Fe_2O_3/N,Fe codoped TiO_2, it is N and Fe which are responsible for high electron and hole mobility despite the mismatch of energy bands of γ-Fe_2O_3 and TiO_2 (Fig. 11.1). Hence, pathways A and B (Table 11.3) lead to the degradation of

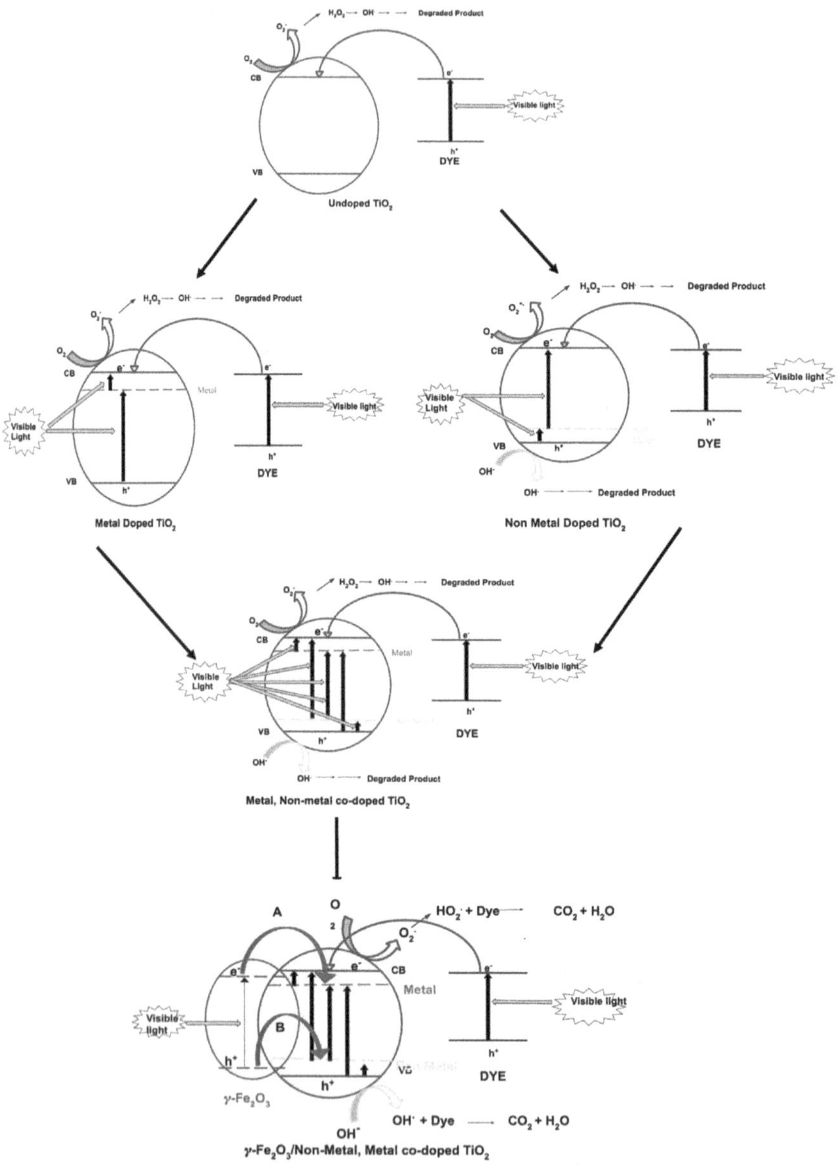

FIG. 11.1 Mechanistic pathways for the degradation of dye with undoped, metal-doped, nonmetal-doped, non-metal/metal-codoped, and γ-Fe_2O_3/nonmetal, metal codoped TiO_2.

TABLE 11.3 Probable reaction pathways for the electrons and holes in γ-Fe_2O_3/N,Fe codoped TiO_2.

Pathway A	Pathway B
$O_2 + e^- \rightarrow O_2^{\cdot -}$	$(H_2O)_{ads} + h^+ \rightarrow H^+ + {}^{\cdot}OH$
$O_2^{\cdot -} + H^+ \rightarrow HO_2^{\cdot}$	$(^-OH)_{ads} + h^+ \rightarrow {}^{\cdot}OH$
$HO_2^{\cdot} + dye \rightarrow \rightarrow \rightarrow CO_2 + H_2O$	${}^{\cdot}OH + dye \rightarrow \rightarrow \rightarrow CO_2 + H_2O$

reactive dyes with γ-Fe_2O_3/N,Fe codoped TiO_2. The recycled γ-Fe_2O_3/N,Fe codoped TiO_2 also demonstrate good repeatability of photoactivity (Kaur et al., 2015).

4. Ionic liquid/deep eutectic solvents as hydrolyzing and templating agents for synthesis of modified titania

Recently, ILs also known as "designer solvents" have proved to be the potential candidate for the synthesis of the micro structures and nanostructures due to the interplay between Coulomb, van der Waals interactions, and hydrogen bonding with the colloidal particles. However, despite their widespread application in organic chemistry, the utility of ILs in the synthesis of metal oxides, especially TiO_2, has not been reported extensively.

For the synthesis of visible light–responsive titania using ILs as solvent media, the highest occupied molecular orbital (HOMO) and lowest occupied molecular orbital (LUMO) of IL should preferably be in accordance with the energy levels of conduction and valence band of TiO_2 as shown in Fig. 11.2. This fact is quite evident from the work of Hu et al., where visible

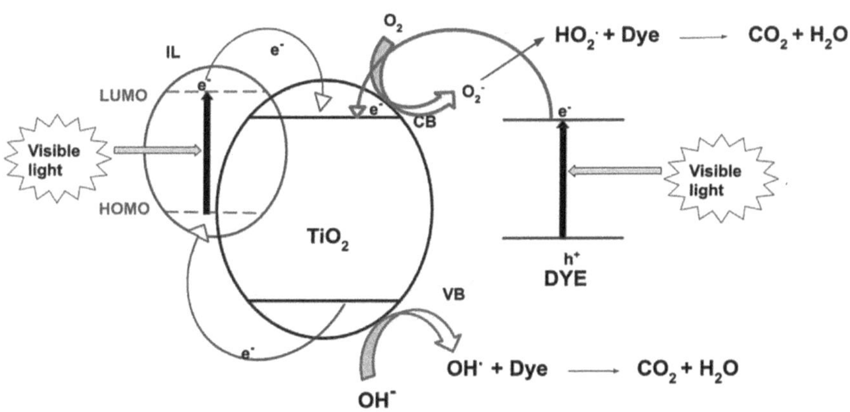

FIG. 11.2 Mechanistic pathway for degradation of dye with IL-TiO_2 nanocomposite.

light–responsive TiO_2 was synthesized using [bmim][OH](Hu et al., 2010). The mechanism for the enhanced photocatalytic performance by IL-modified TiO_2 composites under the exposure to visible light can be explained as follows:

- **i** The electrons from VB of TiO_2 are excited to HOMO of IL resulting into the formation of holes in the VB of TiO_2 under visible light irradiations.
- **ii** The electrons from HOMO of IL are then transferred to LUMO of IL under visible light irradiations which are further injected into the CB of TiO_2 since its energy level lies below the LUMO of the IL.
- **iii** Highly active oxygen radicals are formed when the above photogenerated electrons of TiO_2 react with O_2.
- **iv** The photogenerated holes react with the hydroxyl ion to form highly active hydroxyl radical (·OH) which are responsible for the degradation of dyes.

Even though IL exhibit superior properties such as low melting point and vapor pressure, and the ability to modify the cation and anion structures, they suffer from certain disadvantages, such as high cost and toxicity. A new generation of greener ILs, DESs, has emerged as promising materials and solvents (Wang & et al., 2018). Shahi et al. have synthesized titania nanoparticles using low cost, environmentally benign choline chloride/p-toluene sulfonic acid DES. They found that photocatalytic efficiency of anatase–rutile mixture is higher (98% in 3 h) than commercially available Degussa P-25(Shahi et al., 2017). The development of DES-mediated synthesis of TiO_2 nanomaterials is a growing field and yet to be explored.

5. Facet engineering for the highly reactive titania

Titania has been experimented with controlled particle size, phase, morphology, and electronic band structure to obtain high-quality materials for large number of applications. However, along with all these important features, the surface chemistry concerning atomic arrangement, coordination at the surface, and free energy of single crystal which depend on its shape and varies with the crystallographic orientation at various stabilized crystal facets has recently been reported to be responsible for higher photo efficiency. The reason for the different reactivity of various facets has been attributed to the surface energy of each facet which follows the order: {111}-facet (1.61 J/m^2)> {110}-facet (1.09 J/m^2)> {001}-facet (0.90 J/m^2)> {010}-facet (0.53 J/m^2)> {101}-facet (0.44 J/m^2). The active sites that are considered for the reason of high photocatalytic activity are surface features such as oxygen vacancies and distribution of these vacancies on metal oxides. The different surface energies of various facets of TiO_2 crystals can be altered by decreasing

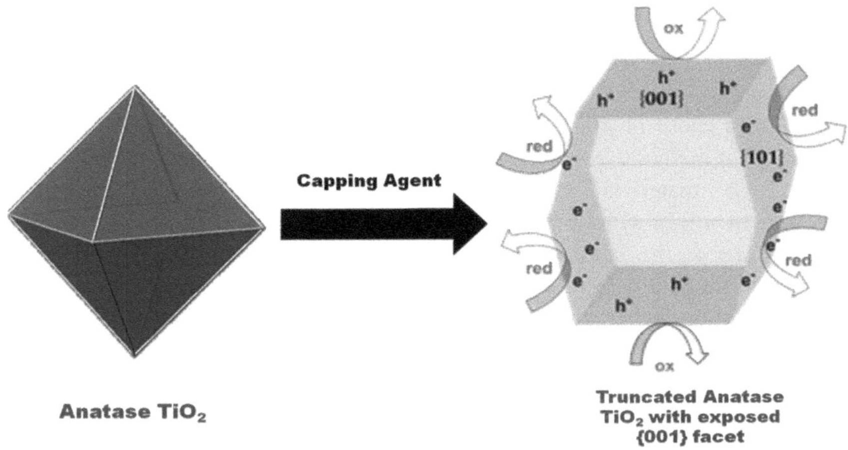

FIG. 11.3 Formation of active sites for the degradation of dyes in Faceted TiO_2.

the selectivity of adsorption of specific inorganic ions which indirectly depends on two factors. The first being the density of undercoordinated Ti atoms, while the second is the distance between two adjacent undercoordinated Ti atoms present on the surface. Hence, the choice of capping agent is more important in deciding the exposed facet (Fig. 11.3) (Xiong et al., 2018).

Studies are being carried out nowadays to synthesize nanosized truncated octahedral bipyramid with exposed reactive facet under nonacidic conditions particularly avoiding use of hydrofluoric acid. Thus, a greener and cost-effective route is highly desirable to obtain titania with higher percentage of exposed reactive facet with enhanced physicochemical properties. ILs/DESs outshine among all the green solvents as the potential candidate acting as hydrolyzing and templating agents for the synthesis of metal oxides and also reported to play an important role in exposing reactive facets (Zhang et al., 2009; Xu et al., 2015).

In the studies carried out by Shahi et al. for preparing nanosized pure anatase, rutile and mixed phase titanium dioxide in acidic recoverable and noncorrosive AILs [Bmim] HSO_4 and $[HO_3S(CH_2)_3MIM][CF_3SO_3]$, a comprehensive analysis of effective control of highly exposed reactive {001} facet by carefully varying various parameters like molar concentration and amount of AILs and calcination temperature has been done. While studying the effect of elevation of the amount of [Bmim] HSO_4, a high percentage of 47.5% of {001} facet is obtained under hydrothermal conditions. This is attributed to the adsorption/desorption of $[BMIM]^+$ on the {001} plane resulting in the exposure of {001} facet. The obtained highly reactive exposed facet was found to be responsible for exceptionally good photocatalytic activity under UV irradiations for the degradation of methyl orange (Sandhu et al., 2018).

6. Summary and outlook

The treatment of pollutants present in waterbodies due to high concentration of organic pollutants from pharmaceutical, cosmetic, agriculture, and textile industries is a cause of great concern in present times. Drawbacks faced during conventional methods for treatment of industrial wastewater could be overcome using TiO_2 materials with high photocatalytic or photoelectrocatalytic efficiency. Until now, treatment of wastewater using modified titania has been mostly done at the laboratory scale. However, industrial implementation of photocatalytic treatment of wastewater using coupled and codoped TiO_2 under solar irradiation is yet to be fully explored. Another potential environmental application of titania toward generating cleaner energy is its utilization in photocatalytic hydrogen production. Thus, even though extensive research has been done toward utilization of doped and undoped titania, its full potential is yet to be tapped with focus on practical implementations.

References

Borolo, G., Livraghi, S., Chiesa, M., Paganini, M.C., Giamello, E., 2012. J. Phys. Chem. C 116, 20887–20894.

Elgendy, K., Elmehasseb, I., Kandil, S., 2019. Synthesis and characterization of rare earth metal and non-metal co-doped TiO_2 nanostructure for photocatalytic degradation of Metronidazole. J. Mater. Environ. Sci. 2508 (1), 60–67.

Hu, S., Wang, A., Li, X., Wang, Y., Löwe, H., 2010. Hydrothermal synthesis of ionic liquid [bmim] OH-modified TiO_2 nanoparticles with enhanced photocatalytic activity under visible light. Chem. Asian J. 5, 1171–1177.

Humayun, M., Raziq, F., Khan, A., Luo, W., 2018. Modification strategies of TiO_2 for potential applications in photocatalysis: a critical review. Green Chem. Lett. Rev. 11 (2), 86–102.

Joseph, A.I.J., Thiripuranthagan, S., 2017. Non-metal doped titania photocatalysts for the degradation of neonicotinoid insecticides under visible light irradiation. J. Nanosci. Nanotechnol. 18 (5), 3158–3164.

Jyothi, M.S., Nayak, V., Reddy, K.R., Naveen, S., Raghu, A.V., 2019. Non-metal (Oxygen, Sulphur, Nitrogen, Boron and Phosphorus)-Doped Metal Oxide Hybrid Nanostructures as Highly Efficient Photocatalysts for Water Treatment and Hydrogen Generation, pp. 83–105.

Kaur, N., Singh, V., 2017. Current status and future challenges in ionic liquid, functional ionic liquid and deep eutectic solvent mediated synthesis of nanostructured TiO_2 : a review. New J. Chem. 41, 2844–2868.

Kaur, N., Kaur Shahi, S., Singh, V., 2015. Synthesis, characterization and photocatalytic activity of magnetically separable γ-Fe_2O_3/N,Fe codoped TiO_2 heterojunction for degradation of Reactive Blue 4 dye. RSC Adv. 5 (76), 61623–61630.

Kaur, N., Shahi, S.K., Singh, V., 2016a. Choline chloride assisted synthesis of N and metal codoped TiO_2 and their photocatalytic activity under visible light. Photochem. Photobiol. 92 (1), 69–75.

Kaur, N., Kaur, S., Singh, V., 2016b. Preparation, characterization and photocatalytic degradation kinetics of Reactive Red dye 198 using N, Fe codoped TiO_2 nanoparticles under visible light. Desalin. Water Treat. 57 (20), 9237–9246.

Kumaravel, V., Mathew, S., Bartlett, J., Pillai, S.C., 2019. Photocatalytic hydrogen production using metal doped TiO_2 : a review of recent advances. Appl. Catal. B Environ. 244, 1021–1064.

Nagaraju, G., Shubha, J.P., Manjunath, K., Dupont, J., 2018. Science Direct Ionothermal synthesis of TiO_2 nanoparticles for enhanced photocatalytic H_2 generation. Int. J. Hydrogen Energy 43 (8), 4028–4035.

Nasirian, M., Lin, Y.P., Bustillo-Lecompte, C.F., Mehrvar, M., 2018. Enhancement of photocatalytic activity of titanium dioxide using non-metal doping methods under visible light: a review. Int. J. Environ. Sci. Technol. 15 (9), 2009–2032.

Reddy, D.R., Dinesh, G.K., Anandan, S., Sivasankar, T., 2016. Sonophotocatalytic treatment of Naphthol Blue Black dye and real textile wastewater using synthesized Fe doped TiO_2. Chem. Eng. Process: Process Intensification 99, 10–18.

Ren, L., Zhou, W., Sun, B., et al., 2019. Defects-engineering of magnetic γ-Fe_2O_3 ultrathin nanosheets/mesoporous black TiO_2 hollow sphere heterojunctions for efficient charge separation and the solar-driven photocatalytic mechanism of tetracycline degradation. Appl. Catal. B Environ. 240, 319–328.

Sandhu, S., Shahi, S.K., Singh, V., 2018. Acidic ionic liquids: an alternative to HF for {001} reactive facet controlled synthesis of anatase titania. New J. Chem. 42, 12762–12765.

Shahi, S.K., Kaur, N., Kaur, A., Singh, V., 2015. Green synthesis of photoactive nanocrystalline anatase TiO_2 in recyclable and recoverable acidic ionic liquid [Bmim] HSO_4. J. Mater. Sci. 50 (6), 2443–2450.

Shahi, S.K., Kaur, N., Singh, V., 2016. Fabrication of phase and morphology controlled pure rutile and rutile/anatase TiO_2 nanostructures in functional ionic liquid/water. Appl. Surf. Sci. 360, 953–960.

Shahi, S.K., Kaur, N., Sandhu, S., Shahi, J.S., Singh, V., 2017. Influences of a new templating agent on the synthesis of coral-like TiO_2 nanoparticles and their photocatalytic activity. J. Sci.: Adv. Mater. Devices 2 (3), 347–353.

Ünlü, B., Çakar, S., Özacar, M., 2018. The effects of metal doped TiO_2 and dithizone-metal complexes on DSSCs performance. Sol. Energy 166, 441–449.

Wang, Q., et al., 2018. Controllable green synthesis of crassula peforata-like TiO_2 with high photocatalytic activity based on deep eutectic solvent (DES). Chem. Eng. J. 348, 811–819.

Wang, M., Deng, K., Lü, W., Deng, X., et al., 2018. Rational design of multifunctional Fe@γ-Fe_2O_3@ H-TiO_2 nanocomposites with enhanced magnetic and photoconversion effects for wide applications: from photocatalysis to imaging-guided photothermal cancer therapy. Adv. Mater. 30 (13), 1–9.

Xiong, Z., Lei, Z., Li, Y., Dong, L., Zhao, Y., Zhang, J., 2018. A review on modification of facet-engineered TiO_2 for photocatalytic CO_2 reduction. J. Photochem. Photobiol. C Photochem. Rev. 36, 24–47.

Xu, Q., Yu, J., Zhang, J., Zhang, J., Liu, G., 2015. Cubic anatase TiO_2 nanocrystals with enhanced photocatalytic CO_2 reduction activity. Chem. Commun. 51, 7950–7953.

Zhang, D.Q., Li, G.S., Yang, X.F., Yu, J.C., 2009. A micrometer-size TiO_2 single-crystal photocatalyst with remarkable 80% level of reactive facets. Chem. Commun. 29, 4381–4383.

Index

Note: 'Page numbers followed by "f" indicate figures and "t" indicates tables.'

Kolibri 360 GmbH
Ferdinand-Jühlke-Straße 7
99095 Erfurt, Deutschland
produktsicherheit@kolibri360.de